Pre-Algebra

Instruction Manual

By Steven P. Demme

Math·U·See®

1·888·854·MATH (6284) - MathUSee.com
Sales@MathUSee.com

Pre-Algebra Instruction Manual

©2009 Math-U-See, Inc.
Published and distributed by Demme Learning

mathusee.com

1-888-854-6284 or +1 717-283-1448 | demmelearning.com
Lancaster, Pennsylvania USA

ISBN 978-1-60826-027-0
Revision Code 0616-B

Printed in the United States of America by Bindery Associates LLC
 2 3 4 5 6 7 8 9 10

For information regarding CPSIA on this printed material call: 1-888-854-6284
and provide reference #0616-110618

Contents

Curriculum Sequence 4

Application and Enrichment Topics 5

How To Use 6

Lesson 01 **Negative Numbers, Addition** 11

Lesson 02 **Negative Numbers, Subtraction** 13

Lesson 03 **Negative Numbers, Multiplication** 15

Lesson 04 **Negative Numbers, Division** 17

Lesson 05 **Exponents** 19

Lesson 06 **Place Value** 21

Lesson 07 **Negative Numbers with Exponents** 25

Lesson 08 **Roots and Radicals** 27

Lesson 09 **Solve for an Unknown** 31

Lesson 10 **Pythagorean Theorem** 37

Lesson 11 **Associative and Commutative Properties** 41

Lesson 12 **Distributive Property** 45

Lesson 13 **Solve for an Unknown** 47

Lesson 14 **Solve for an Unknown** 51

Lesson 15 **Surface Area of Solids** 55

Lesson 16 **Convert Celsius to Fahrenheit** 59

Lesson 17 **Convert Fahrenheit to Celsius** 63

Lesson 18 **Absolute Value** 65

Lesson 19 **Ratio and Proportion** 67

Lesson 20 **Similar Polygons** 71

Lesson 21 **Least Common Multiple** 75

Lesson 22 **Greatest Common Factor** 79

Lesson 23 **Polynomials, Addition** 81

Lesson 24 **Volume of a Cylinder** 85

Lesson 25 **Polynomials, Multiplication** 89

Lesson 26 **Adding and Subtracting Time** 93

Lesson 27 **Volume of a Pyramid and a Cone** 95

Lesson 28 **Military Time, Addition and Subtraction** 97

Lesson 29 **Measurement, Addition and Subtraction** 99

Lesson 30 **Irrational Numbers** 103

Student Solutions 107

Application and Enrichment Solutions 215

Test Solutions 229

Symbols and Tables 255

Glossary 257

Secondary Levels Master Index 263

Pre-Algebra Index 265

Curriculum Sequence

$\int$	**Calculus**
$\cos$	**PreCalculus** with Trigonometry
xy	**Algebra 2**
Δ	**Geometry**
x^2	**Algebra 1**
X	**Pre-Algebra**
ζ	**Zeta** Decimals and Percents
ε	**Epsilon** Fractions
δ	**Delta** Division
γ	**Gamma** Multiplication
β	**Beta** Multiple-Digit Addition and Subtraction
α	**Alpha** Single-Digit Addition and Subtraction
P	**Primer** Introducing Math

Math-U-See is a complete, K-12 math curriculum that uses manipulatives to illustrate and teach math concepts. We strive toward "Building Understanding" by using a mastery-based approach suitable for all levels and learning preferences. While each book concentrates on a specific theme, other math topics are introduced where appropriate. Subsequent books continuously review and integrate topics and concepts presented in previous levels.

Where to Start
Because Math-U-See is mastery-based, students may start at any level. We use the Greek alphabet to show the sequence of concepts taught rather than the grade level. Go to MathUSee.com for more placement help.

Each level builds on previously learned skills to prepare a solid foundation so the student is then ready to apply these concepts to algebra and other upper-level courses.

Major concepts and skills for Pre-Algebra:

- Operations with integers
- Exponents, roots, and radicals
- Pythagorean theorem
- Solving for an unknown
- Commutative, Associative, and Distributive Properties
- Volume and surface area of solids
- Ratio and proportion
- Addition and multiplication of polynomials

Additional concepts and skills for Pre-Algebra:

- Temperature conversions (Fahrenheit and Celsius)
- Least common multiple and greatest common factor
- Military time
- Adding and subtracting measurements

Find more information and products at MathUSee.com

APPLICATION AND ENRICHMENT TOPICS

Here are the topics for the special challenge lessons included in the student workbook. You will find one Application and Enrichment page after the last systematic review page for each regular lesson. Instructions for the Application and Enrichment pages are included in the student text.

LESSON	TOPIC
01	Challenge word problems
02	Graphs; word problems
03	Changing dimensions of a rectangle; challenge area problems
04	Challenge word problems; diameter/area relationship of circle
05	Exponential changes; area of trapezoid
06	Patterns and charts
07	Sequences; Fibonacci numbers; three–dimensional patterns
08	Word problem skills; solving word problems with variables only
09	Performing basic operations with variables only
10	Word problems with Pythagorean theorem; more operations with variables only
11	More on finding patterns
12	Rectangles—same area/different perimeter and same perimeter/different area
13	Solving equations with variables only; estimating irrational square roots
14	Using equations to solve word problems
15	Practical word problems on finding surface area and purchasing materials
16	Exploring five possible regular polyhedrons
17	Exploring patterns in Pascal's triangle
18	Using algebra to solve word problems
19	Word problems involving ratio and proportion; inverse ratio problems
20	Proportion with maps
21	Prime numbers in Pascal's triangle
22	Advanced divisibility rules
23	Using polynomial addition to find perimeter
24	Practical applications using linear, square, and cubic measures in different units
25	Using polynomial multiplication to find area
26	Percent of change
27	Comparison of results using median and mean
28	Percent of error
29	Metric measure
30	Categories of numbers

HOW TO USE

Introduction

Welcome to *Pre-Algebra*. I believe you will have a positive experience with the unique Math-U-See approach to teaching math. These first few pages explain the essence of this methodology, which has worked for thousands of students and teachers. I hope you will take a few minutes and read through these steps carefully.

I am assuming your student has a thorough grasp of the four basic operations (addition, subtraction, multiplication, and division), along with a mastery of fractions, decimals, and percents. If not, you may wish to spend some time reviewing these concepts before beginning *Pre-Algebra*.

If you are using the program properly and still need additional help, visit us at MathUSee.com or call us at 888-854-6284. –**Steve Demme**

The Goal of Math-U-See

The underlying assumption or premise of Math-U-See is that the reason we study math is to apply math in everyday situations. Our goal is to help produce confident problem solvers who enjoy the study of math. These are students who learn their math facts, rules, and formulas and are able to use this knowledge to solve word problems and real-life applications. Therefore, the study of math is much more than simply committing to memory a list of facts. It includes memorization, but it also encompasses learning the underlying concepts of math that are critical to successful problem solving.

Support and Resources

Math-U-See has a number of resources to help you in the educational process. Many of our customer service representatives have been with us for over 10 years. They are able to answer your questions, help you place your student in the appropriate level, and provide knowledgeable support throughout the school year.

Visit MathUSee.com to use our many online resources, find out when we will be in your neighborhood, and connect with us on social media.

More than Memorization

Many people confuse memorization with understanding. Once while I was teaching seven junior high students, I asked how many pieces they would each receive if there were fourteen pieces. The students' response was, "What do we do: add, subtract, multiply, or divide?" Knowing how to divide is important; understanding when to divide is equally important.

THE SUGGESTED 4-STEP MATH-U-SEE APPROACH

In order to train students to be confident problem solvers, here are the four steps that I suggest you use to get the most from the Math·U·See curriculum.

Step 1. Prepare for the lesson
Step 2. Present the new topic
Step 3. Practice for mastery
Step 4. Progress after mastery

Step 1. Prepare for the lesson

Watch the DVD to learn the new concept and see how to demonstrate this concept with the manipulatives when applicable. Study the written explanations and examples in the instruction manual. Many students watch the DVD along with their instructor. Older students in the secondary level who have taken responsibility to study math themselves will do well to watch the DVD and read through the instruction manual.

Step 2. Present the new topic

Present the new concept to your student. Have the student watch the DVD with you, if you think it would be helpful. Older students may watch the DVD on their own.

a. **Build:** Demonstrate how to use the manipulatives to solve the problem, if applicable. As students mature, they learn to think abstractly. However, we will still be using the manipulatives for much of Pre-Algebra.

b. **Write:** Record the step-by-step solutions on paper as you work them through with manipulatives.

c. **Say:** Explain the *why* and *what* of math as you build and write.

Do as many problems as you feel are necessary until the student is comfortable with the new material. One of the joys of teaching is hearing a student say, *"Now I get it!"* or *"Now I see it!"*

Step 3. Practice for mastery

Using the examples and the lesson practice problems from the student text, have the students practice the new concept until they understand it. It is one thing for students to watch someone else do a problem; it is quite another to do the same problem themselves. Do enough examples together until they can do them without assistance.

Do as many of the lesson practice pages as necessary (not all pages may be needed) until the students remember the new material and gain understanding. Give special attention to the word problems, which are designed to apply the concept being taught in the lesson.

Step 4. Progress after mastery

Once mastery of the new concept is demonstrated, proceed to the systematic review pages for that lesson. Mastery can be demonstrated by having each student teach the new material back to you. The goal is not to fill in worksheets but to be able to teach back what has been learned.

The systematic review worksheets review the new material as well as provide practice of the math concepts previously studied. Remediate missed problems as they arise to ensure continued mastery.

After the last systematic review page in each lesson, you will find an Application and Enrichment page. These are optional but highly recommended for students who will be taking advanced math or science classes. These challenging lessons are a good way for all students to hone their problem-solving skills.

Proceed to the lesson tests. These were designed to be an assessment tool to help determine mastery, but they may also be used as extra worksheets.

Your students will be ready for the next lesson only after demonstrating mastery of the new concept and continued mastery of concepts found in the systematic review worksheets.

Confucius was reputed to have said, "Tell me, I forget; show me, I understand; let me do it, I will remember." To this we add, **"Let me teach it, and I will have achieved mastery!"**

Length of a Lesson

How long should a lesson take? This will vary from student to student and from topic to topic. You may spend a day on a new topic, or you may spend several days. There are so many factors that influence this process that it is impossible to predict the length of time from one lesson to another. I have spent three days on a lesson, and I have also invested three weeks in a lesson. This occurred in the same book with the same student. If you move from lesson to lesson too quickly without the student demonstrating mastery, he will become overwhelmed and discouraged as he is exposed to more new material without having learned the previous topics. However, if you move too slowly, your student may become bored and lose interest in math. I believe that, as you regularly spend time working along with your student, you will sense when is the right time to take the lesson test and progress through the book.

By following the four steps outlined above, you will have a much greater opportunity to succeed. Math must be taught sequentially, as it builds line upon line and precept upon precept on previously-learned material. I hope you will try this methodology and move at your student's pace. As you do, I think you will be helping to create a confident problem solver who enjoys the study of math.

LESSON 1

Negative Numbers, Addition

A *negative number* is a number that is less than zero. It is shown by turning the blocks upside down so the cavities are showing. This way you can see that you are "in the hole," or that you owe a certain amount. I have found success by introducing this subject with positive numbers representing "having" and negative numbers representing "owing," as shown in Figure 1.

Figure 1

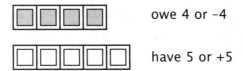

owe 4 or −4

have 5 or +5

Up to this point, all numbers were understood to be positive. Adding 10 + 2 is the same as $(+10) + (+2) = (+12)$. When working with negative numbers, it is a good idea to use parentheses to keep the signs distinct and separate. Even with subtraction, the student was always working with positive numbers. Finding 10 − 2 is the same as $(+10) − (+2) = (+8)$.

When you add two positive numbers, the sum will be a positive number.

Example 1
Solve: $(+5) + (+3)$

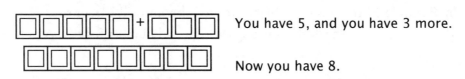

You have 5, and you have 3 more.

Now you have 8.

When you add two negative numbers, the sum will be a negative number.

Example 2
Solve: (–4) + (–2)

 You owe 4, and you owe 2 more.
Now you owe 6.

When you add two numbers with different signs, one positive and one negative, the sign of the answer will be the same as the greater number. The answer will be the difference between the numbers. Study Examples 3 and 4. The blocks really make this clear. Place the smaller block on top of the larger block. You can see that the difference is a positive (raised) two. The way you read it—saying "have" and "owe"—helps to clarify this as well.

Example 3
Solve: (–3) + (+5)

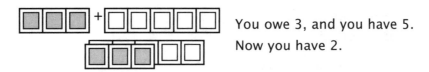 You owe 3, and you have 5.
Now you have 2.

Example 4
Solve: (+4) + (–7)

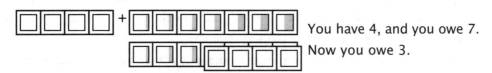 You have 4, and you owe 7.
Now you owe 3.

In terms of a rule, when adding numbers with different signs, take the sign from the greater number and then subtract the numbers to find the difference. In Example 3, the answer will be positive because five is greater than three. Then subtract 5 – 3 to find 2. Putting the sign and number together yields +2.

Negative Numbers, Subtraction

In lesson 1 you added numbers with the same signs and with different signs. By switching the signs between the number and the operation, you can make subtraction into a familiar equation. Consider $(+5) + (-3)$, which you have already solved. This is the same as $(+5) - (+3)$. If you take away the positive signs, which were always just assumed, then this is a typical subtraction problem. The answer in both cases is $(+2)$, or 2. Adding a negative number is the same as subtracting a positive number.

Example 1

$$(+5) + (-3) = (+5) - (+3) = +2 \text{ or } 2$$

A negative number may also be thought of as the opposite of a positive number. When the negative sign is placed before a number, it makes the sign of that number the opposite of what it was before. Think of flipping the block upside down every time a negative sign is introduced. To illustrate this, take a three-block and hold it with the solid, or positive side, facing up. It should look like Figure 1.

Figure 1

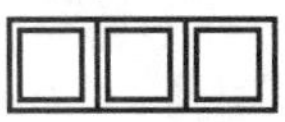

+3

Flip it over to show the opposite of that number. Now it is hollow and has the opposite sign (Figure 2).

Figure 2

 (–3)

Flip it over again to show the opposite of that number; now it is solid again and positive (Figure 3).

Figure 3

–(–3) = +3

The opposite of this is shown with another negative sign (Figure 4).

Figure 4

–[–(–3)] = –3

Because of this concept of opposites, subtracting a negative number is the same as adding a positive number. The negative sign in the operation makes the negative sign on the number positive. See Examples 2 and 3. Notice how it is easier to solve a problem if the operation sign between the numbers is positive.

Example 2

(+7) – (–3) = (+7) + (+3) = +10, or 10

Example 3

(–9) – (–5) = (–9) + (+5) = –4

Example 4

(+10) – (+3) = (+10) + (–3) = +7 or 7

Example 5

(–6) – (–5) = (–6) + (+5) = –1

Negative Numbers, Multiplication

Multiplication is fast adding of the same number. The problem (3) x (–2) is a way of writing (–2) counted three times, or (–2) + (–2) + (–2), or (–6). Think of it as borrowing $2 from someone for three days in a row. After three days, you will owe $6.

Example 1
(–6)(+3) = (–18)

Example 2
(+7)(–6) = (–42)

Once you understand multiplying a negative number by a positive number, consider what you would have if you were multiplying a negative number by a negative number. It will be the opposite of what you just learned, so you are back to being positive. There are only two options for a number: either it is negative, or it is positive.

Since you first learned about multiplication, you always multiplied positive numbers by positive numbers. The following examples will help you understand how to multiply a negative number times a negative number.

Example 3
(+3)(+7) = (+21)

Example 4
(–3)(+7) = (–21)

Example 5

$(+3)(-7) = (-21)$

The only option remaining is shown in Example 6.

Example 6

$(-3)(-7) = (+21)$

Think of negative anything as the opposite of what it was. When multiplying two negative numbers, the product is a positive number. Here are a several more ways of thinking of this.

In the English language, a double negative is a positive. I used to ask students if they were going to the local town fair. If they replied that they weren't not going. I would respond by saying that I would see them there. In response to their puzzled expressions I would explain that if they were "not, not going," then they were going.

Another way to think of it is using the idea of opposites as discussed in the previous lesson. Recall that $-(-21)$ is the same as $+21$. Using brackets for clarification, you can write $(-3)(-7)$ as $-[(3)(-7)]$ by moving the negative sign in front of the 3 outside of the brackets. After multiplying $(3)(-7)$, you have (-21) inside the brackets. Then, putting it all together, you have $-[-21]$, which is $(+21)$.

Example 7

$(-12)(-5) = (+60)$

Have you observed the pattern that, if you have two negative signs, you have a positive? The same holds for four negative signs. Whenever you have an even number of negative signs, the answer is positive, and an odd number of negative signs produces a negative number.

Figure 1

$$(-12) = (-12)$$
$$-(-12) = (+12)$$
$$-[-(-12)] = (-12)$$
$$-\{-[-(-12)]\} = (+12)$$

LESSON 4

Negative Numbers, Division
Integers and Number Line

Division is the opposite, or inverse, of multiplication. Think of each of these problems as a multiplication problem. The multiplication problem $(+3)(-7) = (-21)$ may be rewritten as a division problem in one of three ways.

Figure 1
$(+3)(-7) = (-21)$ may be renamed as:

$$+3 \overline{\smash{\big)}{-21}} \;=\; -7 \quad \text{or} \quad (-21) \div (+3) = (-7) \quad \text{or} \quad \frac{-21}{+3} = -7$$

There are four possibilities.

Example 1

$$\frac{-21}{+3} = -7 \qquad\qquad +3\,\overline{\smash{\big)}{-21}} = -7 \qquad \text{Recall that } (+3) \text{ times } (-7) \text{ equals } (-21).$$

Example 2

$$\frac{-21}{-3} = 7 \qquad\qquad -3\,\overline{\smash{\big)}{-21}} = +7 \qquad \text{Recall that } (-3) \text{ times } (+7) \text{ equals } (-21).$$

Example 3

$$\frac{+21}{-3} = -7 \qquad\qquad -3\,\overline{\smash{\big)}{+21}} = -7 \qquad \text{Recall that } (-3) \text{ times } (-7) \text{ equals } (+21).$$

Example 4

$$\frac{+21}{+3} = +7 \qquad +3 \overline{\smash{\big)}\,+21}^{\,+7} \qquad \text{Recall that (+3) times (+7) equals (+21).}$$

You do not have to come up with new rules for division if you thoroughly understand multiplication with negative numbers.

Example 5

$$\frac{-42}{-7} = +6 \qquad -7 \overline{\smash{\big)}\,-42}^{\,+6} \qquad \text{(+6) times (−7) equals (−42).}$$

Example 6

$$\frac{-75}{-15} = +5 \qquad -15 \overline{\smash{\big)}\,-75}^{\,+5} \qquad \text{(+5) times (−15) equals (−75).}$$

Integers

Whole numbers begin with zero and continue by one. Zero is not considered a positive number, but all of the other whole numbers are positive and are shown as 0, 1, 2, 3 . . . If you include the negative numbers (−1, −2, −3 . . .) with the whole numbers, the resulting set is called *integers*. They are shown on the *number line* in Figure 1. The arrows indicate that the number line extends infinitely in both directions.

Figure 1

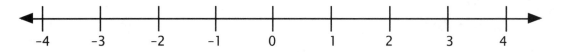

LESSON 5

Exponents

Multiplication is fast adding of the same number. To take this a step further, fast multiplying of the same number is raising to a *power*, or an *exponent*. You can think of exponents in several ways. Picture a square that is 10 over and 10 up. It is 10 two ways (over and up); it is also a square, or "10 squared." By definition, the two stands for how many times 10 is multiplied by itself. Ten is used as a factor twice. Another way to state it is "10 to the two power" or "10 to the power of two." Any number may be expressed as a square, such as two squared or three squared. Figure 1 shows the many ways to express the same concept.

Figure 1

5 over and 5 up	3 over and 3 up
5 used as a factor two times	3 used as a factor two times
5 to the second power	3 to the second power
5 squared	3 squared
5 to the two power	3 to the two power
5 raised to the power of two	3 raised to the power of two

$$5^2 = 5 \cdot 5 = 25 \qquad\qquad 3^2 = 3 \cdot 3 = 9$$

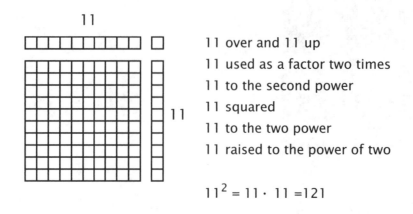

11

11 over and 11 up

11 used as a factor two times

11 to the second power

11 squared

11 to the two power

11 raised to the power of two

$$11^2 = 11 \cdot 11 = 121$$

While it is difficult to show greater exponents with the manipulatives, you can have numbers raised to other powers besides two. For example, $2 \times 2 \times 2 = 2^3$, which is 8, and $3 \cdot 3 \cdot 3 \cdot 3 = 3^4$. It helps to say, "Three times itself four times is the same as 3^4, which is 81."

Example 1
$$7^2 = 7 \cdot 7 = 49$$

Example 2
$$10 \times 10 \times 10 \times 10 \times 10 = 10^5 = 100,000$$

When a number is raised to a power of three, such as 4^3, it is often read as four to the third power, or four *cubed*. You can make a figure that is a cube with dimensions 4 by 4 by 4.

Figure 2

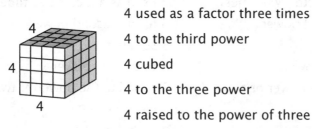

4 over and 4 up and 4 high

4 used as a factor three times

4 to the third power

4 cubed

4 to the three power

4 raised to the power of three

$$4^3 = 4 \cdot 4 \cdot 4 = 64$$

Place Value
with Expanded and Exponential Notation

In the base ten, or decimal, system, all numbers are composed of digits, 0 to 9, and place values, such as units, tens, hundreds, and so on. There are many ways to decompose a number into those components. We will explore three of them. The first, which should be review, is place value notation. The number 156 can be rewitten as $100 + 50 + 6$. The second is expanded notation: $156 = 1 \times 100 + 5 \times 10 + 6 \times 1$. This shows the individual digits times each place value. The third way is exponential notation. Using what you know about exponents, you can express all the place values in the expanded notation as 10 to a power.

Figure 1

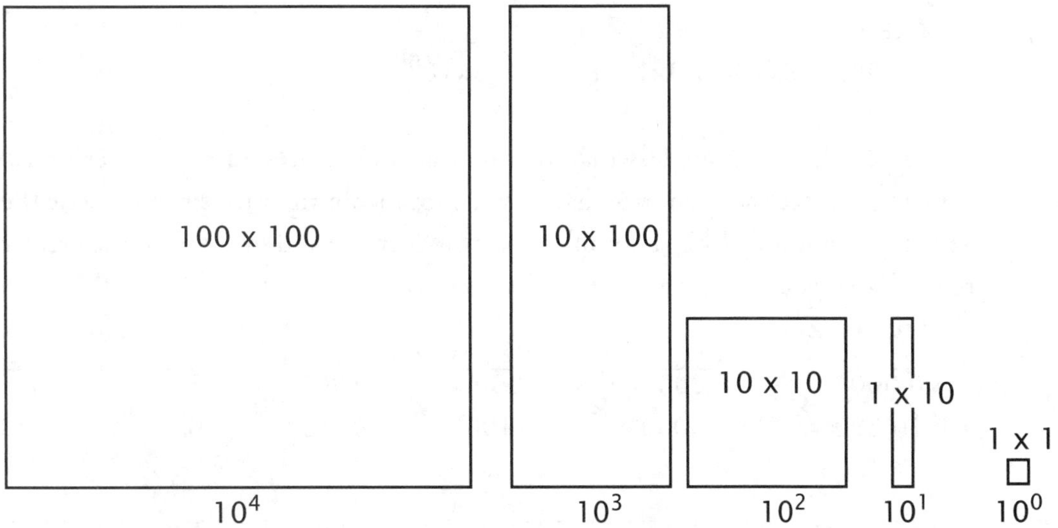

100 x 100		10 x 100		10 x 10	1 x 10	1 x 1
10^4		10^3		10^2	10^1	10^0

_____ ,	_____	_____	_____ ,	_____	_____	_____
1,000,000	100,000	10,000	1,000	100	10	1
10^6	10^5	10^4	10^3	10^2	10^1	10^0

Figure 1 on the previous page, shows several of the place values, beginning with the units place and continuing through the 10,000 place. The first three places are easily shown with the blocks, but 1,000 poses a challenge. One way to make it is with a cube that is 10 by 10 by 10. However, if you follow this progression, it is impossible to show 10,000, so leave 1,000 in two dimensions as 10 by 100. Ten thousand is 100 by 100.

Notice the shapes as you move from right to left: square, rectangle, square, rectangle, square . . . Place values with even exponents form squares, and place values with odd exponents form rectangles.

Notice how the place value corresponds to the powers of 10. (You will learn how $10^0 = 1$ in *Algebra 1*.) Also notice how the exponent corresponds to the number of zeros in the place value. Thus, 10^3 is 1,000, which has three zeros, and 10^1 is 10, which has one zero. In Examples 1–3, the numbers are rewritten first in expanded notation and then in exponential notation.

Example 1
$245 = 2\times100 + 4\times10 + 5\times1 = 2\times10^2 + 4\times10^1 + 5\times10^0$

Example 2
$1,759 = 1\times1,000 + 7\times100 + 5\times10 + 9\times1$
$= 1\times10^3 + 7\times10^2 + 5\times10^1 + 9\times10^0$

Example 3
$803 = 8\times100 + 3\times1 = 8\times10^2 + 3\times10^0$

Place values that are less than one may also be expressed as a power of 10. Looking at place values, notice that, as you move from right to left to increase the values, you multiply by a factor of 10. As you decrease and move from the greater to the lesser values, or from left to right, you divide by a factor of 10.

Figure 2

$\overline{100,000}$		$\overline{10,000}$		$\overline{1,000}$		$\overline{100}$		$\overline{10}$		$\overline{1}$
$10\cdot10,000$	↙	$10\cdot1,000$	↙	$10\cdot100$	↙	$10\cdot10$	↙	$10\cdot1$	↙	

$\overline{100,000}$		$\overline{10,000}$		$\overline{1,000}$		$\overline{100}$		$\overline{10}$		$\overline{1}$
	↘	$100,000\div10$	↘	$10,000\div10$	↘	$1,000\div10$	↘	$100\div10$	↘	$10\div10$

Notice that the place values to the right of the decimal point are tenths, hundredths, and thousandths. A tenth is written as 1 over 10 to the one power; a hundredth is 1 over 100, or 1 over 10 squared, or 1 over 10 to the second power; and a thousandth is 1 over 10 to the third power.

Figure 3

$$\overline{100} \quad \overline{10} \quad \overline{1} \cdot \overline{\frac{1}{10}} \quad \overline{\frac{1}{100}} \quad \overline{\frac{1}{1,000}}$$

$$10^2 \quad 10^1 \quad 10^0 \quad \frac{1}{10^1} \quad \frac{1}{10^2} \quad \frac{1}{10^3}$$

Example 4

Express 34.762 in expanded and exponential notation.

$$34.762 = 3 \times 10 + 4 \times 1 + 7 \times \frac{1}{10} + 6 \times \frac{1}{100} + 2 \times \frac{1}{1,000}$$

$$34.762 = 3 \times 10^1 + 4 \times 10^0 + 7 \times \frac{1}{10^1} + 6 \times \frac{1}{10^2} + 2 \times \frac{1}{10^3}$$

One of the best ways to understand decimals is by using money. Think of dollars as units, dimes as tenths, and pennies as hundredths. The amount $2.39 would be read as "two dollars and thirty-nine cents."

Figure 4

$2.39 = 2 dollars and 3 dimes and 9 pennies =
2.00 + 0.30 + 0.09 = $2.39

In terms of dollars, 1/10 of a dollar is a dime, and 1/100 of a dollar is a penny.

$$\$2.39 = 2 \times 1 + 3 \times \frac{1}{10} + 9 \times \frac{1}{100}$$

Negative Numbers with Exponents

You know that raising a number to a power, or an exponent, is fast multiplying of the same number. The expression 3 x 3 may be rewritten using a superscript: 3^2. This means "three used as a factor two times," or 3 x 3. The product is 9.

You have also learned that a negative number times another negative number produces a positive product. This means that (-3) x (-3) = +9. Thus, $(-3)^2$, which is the same as (-3) x (-3), is also equal to +9.

Example 1
$(-5)^2$ is (-5) used as a factor two times, or (-5) x (-5), which is +25.

Example 2
$(-4)^3$ is (-4)(-4)(-4) = -64

Notice that using the parentheses in Example 1 make it clear that (-5) is being squared. The parentheses show that the number 5 *as well as the minus sign* is being squared. This is very important because the next part of this lesson deals with negative numbers raised to a power where the sign is not inside the parentheses. In other words, -5^2 is not the same as $(-5)^2$. In -5^2 only the five is squared, and then the minus sign is taken into consideration. Writing it out with parentheses, -5^2 is the same as $(-)(5)(5) = (-)(25) = -25$.

Example 3
-3^2 is (3) used as a factor two times, or -(3 x 3), which is -(9) or -9.

Example 4

$-(10)^2$ is (10) used as a factor two times, or (10) x (10), which is +100. Then add the minus sign, giving a product of –100.

Example 5

$-(7)^3$ is $-(7)(7)(7) = -(343) = -343$

Another combination is $-(-2)^4$. Written out, this is $(-)(-2)(-2)(-2)(-2) = (-)(16) = -16$. The (-2) is raised to the power of 4; then the sign is added.

Example 6

$-(-1)^3 = -(-1)(-1)(-1) = -(-1) = +1$ or 1

Example 7

$-(-1)^4 = -(-1)(-1)(-1)(-1) = -(+1) = -1$

Example 8

$$-\left(\frac{3}{7}\right)^2 = -\left(\frac{9}{49}\right) = -\frac{9}{49}$$

Example 9

$$-\left(\frac{-2}{9}\right)^2 = -\left(\frac{4}{81}\right) = -\frac{4}{81}$$

LESSON 8

Roots and Radicals

The opposite of finding a square is finding the *roots,* or like factors. Students may find the word *roots* confusing at first because the concept is new and seemingly unrelated. You may prefer to call these the factors of a square, or the square's like factor. When considering 13 squared, build a square with the sides or factors both the same (in this case, 13). The product is 169. The inverse is to start with 169 and find the factors of this square. Build a square and see how long the sides are (in this case, 13). Look at Example 1. The symbol for the square root is $\sqrt{}$.

When you read $\sqrt{169}$, say, "What are the like factors of a square with a product, or area, of 169?" or "What is the square root of 169?" or "What is the square's like factors?"

Example 1

Find the square root, or the square's like factors, with a product of 169.

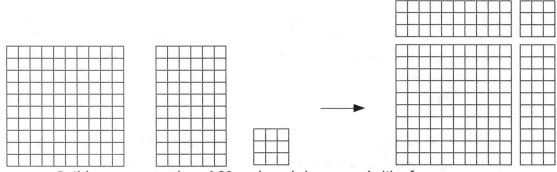

Build a square to show 169 and read the square's like factors.

$\sqrt{169} = 13$

Example 2

Find the square root, or the square's like factors, with a product of 144.

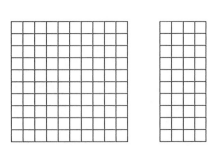 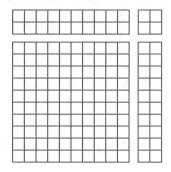

Build a square to show 144 and read the square's like factors.

$$\sqrt{144} = 12$$

Example 3

Find the square root, or the square's like factors, with a product of 225.

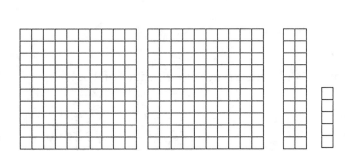 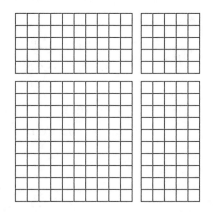

Build a square to show 225 and read the square's like factors.

$$\sqrt{225} = 15$$

Notice that the like factor could also be (-15), since $(-15)^2 = +225$. Thus, when finding the square root of a positive number, the answer could be positive or negative. To indicate the negative factor, you must show this by putting a negative sign in front of the radical.

Example 4

$$\sqrt{16} = 4, \text{ and } -\sqrt{16} = -4$$

Here are a few additional examples.

Example 5
Find: $\sqrt{10^2}$

$10^2 = 100$, and $\sqrt{100} = 10$, so $\sqrt{10^2} = 10$

Example 6
Find: $-\sqrt{6^2}$

$6^2 = 36$, and $\sqrt{36} = 6$, so $-\sqrt{6^2} = -6$

Example 7
Find: $\sqrt{X^2}$

$(X)^2 = X^2$, and $\sqrt{X^2} = X$, so $\sqrt{X^2} = X$

You may also see $\pm\sqrt{4}$, which is read, "plus or minus the square root of four." The answer is ± 2.

LESSON 9

Solve for an Unknown
with Additive Inverse

In previous books, you have solved simple algebraic equations. Now you'll be working with more complex equations that still require a defined strategy. There are two key components essential to developing your strategy in this lesson.

1. Adding the same amount to both sides of an equation
2. Making zero with the additive inverse

Adding the same amount

$$3 = 3$$

Adding 2 to both sides of the equation:

$$3 + 2 = 3 + 2$$

$$5 = 5$$

As long as you add the same amount to both sides of an equation, the integrity of the equation remains intact.

The goal is eventually to subtract the same amount from both sides; however, you actually begin by doing the inverse: adding the same amount to both sides. Remember from what you know about negative numbers that adding a negative number is the same as subtracting a positive number.

When using the manipulatives, the hollow sides of the blocks represent negative numbers, which will be shown in the illustrations with gray shading. The positive numbers are shown with no shading. The blue flat pieces represent (+X) and will be unshaded. The gray flat pieces stand for (–X) and are shaded the same as the negative numbers. This may be the first time you have seen something concrete representing an unknown or a variable. For students who learn visually, this is a great asset. The reason the value is unknown is because it is impossible to count any delineations on the bar; it is smooth. You don't know what number it represents because you cannot count the ridges. Therefore, you say it is unknown and give it a letter. This is algebra: when you don't know the number, represent it with a letter. Avoid letters that are similar to numbers, like a lowercase *l*, which looks like one, or an *O*, which could be confused with zero.

Figure 1

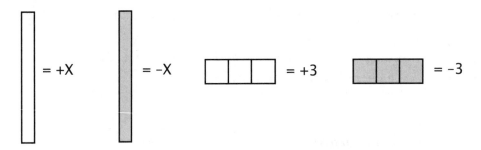

In Example 1 on the next page, Xs, or unknowns, are added to the equation. You know that $1 \cdot X$ is the same as X, since anything times one is equal to itself. If the student is more comfortable with 1X than X, that is fine. Consider the equations on the next page to help with understanding unknowns.

If a student has trouble with this concept, here are some tips. Instead of referring to the unknown as X, call it "something." Thus, 2X is "two somethings," and 2X – X is "two somethings minus one something." Alternately, temporarily replace the X with a number, such as 10. Then 2X is $2 \cdot 10$ or 20, and X is 10. This means that 2X – X could be 20 – 10 = 10; then, switching back, 2X – X = X.

Example 1

Begin with the given equation.

$$X = 3$$

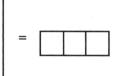

Add X to both sides.

$$2X = X + 3$$

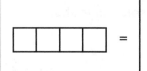 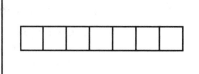

Add 4 to both sides.

$$2X + 4 = X + 7$$

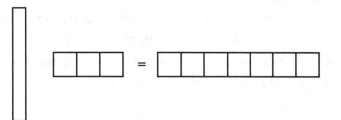

Example 2

$$X + 3 = 7$$

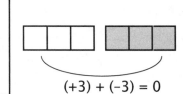

$$X + 3 - 3 = 7 - 3$$
$$\text{or}$$
$$X + 3 = 7$$
$$\underline{-\ 3 \quad -\ 3}$$
$$X = 4$$

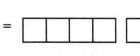 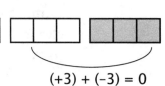

$(+3) + (-3) = 0$ $(+3) + (-3) = 0$

You can add (−3) to both sides or subtract (+3) from both sides. It will have the same effect, since you can add the same amount to both sides without changing the equation. When this is completed, the result is X = 4.

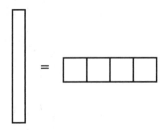

Additive inverse

Example 3

Solve for X. (The solution is shown written in two different ways.)

X − 2 = 5

$$
\begin{array}{rl}
X - 2 &= 5 \\
+2 &+2 \\
\hline
X &= 7
\end{array}
\qquad
\begin{array}{rl}
X - 2 &= 5 \\
X - 2 + 2 &= 5 + 2 \\
X &= 7
\end{array}
$$

When adding, use the **ADDitive inverse** to make zero and eliminate the negative two so that the number is on the opposite side from the unknown.

Always check your work by replacing the letter (the unknown) with the number (the solution). Remember the primary objective: to find the value of the unknown that will satisfy the equation. Placing the solution in parentheses helps clarify the objective and prevents mistakes.

Check:
$$
\begin{array}{rl}
X - 2 &= 5 \\
(7) - 2 &= 5 \\
5 &= 5
\end{array}
$$

Example 4

Solve for X.

2X = X + 12

$$
\begin{array}{rl}
2X &= X + 12 \\
-X &-X \\
\hline
X &= 12
\end{array}
\qquad
\begin{array}{rl}
2X &= X + 12 \\
2X - X &= X - X + 12 \quad \text{Subtract X from both sides.} \\
X &= 12
\end{array}
$$

Check the answer by putting it into the original equation.

$$2(12) = (12) + 12$$
$$24 = 24$$

SUMMARY

The objective of solving for an unknown is finding out what number gives both sides of the equation the same value. To solve, the *variable,* or unknown, must be on one side of the equation, and the numerical value must be on the other. Generally this is done by simplifying both sides until X equals something. You began this lesson by doing the opposite: making a simple equation more complex by adding the same amount to both sides of an equation.

Strategy 1

In Example 5, 3X is greater than 2X. If you were to subtract 2X from both sides, then you would have a positive X remaining on the left side of the equation. Then you would need to position the numbers on the right-hand side by adding the inverse of (–3), which is (+3), to both sides.

Example 5

Solve for X.

$$3X - 3 = 2X + 5$$

$$3X - 3 = 2X + 5$$
$$3X - 2X - 3 = 2X - 2X + 5 \qquad \text{Subtract 2X from both sides.}$$
$$X - 3 + 3 = 5 + 3 \qquad \text{Add 3 to both sides.}$$
$$X = 8$$

Check the answer by putting it into the original equation.

$$3(8) - 3 = 2(8) + 5$$
$$21 = 21$$

Strategy 2

If you were to add 2X to both sides, you would have a positive X remaining on the right-hand side of the equation. Then you would need to get the numbers on the left-hand side by adding the inverse of (+2), which is (−2), to both sides.

Example 6

$-2X + 8 = -X + 2$; solve for X.

$$-2X + 8 = -X + 2$$
$$-2X + 8 + 2X = -X + 2 + 2X \qquad \text{Add 2X to both sides.}$$
$$8 - 2 = X + 2 - 2 \qquad \text{Add −2 to both sides.}$$
$$6 = X$$

Check the answer by putting it into the original equation.
$$-2(6) + 8 = -(6) + 2$$
$$-4 = -4$$

Strategy 3

In this problem, the greater unknown is on the right-hand side, so begin by adding (−3A) to both sides; then add (+5) to both sides so the numerical value is opposite the unknown.

Example 7

$3A - 3 = 4A - 5$; solve for A.

$$3A - 3 = 4A - 5$$
$$3A - 3 + (-3A) = 4A - 5 + (-3A) \qquad \text{Add −3A to both sides.}$$
$$-3 + 5 = A - 5 + 5 \qquad \text{Add +5 to both sides.}$$
$$2 = A$$

Check the answer by putting it into the original equation.
$$3(2) - 3 = 4(2) - 5$$
$$3 = 3$$

Pythagorean Theorem

Pythagoras was an ancient Greek mathematician and philosopher who lived in the sixth century BC. He was credited with formulating the Pythagorean theorem, even though evidence of it was known in other ancient cultures. It is usually stated, "The square of the length of the hypotenuse of a right triangle is equal to the sum of the squares of the lengths of the other two sides." It has also been expressed as $A^2 + B^2 = C^2$.

To understand this theorem, begin by defining a right triangle and its parts. A triangle has three straight sides and three angles. A *right triangle* has one angle with a measure of 90°, called a right angle. A triangle that has a right angle is called a right triangle. The side opposite the right angle is called the **hypotenuse.** The two sides that form the right angle are called the **legs.** You remember this because *leg* starts with the letter *L*, and a capital *L* is a right angle.

Figure 1 is an example of a right triangle. The box in the corner denotes a right angle, or a 90° angle. Four right angles in the center of a circle divide the circle into four equal parts, so a circle has 360° (4 x 90° = 360°), as shown in Figure 2.

Figure 1

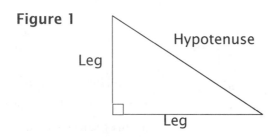

Hypotenuse

Leg

Leg

Figure 2

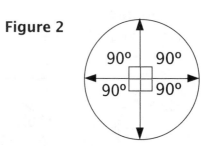

90° 90°
90° 90°

The most famous right triangle that illustrates this formula is the 3–4–5 right triangle, with the legs measuring three units and four units and the hypotenuse measuring five units. Figure 3 shows the formula worked out beside it. This right triangle is illustrated in Figure 4.

Figure 3

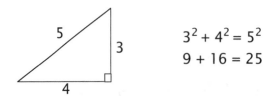

$$3^2 + 4^2 = 5^2$$
$$9 + 16 = 25$$

Figure 4

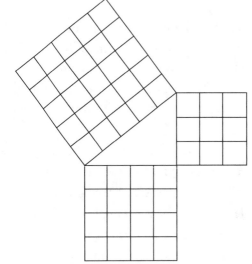

This theorem holds true for all right triangles. The converse is also true. If the length of one leg squared plus the length of the other leg squared equals the length of the hypotenuse squared, the triangle is a right triangle. Study the following examples.

Example 1

Is this a right triangle?

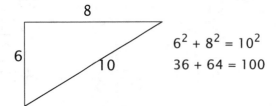

$$6^2 + 8^2 = 10^2$$
$$36 + 64 = 100$$

It is a right triangle because the Pythagorean theorem holds true.

Example 2

Find the length of the missing side X.

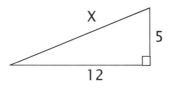

$5^2 + 12^2 = X^2$ It is a right triangle.

$25 + 144 = X^2$ Use the Pythagorean theorem.

$169 = X^2$ Find the square root

$13 = X$ of both sides.

Example 3

Find the length of the missing side L.

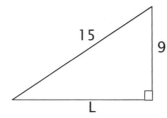

$9^2 + L^2 = 15^2$ Use the Pythagorean theorem.

$81 + L^2 = 225$ Subtract 81 from both sides.

$L^2 = 144$ Find the square root

$L = 12$ of both sides.

Example 4

Find the length of the missing side L.

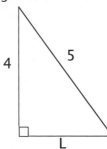

$4^2 + L^2 = 5^2$ Use the Pythagorean theorem.

$16 + L^2 = 25$ Subtract 16 from both sides.

$L^2 = 9$ Find the square root

$L = 3$ of both sides.

Associative and Commutative Properties

Commutative Property

You have been using the *Commutative Property* since first learning addition. You can think of it as the "commute"ative property. Suppose you commute to work seven miles. When you come home, the distance is still seven miles. Regardless of the direction in which you are going, it is the same distance. This is true for addition and multiplication. Changing the order, or direction, does not affect the sum (addition) or the product (multiplication). Study the following examples and notice that the Commutative Property does not apply to subtraction or division, where order and direction are critical.

Example 1

$5 + 7 = 12$ $7 + 5 = 12$ $12 = 12$

Example 2

$5 \times 7 = 35$ $7 \times 5 = 35$ $35 = 35$

Example 3

$7 - 5 = (+2)$ $5 - 7 = (-2)$ $(+2) \neq (-2)$

$\neq$ means "not equal to"

Example 4

$7 \div 5 = 7/5$ $5 \div 7 = 5/7$ $7/5 \neq 5/7$

Associative Property

You can remember the meaning of the *Associative Property* by thinking about the people with whom you associate, or are grouped. Read through the following examples and notice how the grouping by parentheses affects the answer.

Example 5

$$(3 + 5) + 7 = 3 + (5 + 7)$$
$$8 + 7 = 3 + 12$$
$$15 = 15$$

Example 6

$$(2 \times 3) \times 4 = 2 \times (3 \times 4)$$
$$6 \times 4 = 2 \times 12$$
$$24 = 24$$

Example 7

$$(10 - 7) - 4 = 10 - (7 - 4)$$
$$3 - 4 = 10 - 3$$
$$-1 \neq 7$$

Example 8

$$(8 \div 4) \div 2 = 8 \div (4 \div 2)$$
$$2 \div 2 = 8 \div 2$$
$$1 \neq 4$$

Like the Commutative Property, the Associative Property is true for addition and multiplication but not for subtraction and division.

The Commutative and Associative Properties are also used when combining like terms in algebra. Remember that to combine you must have the same kind. Therefore, you can add Xs to Xs and As to As, but you can't add, or combine Xs and As since they are not the same kind.

Example 9

Simplify.   2X + 4A + 3A + 7X

                 (4A + 3A) + (2X + 7X)

                 7A + 9X

Example 10

Simplify.   5A – 3X – 2A + 9A

                 (5A – 2A + 9A) – 3X

                 12A – 3X

In some problems you will see an A or an X without a number in the front. A number in front of an unknown is called the *coefficient*. If the coefficient is one, it does not have to appear, since $1 \cdot A$ is equal to A and $1 \cdot X$ is the same as X. In Examples 11 and 12 there are several instances of unknowns without a "visible" coefficient. When you see this, remember that the one is present, even though it is not written. You can also assume a number is positive if there is no sign.

Example 11

Simplify.   X + 4A – A + 3X – 2X

                 (X + 3X – 2X) + (4A – A)

                 2X + 3A

Example 12

Simplify.   4B + 2X + 3 – 4X

                 4B + (2X – 4X) + 3

                 4B – 2X + 3

Notice that you have three "kinds" in Example 12: Bs, Xs, and units. You can combine the Xs with each other but not the Bs with the units.

Although subtraction is not commutative, you can rewrite subtraction as addition of a negative number. This gives you more flexibility in using the Commutative Property.

Example 13

    7 – 2 = 7 + (-2) = (-2) + 7

Distributive Property

In the following picture and corresponding equation, notice how two is distributed between five and three and their relationship with the eight (5 + 3).

$$(2)(8) = (2)(5) + (2)(3)$$
$$\rightarrow \uparrow \qquad \rightarrow \uparrow \qquad \rightarrow \uparrow$$
$$(2)(5 + 3) = (2)(5) + (2)(3)$$
$$\rightarrow \uparrow \qquad \rightarrow \uparrow \qquad \rightarrow \uparrow$$

$$16 = 10 + 6$$
$$16 = 16$$

$$(2)(5 + 3) = (2)(5) + (2)(3)$$

The ***Distributive Property of Multiplication over Addition*** is very useful in multiplying problems with greater numbers, such as 4 times 12. Decompose 12 as 10 + 2; (4)(12) = (4)(10 + 2) = (4)(10) + (4)(2), which yields 40 + 8, or 48.

Example 1
Distribute: (8)(D + 3)   (8)(D + 3) = (8)(D) + (8)(3) = 8D + 24

Example 2
Distribute: (9)(9 + K)   (9)(9 + K) = (9)(9) + (9)(K) = 81 + 9K

You can also go backward using the same principle. In the equation below, what was distributed between the A + B? As you read this problem, think, "What times A + B is equal to 3A + 3B?" The answer is three.

$$(\quad)(A + B) = 3A + 3B.$$

Take it a step further. If given half of the equation, find the other half.

$$7X + 7M = (\quad)(\quad + \quad).$$

The answer is $7X + 7M = (7)(X + M)$. This is often referred to as finding the *common factor* (in this case, seven) and factoring it out of the equation. It is simply the Distributive Property in reverse.

Example 3
Find the common factor $\qquad$ $(B)(X) + (B)(6) = (B)(X + 6)$
and rewrite (B)(X) + (B)(6).

Example 4
Find the common factor $\qquad$ $(5)(D) + (5)(4) = (5)(D + 4)$
and rewrite (5)(D) + (5)(4).

Throughout the rest of the book you will be using both of these skills. Sometimes you will distribute, and at other times you will divide by a common factor.

Solve for an Unknown
with Multiplicative Inverse

What do you multiply by 3/4 to make a product of one? The answer is 4/3 because it is the reciprocal of 3/4; 3/4 x 4/3 = 1. What do you multiply by five to make one? The answer is 1/5 because 1/5 is the reciprocal of 5/1, and 1/5 x 5/1 = 1. The *multiplicative inverse* of 1/5 is 5/1, since it is the inverse used in multiplying to produce one.

The additive inverse of 3/7 is –3/7, but the multiplicative inverse of 3/7 is 7/3. Use the additive inverse to produce a sum of zero and use the multiplicative inverse to make a product of one. This concept of making one is necessary when solving an equation with a *coefficient* (a number in front of the variable or unknown), as the 3 in 3X or the 5 in 5Y.

The primary objective when solving for an unknown is to get the variable, or unknown, on one side of the equation and the number on the other side, as in X = 3, or G = 1/4, or 500 = K.

Example 1

We want to multiply both sides by 1/3, the reciprocal of 3, to get ⇒

$$3X = 12$$
$$\frac{1}{3} \times 3X = \frac{1}{3} \times 12$$
$$X = 4$$

When multiplying, use the MULTIPLicative inverse to make one so that the number is on one side and the unknown is on the other.

Example 2
Multiply
both sides by 3/2, the
reciprocal of 2/3, to get $\Rightarrow$

$$\frac{2}{3}X = 14$$

$$\frac{3}{2} \times \frac{2}{3}X = \frac{3}{2} \times 14$$

$$X = 21$$

When multiplying, use the reciprocal to make one so the number is on one side and the unknown is on the other.

Always check your work by replacing the letter (the unknown) with the number (the solution). Remember the primary objective: to find the value of X (or whatever letter represents the unknown) that will satisfy the equation. Here are checks for the previous examples. The solutions are shown in parentheses.

Example 1
check:

$$3X = 12$$
$$3(4) = 12$$
$$12 = 12$$

Example 2
check:

$$2/3 \, X = 14$$
$$2/3 \, (21) = 14$$
$$14 = 14$$

MORE TIPS

One of the first steps in solving an equation of this sort is to combine all the terms that are alike. Then, after the equation has been simplified, place all the unknowns on one side and all the numbers on the other by adding the same amount to both sides. Study the examples on the next page.

Example 3

$$7 + 3X - 6 + 4X - 2X + 15 = 7X - 2 + 8 - 4X$$

$5X + 16 = 3X + 6$	Combine like terms.
$\underline{-3X - 16 \quad -3X - 16}$	Add the same amount to both
$2X \quad = \quad -10$	sides to place the unknowns on
$X = -5$	one side and the numbers on
	the other.
	Multiply by the reciprocal.

Check: $7 + 3(-5) - 6 + 4(-5) - 2(-5) + 15 = 7(-5) - 2 + 8 - 4(-5)$

$7 - 15 - 6 - 20 + 10 + 15 = -35 - 2 + 8 + 20$

$32 - 41 = 28 - 37$

$-9 = -9$

Example 4

$$13 - 4X - 11 + 6X - X + 25 = 9X - 1 + 16 - 5X$$

$X + 27 = 4X + 15$	Combine like terms.
$\underline{-X - 15 \quad - X - 15}$	Add the same amount to both
$12 = 3X$	sides to place the unknowns on
	the right and the numbers on
$\frac{1}{3} \times 12 = \frac{1}{3} \times 3X$	the left. I placed the unknowns where they will be positive.
$4 = X$	Multiply by the reciprocal.

Check: $13 - 4(4) - 11 + 6(4) - (4) + 25 = 9(4) - 1 + 16 - 5(4)$

$13 - 16 - 11 + 24 - 4 + 25 = 36 - 1 + 16 - 20$

$62 - 31 = 52 - 21$

$31 = 31$

When solving an equation, you may finish with the unknown on either side of the equal sign. It is easier to place the unknowns where they will be positive, as in Example 4.

Solve for an Unknown
with Order of Operations

Using these three skills, you can now solve more complex equations. Before you do, however, you must first study the order of operations. The operations are addition, subtraction, multiplication, division, and exponents. The order in which you solve them makes a significant difference in many problems. You could solve the following equation in several ways. $3 + 4 \times 2 - 5 =$

$$
\begin{array}{lll}
(3 + 4) \times 2 - 5 = & 3 + (4 \times 2) - 5 = & (3 + 4) \times (2 - 5) = \\
\quad 7 \times 2 - 5 = & \quad 3 + 8 - 5 = & \quad 7 \times (-3) = -21 \\
\quad\quad 14 - 5 = 9 & \quad\quad 11 - 5 = 6 &
\end{array}
$$

Depending on which operation is performed first, there can be three different solutions to the same problem. The order you follow is very important when performing operations.

To help you remember this order, you can think about the (fictitious) parachute expert, my dear Aunt Sally: *PARA*chute (for parentheses) *EXP*ert (for exponents) *M*y (for multiplication) *D*ear (for division) *A*unt (for addition) *S*ally (for subtraction). When you learn math, you proceed up the levels from one to four:

Level 1 Counting
Level 2 Addition and subtraction
Level 3 Multiplication and division
Level 4 Exponents

When you think about the four levels of math, order of operations is in reverse order, from level four to level two. Understanding this and remembering the "Parachute Expert, My Dear Aunt Sally" will help you learn the order of operations. Try the following example.

Example 1

$(7 + 1)^2 \div 4 + 7 \times 3 - 6$	Step 1. Parentheses
$(8)^2 \div 4 + 7 \times 3 - 6$	Step 2. Exponents (level 4)
$64 \div 4 + 7 \times 3 - 6$	Step 3. Multiply and divide (level 3)
$16 + 21 - 6$	Step 4. Add and subtract (level 2)
31	

When you have two operations from the same level next to each other in the equation, proceed from left to right and compute them in the order in which they appear in the problem. For example, if you have $8 \div 2 \times 12 \div 3$, since multiplication and division are on the same level, start with 8 divided by 2; then take this answer times 12 and divide by 3 for your solution.

Example 2

$2 \times 4 + 3^2 - 9 + 17$	Step 1. Parentheses (none)
$2 \times 4 + 3^2 - 9 + 17$	Step 2. Exponents (level 4)
$2 \times 4 + 9 - 9 + 17$	Step 3. Multiply and divide (level 3)
$8 + 9 - 9 + 17$	Step 4. Add and subtract (level 2)
25	

Example 3

$5 (3 \times 4) - 18 + (5 - 7)^2 + 12$	Step 1. Parentheses
$5 (12) - 18 + (-2)^2 + 12$	Step 2. Exponents (level 4)
$5(12) - 18 + 4 + 12$	Step 3. Multiply and divide (level 3)
$60 - 18 + 4 + 12$	Step 4. Add/subtract left to right (level 2)
58	

Example 4

$5^2 - (3 + 1)^2 \times 6 \div 2 + 9$	Step 1. Parentheses
$5^2 - 4^2 \times 6 \div 2 + 9$	Step 2. Exponents (level 4)
$25 - 16 \times 6 \div 2 + 9$	Step 3. Multiply (level 3)
$25 - 96 \div 2 + 9$	Step 3. Divide (level 3)
$25 - 48 + 9$	Step 4. Add/subtract left to right (level 2)
-14	

Remember that when you have two operations from the same level in the equation, you proceed from left to right and complete them in the order in which they appear in the expression.

LESSON 15

Surface Area of Solids

Surface area is the combined area of the surfaces of a three-dimensional shape. An example is a cardboard box. The surface area of the box is the area of all the sides of the box, including the top and bottom, added together. It differs from volume, which describes in cubic units the amount of space contained inside the box. Take apart the box, and you can see how much cardboard makes up the surface area of the box.

Another good way to explain this subject is to observe the room you are in and count the walls (four); then add the ceiling and floor (two) to get the total number of flat surfaces in the room (six). In a *rectangular solid*, these flat surfaces are called *faces*. A *cube* (a rectangular solid with all the sides the same length, or a three-dimensional square) also has six faces.

In Example 2, a *pyramid* is introduced. If the base of a pyramid is a square or a rectangle, the pyramid has five faces. If the base is a triangle, the pyramid has four faces. Have the student build these shapes out of paper. Included in this lesson are templates of a cube, a rectangular solid, and two pyramids to trace. After you trace the templates, cut the paper along the outside lines; then fold on the heavy lines to form the three-dimensional shapes.

Finding surface area is finding the area of each of the surfaces, or faces, and then adding them together. You will use this skill in word problems to determine the amount of paint needed to paint a room, wallpaper needed for the walls, or siding needed for the outside of the house.

Example 1

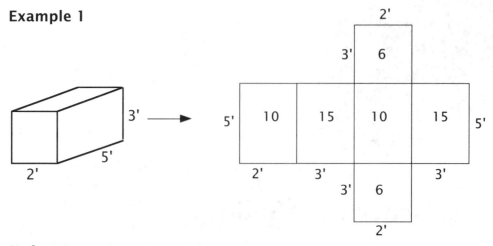

Surface Area =

10 + 10 + 15 + 15 + 6 + 6 = 62 square feet, or ft^2

Find the area of a square or rectangle by multiplying the area of the base times the height, or A = bh.

Example 2

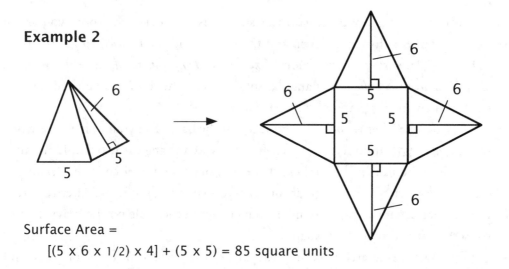

Surface Area =

[(5 x 6 x 1/2) x 4] + (5 x 5) = 85 square units

Each of the four triangles has an area of 15 square units (5 x 6 x 1/2), and the area of the base is 25 square units (5 x 5).

Find the area of a triangle by multiplying 1/2 times the area of the base times the height, or A = 1/2bh.

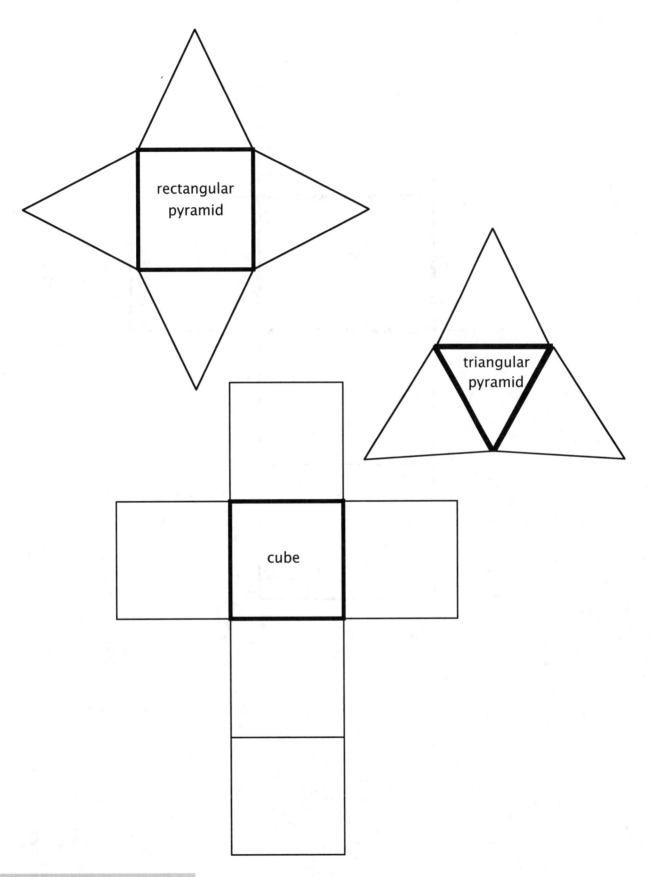

rectangular pyramid

triangular pyramid

cube

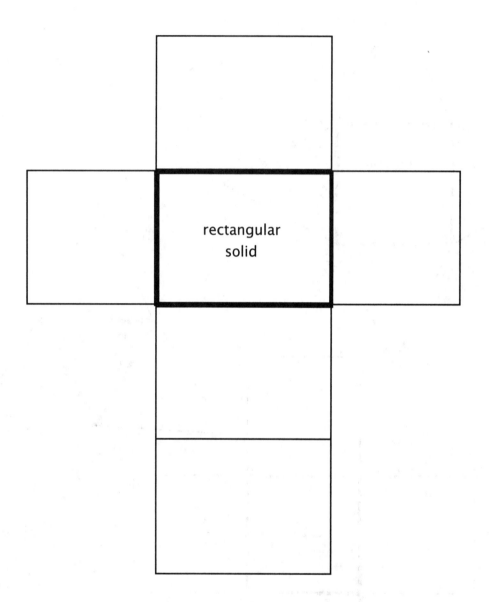

rectangular
solid

LESSON 16

Convert Celsius to Fahrenheit

To convert from *Celsius*, or Centigrade, to *Fahrenheit*, you need to know two key pieces of information: the point at which water freezes, which is 0°C or 32°F; and the fact that, for every 5° that Celsius increases or decreases, there is a corresponding 9° change in Fahrenheit.

These temperatures are important to memorize.	
Freezing point of water	0°C = 32°F
Normal body temperature	37°C = 98.6°F
Boiling point of water	100°C = 212°F

The freezing temperature you simply memorize; the other temperatures you can discern as you move from the freezing point to the boiling point. In Celsius, the change is from 0° to 100°, or a change of 100°. In Fahrenheit, the change is from 32° to 212°, or a change of 180°. The ratio is 100 to 180, which simplifies to a ratio of 5 to 9.

For every five degrees that Celsius increases or decreases, Fahrenheit changes by nine degrees. To find out the equivalent of 5°C in Fahrenheit, begin at the freezing point. This is 0°C and 32°F. Add 5° to 0°C; then add the corresponding 9° to 32°F., giving the answer of 41°F. Look at the chart in Figure 1 on the next page to see this relationship worked out.

Figure 1

0°C	32°F
5°C	41°F
10°C	50°F
15°C	59°F
20°C	68°F
100°C	212°F

Example 1

Convert 15°C to F°.

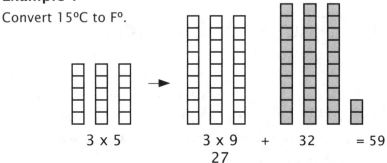

3 x 5   3 x 9   +   32   = 59
                  27

Step 1. Find out how many multiples of 5° are in 15°. 15 ÷ 5 = 3

Step 2. Multiply this number by 9°. 3 x 9° = 27°

Step 3. Add 32° to the number. 27° + 32° = 59°

Example 2

Convert 20°C to F°.

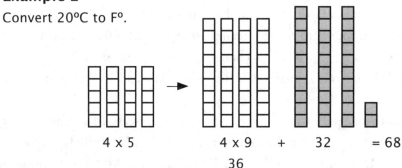

4 x 5   4 x 9   +   32   = 68
                  36

Step 1. Find out how many multiples of 5° are in 20°. 20 ÷ 5 = 4

Step 2. Multiply this number by 9°.   4 x 9° = 36°

Step 3. Add 32° to the number. 36° + 32° = 68°

Example 3
Convert 35ºC to Fº.

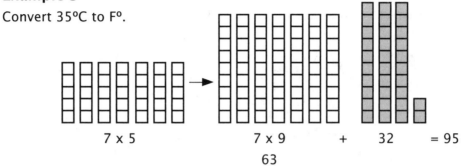

$$7 \times 5 \qquad 7 \times 9 \qquad + \qquad 32 \qquad = 95$$
$$63$$

Step 1. Find out how many multiples of 5º are in 35º. $35 \div 5 = 7$

Step 2. Multiply this number by 9º. $7 \times 9º = 63º$

Step 3. Add 32º to the number. $63º + 32º = 95º$

If you were to write out on one line what is shown in Example 3, it would look like this: $35º \div 5 \times 9 + 32º = 95º$. Recall that ÷ 5 could be written as /5, which is the same as multiplying by 1/5. Therefore, the equation could be written as $35º \times 1/5 \times 9 + 32º = 95º$, and 1/5 x 9 is 9/5. Now you have $35º \times 9/5 + 32º = 95º$. This leads to the following formula:

$$\left(Cº \times \frac{9}{5}\right) + 32º = Fº$$

Using the formula is especially helpful for problems that don't have a whole number answer, but remember the key concept, which is 5°C for every 9°F, plus the freezing temperature. This will always prove helpful for estimating.

Example 4
Convert 100ºC to Fº, using the formula.

$$\left(Cº \times \frac{9}{5}\right) + 32º = Fº$$
$$\left(100º \times \frac{9}{5}\right) + 32º = Fº$$
$$(180º) + 32º = Fº$$
$$212º = Fº$$

Example 5
Convert 37°C to F°, using the formula.

$$\left(C° \times \frac{9}{5}\right) + 32° = F°$$

$$\left(37° \times \frac{9}{5}\right) + 32° = F°$$

$$\left(66.6°\right) + 32° = F°$$

$$98.6° = F°$$

Notice that multiplying by 9/5 is the same as multiplying by 1.8. If you are more comfortable with decimals, you may use this formula: (C° x 1.8) + 32° = F°.

Example 6
(Example 4 with decimals)

$$(100° \times 1.8) + 32° = F°$$

$$(180°) + 32° = F°$$

$$212° = F°$$

Example 7
(Example 5 with decimals)

$$(37° \times 1.8) + 32° = F°$$

$$(66.6°) + 32° = F°$$

$$98.6° = F°$$

LESSON 17

Convert Fahrenheit to Celsius

In this lesson you will learn the reverse of converting Celsius to Fahrenheit. In that process, you divided by five, multiplied by nine, and added 32°. The inverse is to subtract 32°, divide by nine, and multiply by five. To help you recall the formula, remember the boiling points as 100°C and 212°F. Then think, "How could I get from 212° to 100°?" Subtract 32°, which leaves 180°. Then divide 180 by nine to get 20° and multiply this by five to get 100°.

$$\left(F° - 32°\right) \div 9 \times 5 \quad \text{or} \quad \left(C° \times \frac{9}{5}\right) + 32° = F°$$

If you want to use decimals, $5 \div 9$ is 0.56, rounded to the hundredths place. Be aware that your answer will not be as accurate if you work the problem this way, but it will be very close. $(F° - 32°) \times 0.56 = C$. Examples 2 and 4 show both ways.

Example 1
Convert 212°F to C°.

$$\left(212° - 32°\right)\frac{5}{9} = \left(180\right)\frac{5}{9} = 100°C$$

Example 2
Convert 72°F to C°.

$$\left(72° - 32°\right)\frac{5}{9} = \left(40\right)\frac{5}{9} = \frac{200}{9} = 22\frac{2}{9} \text{ or } 22.\overline{22}°C$$

Using decimals, convert 72°F to C°.

$$(F° - 32°) \times 0.56 = C$$
$$(72° - 32°) \times 0.56 = C$$
$$(40°) \times 0.56 = 22.4°C$$

Example 3
Convert 32°F to C°.

$$(32° - 32°)\frac{5}{9} = (0)\frac{5}{9} = 0°C$$

Example 4
Convert 27°F to C°.

$$(27° - 32°)\frac{5}{9} = (-5)\frac{5}{9} = -2\frac{7}{9} \text{ or } -2.\overline{77}°C$$

Using decimals, convert 27°F to C°.

$$(F° - 32°) \times 0.56 = C$$
$$(27° - 32°) \times 0.56 = C$$
$$(-5°) \times 0.56 = -2.8°C$$

Absolute Value

Another concept employed in algebra is **absolute value**. The absolute value of a number tells the *distance* from zero, while a positive or negative sign tells the *direction* from zero. If you are traveling on an interstate highway and go from mile marker 23 to mile marker 34, it is 34 – 23, or 11 miles. In the other direction, it is 23 – 34, or –11 miles. The symbol used to represent absolute value is two parallel lines, or bars, outside the number or numbers in question. You can use these bars to show that |34 – 23| is the same distance as |23 – 34|, or positive 11 miles.

When working a problem with several parts, treat the absolute value symbols just like parentheses and do the part inside them first; then make the result positive.

Example 1
$$|9| = +9, \text{ and } |-9| = +9$$

Example 2
$$|3 - 11^2| = |3 - 121| = |-118| = 118$$

Example 3
$$|4^2 + 6| = |16 + 6| = |22| = 22$$

Example 4
$$-|-28| = -(+28) = -28$$

Example 5
$$-|2^3 - 10| = -|8 - 10| = -|-2| = -(+2) = -2$$

Ratio and Proportion

Most of your study of fractions has involved fractions of one. Now your knowledge of fractions will be applied to *ratios*, which are relationships between whole numbers. Ratios might be used in a classroom. Suppose the ratio of boys to girls is 2 to 3, which can be written as 2:3 or 2/3. Because ratios can be written in fractional form, **ratio**nal numbers refer to numbers that can be represented as fractions.

Two equivalent ratios create a *proportion*. The proportion 2/3 = 8/12 is read as "Two is to three as eight is to twelve." Notice that proportions look exactly like equivalent fractions, but they represent relationships instead of parts of a whole. The proportion shown doesn't read "two thirds equals eight twelfths." It is not same amount, more pieces; rather, it means more pieces but the same ratio. In the example, if the ratio is 2 to 3, then if there are 8 boys, there must be 12 girls. More pieces, same ratio! There are three ways to look at ratios.

Example 1

$$1\frac{1}{2} \times \left(\frac{2}{3} = \frac{8}{12} \right) \times 1\frac{1}{2}$$

(1) Notice the relationship between the numerator and the denominator. If 2 x 1½ is 3 in the original ratio, then it follows that 8 x 1½ will be 12.

Example 2

$$\frac{2}{3} \overset{\times 4}{\underset{\times 4}{\rightrightarrows}} \frac{8}{12}$$

(2) Notice the relationship between numerator and numerator. You see that 4 x 2 is 8. There must be the same relationship in the denominator. It follows that 4 x 3 = 12.

Example 3

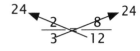 (3) Cross-multiplying the denominators by the opposite numerators yields 12 x 2 and 3 x 8, which are both 24.

You may use any of these relationships to find missing information. Choose the one that fits the given data most efficiently.

Example 4

The ratio of blue cars to red cars is 2 to 5. There are 30 red cars in the parking lot. How many cars are blue?

$$\frac{2}{5} = \frac{B}{30} \qquad \frac{2}{5} \xrightarrow[x6]{} \frac{B}{30} \qquad \frac{2}{5} \underset{x6}{\overset{x6}{\xrightarrow{\quad}}} \frac{12}{30}$$

First, set up the proportion. You can see that 30 is 6 times greater than 5. 6 x 2 = 12; the answer is 12 blue cars.

Example 5

The ratio of cloudy days to sunny days is 3 to 5. If there are 18 cloudy days, how many are sunny?

$$\frac{3}{5} = \frac{18}{S}$$

 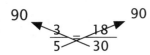

Cross-multiplying yields 5 x 18 = 90, which means that 3S = 90. Dividing 3S and 90 by 3 shows that S = 30 sunny days.

Some problems do not have whole-number answers. Here you can use your skills for solving for an unknown to find the answer.

Example 6

Solve for X.

$$\frac{2}{5} = \frac{X}{11}$$

$$\frac{11}{1} \times \frac{2}{5} = \frac{X}{\cancel{11}} \times \frac{\cancel{11}}{1}$$

$$\frac{22}{5} = X$$

Estimating first shows that 11 is a little more than 2 x 5, so X should be a little more than 2 x 2, or more than 4.

$\dfrac{X}{11}$ is the same as $\dfrac{1}{11} \cdot X$

Multiply both sides by $\dfrac{11}{1}$,

which is the reciprocal of $\dfrac{1}{11}$.

The answer is $\dfrac{22}{5}$ or $4\dfrac{2}{5}$, or 4.4.

Example 7

Solve for A.

$$\frac{3}{A} = \frac{17}{40}$$

Do you see that A ≠ 0?
You cannot divide by zero.
When you estimate, 17 is 5 times greater than 3 (almost 6 times, or 18). Dividing 5 into 40 gives 8, so A should be around 8.

$$\cancel{A} \times \frac{3}{\cancel{A}} = \frac{17}{40} \times A$$

Multiply both sides by A.

$$3 = \frac{17}{40} \times A$$

$$\frac{40}{17} \times 3 = \frac{\cancel{17}}{\cancel{40}} \times A \times \frac{\cancel{40}}{\cancel{17}}$$

$$\frac{120}{17} = A$$

$$7\frac{1}{17} = A$$

Then multiply both sides by the reciprocal, 40/17.

LESSON 20

Similar Polygons

Definition of Similar

Imagine standing beside a smaller (or larger) picture of yourself. You have the same shape as the picture, but you aren't the same size. Now think about a map of your state. It isn't the same exact size, but it is in the same proportion. Polygons may have the same proportions, or same shape, without being the same size. A polygon is a two-dimensional figure with a certain number of straight sides. Rectangles, squares, and triangles are examples of polygons.

Two polygons that are not identical in size but have the same proportions are said to be *similar* (~). Note how the squares are similar but not congruent (exactly the same). The measures of the angles in each square are 90°, but the sides have different lengths. The squares are similar (~).

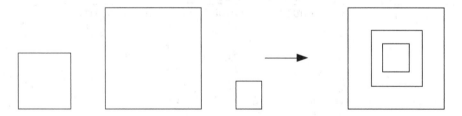

These two triangles are the same shape and have angles with the same measure, but their sides do not have the same length. If the angles had the same measure and the sides had the same length, they would be exactly the same, or *congruent*.

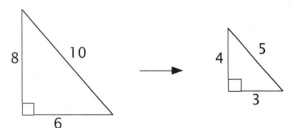

SIMILAR POLYGONS - LESSON 20   **71**

Proportion and Ratio

Looking closely at the corresponding sides, you can write a proportion. You may recall that a proportion shows two or more equivalent ratios. The ratio of the hypotenuse of the small triangle to the hypotenuse of the large triangle is 5 to 10, which may be written as a fraction and subsequently simplified to 1 to 2.

$$\frac{\text{hyp. small triangle}}{\text{hyp. large triangle}} = \frac{5}{10} = \frac{1}{2}$$

The ratio is read as "Five is to ten as one is to two."

You could also write the ratios of the short legs to the short legs and the long legs to the long legs and then put them together to make a proportion. Make sure you always move consistently in the same direction thoroughout the proportions, as in small to large or large to small.

$$\frac{\text{short leg sm. triangle}}{\text{short leg lg. triangle}} = \frac{3}{6} = \frac{1}{2} \qquad \frac{\text{long leg sm. triangle}}{\text{long leg lg. triangle}} = \frac{4}{8} = \frac{1}{2}$$

$$\frac{\text{short leg sm. triangle}}{\text{short leg lg. triangle}} = \frac{\text{hypotenuse sm. triangle}}{\text{hypotenuse lg. triangle}} = \frac{\text{long leg sm. triangle}}{\text{long leg lg. triangle}} = \frac{3}{6} = \frac{5}{10} = \frac{4}{8}$$

Example 1

Find the length of side A.

The rectangles are similar.

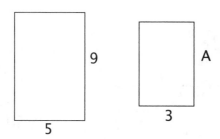

$$\frac{\text{lg. rectangle up dimension}}{\text{sm. rectangle up dimension}} = \frac{\text{lg. rectangle over dimension}}{\text{sm. rectangle over dimension}}$$

$$\frac{9}{A} = \frac{5}{3} \qquad \text{Multiply both sides by A.}$$

$$\frac{A}{1} \times \frac{9}{A} = \frac{5}{3} \times \frac{A}{1} \qquad \frac{9}{1} = \frac{5A}{3} \qquad \text{Multiply both sides by 3/5.}$$

$$\frac{3}{5} \times \frac{9}{1} = \frac{5A}{3} \times \frac{3}{5} \qquad \frac{27}{5} = \frac{A}{1} \qquad \text{A = 5 2/5, or 5.4}$$

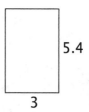

Example 2

Find the length of side H.
The triangles are similar.

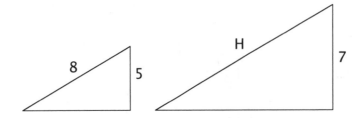

$$\frac{\text{hypotenuse large triangle}}{\text{hypotenuse small triangle}} = \frac{\text{vertical side large triangle}}{\text{vertical side small triangle}}$$

$$\frac{H}{8} = \frac{7}{5} \qquad \text{Multiply both sides by 8.}$$

$$\frac{8}{1} \times \frac{H}{8} = \frac{7}{5} \times \frac{8}{1} \qquad \frac{H}{1} = \frac{56}{5} \qquad H = 56/5 \text{ or } 11 \text{ } 1/5 \text{ or } 11.2$$

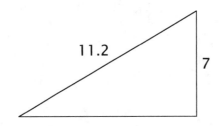

Least Common Multiple
and Prime Factorization

The term *least common multiple* is made up of three words, which you'll examine one at at time. *Multiple* is a result of multiplying, or skip counting. The multiples of six are 6, 12, 18, 24, and so on. *Common* multiple: when considering two or more numbers, these are the multiples that the numbers share, or have in common. The common multiples of four and six are shown in Figure 1.

Figure 1
4: 4, 8, <u>12</u>, 16, 20, <u>24</u>, 28, 32, <u>36</u> . . .
6: 6, <u>12</u>, 18, <u>24</u>, 30, <u>36</u> . . .   12, 24, and 36 are common multiples.

Least common multiple: of these common multiples, 12 is the least.

Example 1
Find the least common multiple of 6 and 8.

The *multiples* of 6 and 8 are shown below.
   6: 6, 12, 18, 24, 30, 36, 42, 48, 54, 60, 66, 72, 78, 84, 90 . . .
   8: 8, 16, 24, 32, 40, 48, 56, 64, 72, 80, 88, 96, 104, 112 . . .

The *common* multiples are shown in bold type.
   6: 6, 12, 18, **24**, 30, 36, 42, **48**, 54, 60, 66, **72**, 78, 84, 90 . . .
   8: 8, 16, **24**, 32, 40, **48**, 56, 64, **72**, 80, 88, 96, 104, 112 . . .

24, 48, and 72 are each common multiples of 6 and 8, and 24 is the *least* common multiple.

There is another way to find the least common multiple (LCM). It involves finding the *prime factors* of each number first. (If you need a review of prime numbers, see the next section.) Using the same problem as in Example 1, the prime factors of six and eight are: 6 = 2 x 3, and 8 = 2 x 2 x 2. The LCM must have all the prime factors of six as well as the prime factors of eight because the LCM is a number common to both. It must also be the least common multiple, so it must have no extra factors included.

Figure 2

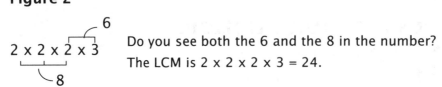

Do you see both the 6 and the 8 in the number?
The LCM is 2 x 2 x 2 x 3 = 24.

Example 2
Find the LCM for 12 and 20.
Find the prime factors for each number.

$$12 = 2 \cdot 2 \cdot 3 \qquad\qquad 20 = 2 \cdot 2 \cdot 5$$

Start with the 2 · 2 · 3 from 12; with
2 · 2 already, we need only the 5 from 20, so the
LCM is 2 · 2 · 3 · 5.

Prime Numbers

For this lesson and the one on finding the greatest common factor (lesson 22), a brief review of prime numbers and how to find them may be necessary.

To illustrate the concept of prime numbers, begin with seven green unit blocks and see if you can make two different rectangles, using all of the blocks in each rectangle. You will soon find out that only one rectangle can be built. Note that a rectangle 1 x 7 is the same as a rectangle that is 7 x 1. Then ask, "What do you call a number for which you can build only one rectangle?" "A prime number." A prime number has only factors (or dimensions) of factors one and itself, or that can only be divided evenly by one and itself. Having seen this relationship with the manipulatives, the definition now makes sense. (The number one is not considered a prime number.)

Example 3

Find all the possible factors (by building rectangles) of 13 and tell whether the number is prime or composite.

The factors are 13 and 1, or 1 and 13, so 13 is a prime number.

The prime factors of a number may be discovered (prime factorization) using a factor tree or repeated division. Here is an example of each method for the numbers 12 and 18.

Example 4

What are the prime factors of 12?

factor tree   repeated division

```
    12                        3
   / \                      2⌐6
  3   4                    2⌐ 12
     / \
    2   2
```

2 x 2 x 3   2 x 2 x 3

Example 5

What are the prime factors of 18?

```
    18                        3
   / \                      3⌐9
  2   9                    2⌐ 18
     / \
    3   3
```

2 x 3 x 3   2 x 3 x 3

LESSON 22

Greatest Common Factor

There are three words to consider in **greatest common factor**: the *factor*, the *common factor*, and the *greatest common factor*.

Factor: The factors of six are 1 x 6 and 2 x 3. These can be written in order as 1, 2, 3, and 6. The factors of 15 are 1 x 15 and 3 x 5. These can be written in order as 1, 3, 5, and 15.

Common factor: the common factors of 6 and 15 are 1 and 3. They are underlined in Figure 1.

Figure 1   6: <u>1</u>, 2, <u>3</u>, 6
                15: <u>1</u>, <u>3</u>, 5, 15

Greatest common factor: the greatest common factor (GCF) of 6 and 15 is 3.

Example 1
Find the GCF of 12 and 18.
The *factors* of 12 and 18 are as follows:
    12: 1, 2, 3, 4, 6, 12
    18: 1, 2, 3, 6, 9, 18
The *common* factors of 12 and 18 are underlined.
    12: <u>1</u>, <u>2</u>, <u>3</u>, 4, <u>6</u>, 12
    18: <u>1</u>, <u>2</u>, <u>3</u>, <u>6</u>, 9, 18

1, 2, 3, and 6 are each common factors of 12 and 18, but 6 is the *greatest* common factor.

You can also find the GCF by using prime factorization. Find the prime factors of the numbers in the problem and then find which factors are common to both. If there is more than one common prime factor, multiply them to find the GCF.

Example 2
Find the GCF of 18 and 27 using prime factorization.

Find the prime factors of both numbers.
    18 = 2 x 3 x 3
    27 = 3 x 3 x 3

Underline the common prime factors of both numbers.
    18 = 2 x 3 x 3
    27 = 3 x 3 x 3

The greatest common factor (GCF) is 3 x 3, or 9.

Polynomials, Addition

Another important concept in algebra is the concept of X. For this section, you will need the algebra inserts. These snap into the backs of the red hundred pieces and the blue ten bars. After you have inserted the blue piece into the back of the ten bar, look at it with the smooth side facing you. The over dimension is obviously one, but what about the up dimension? There are 10 lines on the other side, but this side is smooth. You can't count the up dimension. When you know that it is obviously something, but you aren't sure what, you give it a letter to represent how long it is. A simple definition of algebra is "when you don't know what number to use, choose a letter!" The gray pieces represent negative X, or $(-X)$.

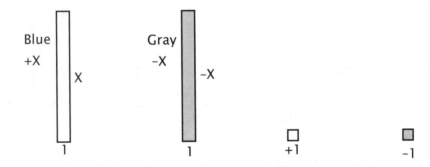

The blue X bar becomes the standard of reference. Using the blue X bar as a ruler, measure the back of the red square with the insert in it (smooth side). Notice the over and up factors are both X. It is X used as a factor two ways, or X squared, or X to the two power. Using your blocks, build $2X^2 + 3X + 4$.

Example 1

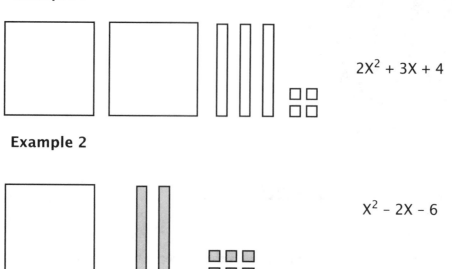

$2X^2 + 3X + 4$

Example 2

$X^2 - 2X - 6$

A mathematical expression with more than one term is called a *polynomial*. An expression with one term is referred to as a *monomial*. If it has only two terms, such as X + 2, it is called a *binomial*. A polynomial with three terms is a *trinomial*.

Algebra can be described as arithmetic in *base X* instead of base 10. Notice how it parallels place value in base 10. Instead of 10, we use a power of X for the place value.

X^6	X^5	X^4	X^3	X^2	X^1	X^0
10^6	10^5	10^4	10^3	10^2	10^1	10^0
1,000,000	100,000	10,000	1,000	100	10	1
million	hundred thousand	ten thousand	thousand	hundred	ten	unit

When adding polynomials, follow the same pattern that you use when adding numbers in base 10. To compare or combine, you must have the same kind. You add units to units, Xs to Xs, and X squares to X squares. Even though you are in base X, the same concept still applies. In algebra, since you don't know what the value of X is, you can't regroup. In some ways, this makes it easier to add polynomials. Study the examples.

Example 3

$$
\begin{array}{r}
2X^2 + 5X + 7 \\
+\ 3X^2 + 8X + 6 \\
\hline
5X^2 + 13X + 13
\end{array}
$$

Example 4

$$
\begin{array}{r}
2X^2 - 5X + 7 \\
+\ 3X^2 + 8X - 8 \\
\hline
5X^2 + 3X - 1
\end{array}
$$

Example 5

$$
\begin{array}{r}
-2X^2 - 5X - 7 \\
+\ 3X^2 - 8X + 8 \\
\hline
X^2 - 13X + 1
\end{array}
$$

Example 6

$$
\begin{array}{r}
2X^2 + 5X + 7 \\
+\ -3X^2 - 8X - 8 \\
\hline
-X^2 - 3X - 1
\end{array}
$$

Volume of a Cylinder

The formula for most volume problems (cube, rectangular solid, or *cylinder*) is *area of the base times the height*. To distinguish this formula from the area of a polygon, which is two-dimensional, use the capitalized *B* for area of the Base. In the formula for the area of polygons, use the lowercase *b* for base.

What is unique about the area of the base of a cylinder is that it is a circle and is found by using the formula πr². (For a review of finding the area of a circle, please turn to the next page.) When you find the area of the base, multiply it by the height of the solid to find the volume.

$$V = Bh$$

Think of finding the volume of a cylinder as finding how much pineapple is in a can of rings, as shown in Figure 1. First you find the area of one pineapple ring. This is the area of the base or B, which is found by computing πr². Then you multiply that number by the number of pineapple rings in the can. This is the height, or h.

Figure 1

Example 1
Find the volume of the cylinder.

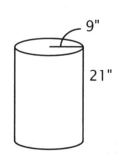

Volume = πr^2 x h

$\approx (3.14)(9^2)$ x 21

$\approx 5,341.14$ cubic inches

This is usually written as 5,341.14 in^3.

Notice the symbol that is used to show that the answer is only an approximation (≈). This is because the number 3.14 is a rounded value for pi and therefore only gives a rounded answer.

Example 2
Find the volume of the cylinder.

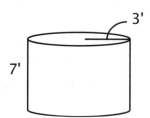

Volume = πr^2 x h

$\approx (3.14)(3^2)$ x 7

≈ 197.82 ft^3

Example 3

Find the volume of the cylinder.

Since the diameter is given,
find half of it to find the radius.

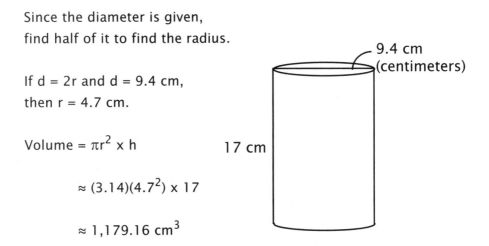

If d = 2r and d = 9.4 cm,
then r = 4.7 cm.

Volume = πr^2 × h

$\approx$ (3.14)(4.7^2) × 17

$\approx$ 1,179.16 cm^3

Review of the Area of a Circle

The formula for the area of a circle is πr^2. ***Pi***, or π, is a symbol for a value that is a little over three. The *r* represents the radius of a circle. The ***radius*** is the distance from the center of the circle to the edge of the circle. The ***diameter*** of a circle is a line all the way across the circle through the center. Take one half of the diameter to find the radius.

Whenever finding area, remember the word SQUAREA, which is a conjoining of *square* and *area*. Area is always computed in square units.

The value of π is normally represented by one of two values, 22/7 or 3.14. Both of these are approximations for a number that extends as a decimal indefinitely. *Pi* = 3.1415927. . . There are some problems where it is advantageous to use the decimal and others where the fraction is more convenient.

Example 4

Find the area of the circle.

Area = πr^2

Area $\approx$ 3.14(6.5)^2

Area $\approx$ 3.14(42.25)

Area $\approx$ 132.67 square inches or in^2

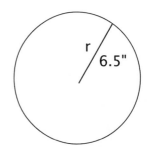

Polynomials, Multiplication

Polynomial multiplication uses the same process as double-digit multiplication. Instead of over $(10 + 2)$ and up $(10 + 3)$, as in Example 1, it is over $(X + 2)$ and up $(X + 3)$, as in Example 2. The area or product is not $10^2 + 50 + 6$ or 156, as in Example 1, but $X^2 + 5X + 6$.

Example 1

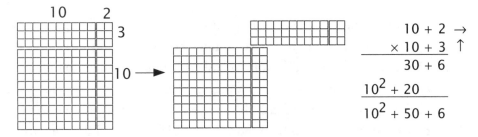

$$
\begin{array}{r}
10 + 2 \rightarrow \\
\times\ 10 + 3 \uparrow \\
\hline
30 + 6 \\
10^2 + 20 \\
\hline
10^2 + 50 + 6
\end{array}
$$

Example 2

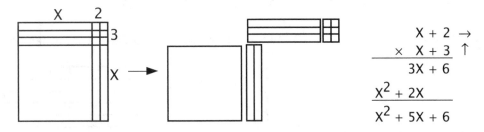

$$
\begin{array}{r}
X + 2 \rightarrow \\
\times\ X + 3 \uparrow \\
\hline
3X + 6 \\
X^2 + 2X \\
\hline
X^2 + 5X + 6
\end{array}
$$

Example 3

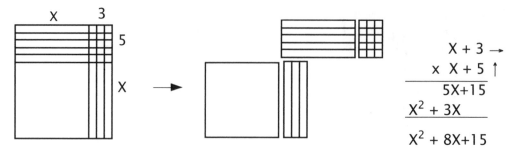

$$X + 3 \rightarrow$$
$$\underline{\times \ X + 5 \uparrow}$$
$$5X + 15$$
$$\underline{X^2 + 3X}$$
$$X^2 + 8X + 15$$

Example 4

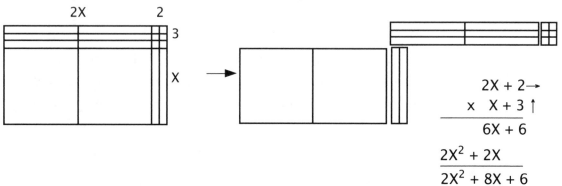

$$2X + 2 \rightarrow$$
$$\underline{\times \ X + 3 \uparrow}$$
$$6X + 6$$
$$\underline{2X^2 + 2X}$$
$$2X^2 + 8X + 6$$

The following aspect of this lesson is very similar to the first part in which you are given a rectangle and are asked to find the factors and product. What if you are given the two binomial factors and asked to multiply them to find the product *without* a picture? Example 5 has a picture where just the lines along the edges, or the length of the sides (factors), are given. If you want to build the rest of the rectangle and fill it in, you have the same problem as you did at the beginning of this lesson.

Example 5

Find the factors and product of the figure given.

When you fill it in, the area looks like this.

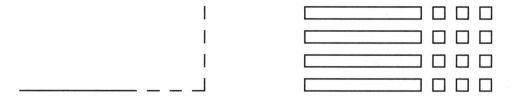

The factors are (X + 3) and (4), and the product is 4X + 12.

Example 6
Find the factors and product of the figure given.

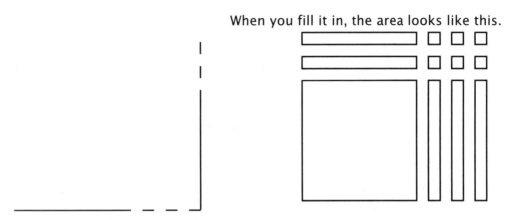

When you fill it in, the area looks like this.

The factors are (X + 3) and (X + 2), and the product is $X^2 + 5X + 6$.

Adding and Subtracting Time
and the "Same Difference Theorem"

Adding Time

When working with time, remember that there are 60 minutes, not 100 minutes, in one hour. This comes into play when regrouping from minutes to hours. If you are adding 50 minutes plus 30 minutes, the answer is 80 minutes, which is the same as one hour and 20 minutes.

Example 1

```
      1
    0:50
+   0:30
    1:20
```

When regrouping from the tens place to the hours place, six tens is the same as one hour.

Example 2

```
    1  1
    0:52
+   0:38
    1:30
```

When regrouping from the units place to the tens place, 10 units is the same as one ten.

Subtracting Time

Subtracting time is the inverse of adding time. If Examples 1 and 2 are clearly understood, then subtracting time is just the opposite. When you regroup from the tens place with regular numbers, you use 10. However, when you regroup from the hours place, you bring over 6 tens because 60 minutes is one hour.

Example 3

$$
\begin{array}{r}
\overset{1}{}\;\overset{60}{} \\
2:15 = \cancel{2}:\cancel{15} = 1:75 \\
-\;0:44 \qquad = 0:44 \\
\hline
1:31
\end{array}
$$

When regrouping from the hours place to the tens place, cross out the two, and you are left with one hour. Then you add the 60 minutes to the 15 minutes. Now you can subtract.

Example 4

$$
\begin{array}{r}
\overset{3}{}\;\overset{60}{} \qquad \overset{7}{}\;\overset{1}{} \\
4:26 = \cancel{4}:26 = 3:86 = 3:\cancel{86} \\
-\;1:39 = \qquad\qquad = 1:39 \\
\hline
= 2:47
\end{array}
$$

Subtracting Time with the "Same Difference Theorem"

You can do Example 4 again but make it easier without all the regrouping. By making the bottom number a whole hour, the subtraction would be much simpler. If you add 0:21 (21 minutes) to the top and bottom numbers, the result is shown in Example 5. The question is always, "What do you have to add to the bottom number to make it a whole number (or whole hour, when working with time)?"

Example 5

$$
\begin{array}{r}
4:26 + 0:21 = 4:47 \\
-\;1:39 + 0:21 = 2:00 \\
\hline
2:47
\end{array}
$$

LESSON 27

Volume of a Pyramid and a Cone

A pyramid has three or more triangular faces, depending on how many sides the base has. In this book you'll be using pyramids with square bases and four triangular faces. The height of the pyramid itself is the *altitude*. The altitude is perpendicular to the base and runs up to the highest point where the faces meet. The points where the edges of the base and faces meet are called *vertices* (the plural of *vertex*). The vertex that is the farthest away from the base is the *apex* of the pyramid. The height of a face is the *slant height*. The volume of a pyramid or a cone is found by multiplying B (the area of the base) by the height (altitude) by 1/3.

$$V = 1/3 \ Bh$$

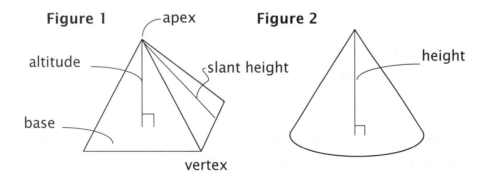

Figure 1   apex

altitude

base

slant height

vertex

Figure 2   height

Example 1

Find the volume of the pyramid with a square base.

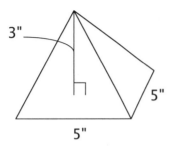

$V = 1/3\ Bh$

$V = 1/3\ (5 \times 5)(3)$

$V = 25$ cubic inches, or $25\ in^3$

Example 2

Find the volume of the cone. All cones have a circular base.

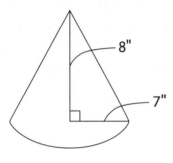

$V = 1/3\ Bh$

$V \approx 1/3\ (3.14)(7^2)(8)$

$V \approx 410.29$ cubic inches, or $410.29\ in^3$

Notice that the volume of a cone is only an approximation, since you must use a rounded value for *pi* to find the area of the base.

LESSON 28

Military Time, Addition and Subtraction

If you were asked to find the difference between 2:30 p.m. and 10:15 a.m., you would be subtracting a greater number from a lesser number in the hours place. To facilitate this, you can change the 2 in 2:30 p.m. to 14 by adding 12 to the hours. In *military time*, 2:30 p.m. is referred to as 1430. Time on the 12-hour clock changes every 12 hours, but military time proceeds from midnight to midnight before turning over. This way of writing and telling time is also used by medical personnel. It is sometimes called 24-hour time.

However you think of it, now you can compute 14:30 minus 10:15 normally. An additional benefit to military time is being able to recognize a.m. as hours that are 11 or less and p.m. as hours from 12 to 23. You always need four digits in military time, so 3:59 a.m. is 0359. For fun, practice converting times from the 12-hour clock to military time and vice versa.

Figure 1	0000 – 12:00 a.m.	1200 – 12:00 p.m.
	0100 – 1:00 a.m.	1300 – 1:00 p.m.
	0200 – 2:00 a.m.	1400 – 2:00 p.m.
	0300 – 3:00 a.m.	1500 – 3:00 p.m.
	0400 – 4:00 a.m.	1600 – 4:00 p.m.
	0500 – 5:00 a.m.	1700 – 5:00 p.m.
	0600 – 6:00 a.m.	1800 – 6:00 p.m.
	0700 – 7:00 a.m.	1900 – 7:00 p.m.
	0800 – 8:00 a.m.	2000 – 8:00 p.m.
	0900 – 9:00 a.m.	2100 – 9:00 p.m.
	1000 – 10:00 a.m.	2200 – 10:00 p.m.
	1100 – 11:00 a.m.	2300 – 11:00 p.m.

Example 1
Convert 4:46 p.m. to military time.

$$\begin{array}{r} 4{:}46 \\ +\ 12{:}00 \\ \hline 16\ 46 \end{array}$$

Example 2
Convert 2358 to standard time.

$$\begin{array}{r} 23\ 58 \\ -\ 12{:}00 \\ \hline 11{:}58 \end{array}$$ Since 2358 is greater than 1200, the time is p.m.

Example 3
Add 0347 to 0915 military time.

$$\begin{array}{r} 1\ 1\ 1 \\ 0\ 9\ 1\ 5 \\ +\ 0\ 3\ 4\ 7 \\ \hline 1\ 3\ 0\ 2 \end{array}$$

4+1+1 in the tens place makes 60 minutes, which is regrouped to form one hour. Compute as usual.

Example 4
Subtract 0347 from 0915.

$$\begin{array}{r} 0\ 9\ 1\ 5 = 0\ 8\ 7\ 5 \\ -\ 0\ 3\ 4\ 7 = 0\ 3\ 4\ 7 \\ \hline 0\ 5\ 2\ 8 \end{array}$$

Regroup one hour and change it to 60 minutes. Add it to 10 minutes. Then regroup 10 from 70 and add it to 5.

Example 5
Subtract 0347 from 0915 military time, using the "same difference theorem."

$$\begin{array}{r} 0915 + 13 = 0928 \\ -\ 0347 + 13 = 0400 \\ \hline 0528 \end{array}$$

Add 13 minutes to 0347 to make 0400.
Add 13 minutes to 0915 to make 0928.
Then subtract.

Measurement, Addition and Subtraction

The "same difference theorem" can be used for adding measurements as well as mixed numbers and time. Each problem is shown both ways—with the longer, traditional method and with Math-U-See's "same difference theorem." If you need a quick review of measurement, consult the Appendix.

Example 1

Johnny stood on Steve's head. If Steve is 6' 5" and Johnny is 4' 11", how tall were they together?

A.
$$
\begin{array}{r}
6'\ 5" \\
+\ 4'\ 11" \\
\hline
10'\ 16"
\end{array}
\quad \Rightarrow\ 16" = 1'\ 4"\ \Rightarrow\quad
\begin{array}{r}
1' \\
6'\ 5" \\
+\ 4'\ 11" \\
\hline
11'\ 4"
\end{array}
$$

Another way to solve this problem is by remembering that 12" = 1'. If you take 1" from the 5" and add it to the 4' 11", the problem is much easier.

B.
$$
\begin{array}{r}
6'\ 5" \\
+\ 4'\ 11" \\
\hline
\end{array}
\qquad
\begin{array}{l}
\text{Take 1" from the 5'.} \\
\text{Add 1 to make a} \\
\text{whole number.}
\end{array}
\qquad
\begin{array}{r}
6'\ 4" \\
+\ 5' \\
\hline
11'\ 4"
\end{array}
$$

Example 2

What is the sum of 5 yd 2 ft and 3 yd 2 ft?

```
                                      1 yd
A.    5 yd  2 ft                      5 yd  2 ft
    + 3 yd  2 ft                    + 3 yd  2 ft
     8 yd  4 ft ⇒ 4' = 1 yd 1 ft ⇒   9 yd  1 ft
```

Another way to solve this problem is by remembering that one yard equals three feet. If you take 1 ft from the 2 ft and add it to the 3 yd 2 ft, the problem is much easier.

```
B.   5 yd  2 ft     Take 1' from the 2'.     5 yd  1 ft
   + 3 yd  2 ft     Add 1' to make a       + 4 yd
                    whole number.            9 yd  1 ft
```

Example 3

Find the total weight of 7 lb 8 oz and 2 lb 11 oz.

```
                                          1 lb
A.    7 lb  8 oz                          7 lb  8 oz
    + 2 lb 11 oz                        + 2 lb 11 oz
     9 lb 19 oz ⇒ 19 oz = 1 lb 3 oz ⇒   10 lb  3 oz
```

Another way to solve this problem is by remembering that one pound equals 16 ounces. If you take 5 oz from the 8 oz and add it to the 2 lbs 11 oz, the problem is much easier.

```
B.   7 lb  8 oz     Take 5 oz from the      7 lb 3 oz
   + 2 lb 11 oz     8 oz. Add 5 oz to     + 3 lb
                    make a whole number.    10 lb 3 oz
```

Using the "Same Difference Theorem" to Subtract Measures

The "same difference theorem" is also useful and efficient for subtracting measurements. Each of the following problems is solved two ways. Make sure you understand the concept in the first method and then see how easily each problem is solved with the "same difference theorem."

Example 4

What is the difference in height between Steve, who is 6'5", and Johnny, who is 4'11"?

A.
$$
\begin{array}{r}
6'\ 5" \\
-\ 4'\ 11" \\
\hline
\end{array}
\quad \Rightarrow \quad
\overset{12"}{\curvearrowright}\ 5'\ 5"
\quad \Rightarrow \quad
\begin{array}{r}
5'\ 17" \\
-\ 4'\ 11" \\
\hline
1'\ 6"
\end{array}
$$

B.
$$
\begin{array}{r}
6'\ 5" \\
-\ 4'\ 11" \\
\hline
\end{array}
\quad
\begin{array}{l}
\text{Add } 1". \\
\text{Add } 1" \text{ to make} \\
\text{whole number.}
\end{array}
\quad
\begin{array}{r}
6'\ 6" \\
-\ 5' \\
\hline
1'\ 6"
\end{array}
$$

Example 5

Find the difference between 4 yd 1 ft and 1 yd 2 ft.

(A)
$$
\begin{array}{r}
4\ \text{yd}\ 1\ \text{ft} \\
-\ 1\ \text{yd}\ 2\ \text{ft} \\
\hline
\end{array}
\quad \Rightarrow \quad
\overset{3\ \text{ft}}{\curvearrowright}\ 3\ \text{yd}\ 1\ \text{ft}
\quad \Rightarrow \quad
\begin{array}{r}
3\ \text{yd}\ 4\ \text{ft} \\
-\ 1\ \text{yd}\ 2\ \text{ft} \\
\hline
2\ \text{yd}\ 2\ \text{ft}
\end{array}
$$

(B)
$$
\begin{array}{r}
4\ \text{yd}\ 1\ \text{ft} \\
-\ 1\ \text{yd}\ 2\ \text{ft} \\
\hline
\end{array}
\quad
\begin{array}{l}
\text{add } 1\ \text{ft} \Rightarrow \\
\text{add } 1\ \text{ft} \Rightarrow
\end{array}
\quad
\begin{array}{r}
4\ \text{yd}\ 2\ \text{ft} \\
-\ 2\ \text{yd} \\
\hline
2\ \text{yd}\ 2\ \text{ft}
\end{array}
$$

Example 6

Find the difference between 7 lb 8 oz and 2 lb 11 oz.

16 oz

(A)
$$\begin{array}{r} 7 \text{ lb } 8 \text{ oz} \\ - 2 \text{ lb } 11 \text{ oz} \\ \hline \end{array}$$

$6 \text{ lb } 8 \text{ oz}$

$$\begin{array}{r} 6 \text{ lb } 24 \text{ oz} \\ - 2 \text{ lb } 11 \text{ oz} \\ \hline 4 \text{ lb } 13 \text{ oz} \end{array}$$

(B)
$$\begin{array}{r} 7 \text{ lb } 8 \text{ oz} \\ - 2 \text{ lb } 11 \text{ oz} \\ \hline \end{array}$$

add 5 oz ⇒
add 5 oz ⇒

$$\begin{array}{r} 7 \text{ lb } 13 \text{ oz} \\ - 3 \text{ lb} \\ \hline 4 \text{ lb } 13 \text{ oz} \end{array}$$

Irrational Numbers
and the Square Root Formula, Real Numbers

A number that cannot be written as a *rational number* (a ratio of two integers) is referred to as an *irrational number*. These create non-repeating, non-terminating decimals. The number $\sqrt{4}$ is rational because it is equal to two, which may be expressed as 2/1, but $\sqrt{3}$ is not a rational number; it is irrational. Use your calculator to find the square root of three. The answer that appears on the screen can be rounded to 1.732. If you multiply 1.732 x 1.732 you get 2.999824, which is close to three, or approximately three, but not exactly three. That is why, when speaking of the square root of three, you generally leave it as $\sqrt{3}$.

Pi (π) is also an irrational number. The expressions 3.14 and 22/7 are approximate values of *pi* but are not exact. Perhaps you've read about people who have memorized the first 100 numbers (out of thousands) in the approximation of *pi*. The numbers 22/7 and 3.14 in and of themselves are rational numbers because *pi* has been rounded to hundredths. Even though the values are used to represent *pi*, *pi* itself is an irrational number. When referring to *pi*, its unique value is represented by the Greek letter π. Problems arise when you try to represent these values with whole numbers.

Square Root Formula

There are very few people that still know how to find the square root of a number without a calculator. There are three unique steps in this process. The following example explains each step.

Example 1

Find the square root: 20,449

$$\sqrt{2 _{\wedge}0,4_{\wedge}49}$$

Beginning at the decimal point, mark every two spaces.

$$\begin{array}{r} 1 \\ \hline \sqrt{2 _{\wedge}0,4_{\wedge}49} \end{array}$$

Find the closest square root of 2. This is 1. Put it on top of the radical sign.

$$\begin{array}{r} 1 \\ \hline \sqrt{2 _{\wedge}0,4_{\wedge}49} \\ 1 \\ \hline 1\ 04 \end{array}$$

Then $1^2 = 1$. Write 1 below the 2, subtract, then bring two numbers down. (Hence the reason for the marks.)

$1 \times 20 = 20$

$$\begin{array}{r} 1 \\ \hline \sqrt{2 _{\wedge}0,4_{\wedge}49} \\ 1 \\ \hline 1\ 04 \end{array}$$

Multiply 1 (the one on top) by 20 and write to the left.

$4 \times 24 = 96$

$$\begin{array}{r} 1\ \ 4 \\ \hline \sqrt{2 _{\wedge}0,4_{\wedge}49} \\ 1 \\ \hline 1\ 04 \\ -\ 96 \\ \hline 8 \end{array}$$

Estimate how many times 20 will divide into 104. Whatever number you estimate, you must add to 20 before you multiply. If you select 5, you have to multiply 5 by 20 + 5, or 25. The answer is 125, which is too much. Try 4. 4 x 24 = 96. Subtract and bring down two more places and repeat the last step again.

$14 \times 20 = 280$

Choose 3.

$3 \times 283 = 849$

$$\begin{array}{r} 1\ 4\ 3 \\ \hline \sqrt{2 _{\wedge}0,4_{\wedge}49} \\ 1 \\ \hline 104 \\ -\ 96 \\ \hline 8\ 49 \\ 8\ 49 \end{array}$$

Estimate how many times 280 (14 x 20) will divide into 849. Try 3. 3 x 283 = 849. Then check your answer.

Check
$$
\begin{array}{r}
143 \\
\times\,143 \\
\hline
429 \\
572 \\
143 \\
\hline
20{,}449
\end{array}
$$

Example 2

Find the square root: 536.8489

$$
\begin{array}{rr}
2\times20+3= & 43 \\
& \times\,3 \\
\hline
& 129 \\
23\times20+1= & 461 \\
& \times\,1 \\
\hline
& 461 \\
231\times20+7= & 4{,}627 \\
& \times\quad7 \\
\hline
& 32{,}389
\end{array}
\qquad
\begin{array}{r}
2\quad3.\;1\;7 \\
\sqrt{5\,{}_\wedge36.84\,{}_\wedge89} \\
4 \\
\hline
136 \\
129 \\
\hline
784 \\
461 \\
\hline
32389 \\
32389 \\
\hline
\end{array}
$$

Check
$$
\begin{array}{r}
23.17 \\
\times\,23.17 \\
\hline
16219 \\
2317 \\
6951 \\
4634 \\
\hline
536.8489
\end{array}
$$

Example 3

Find $\sqrt{3}$ to the hundredths place.

$$1 \times 20 + 7 = \quad 27$$
$$\underline{\times \quad 7}$$
$$189$$

$$17 \times 20 + 3 = \quad 343$$
$$\underline{\times \quad 3}$$
$$1{,}029$$

$$173 \times 20 + 2 = 3{,}462$$
$$\underline{\times \quad 2}$$
$$6{,}924$$

$$
\begin{array}{l}
\phantom{\sqrt{}}1.\ \ 7\ \ 3\ \ 2 \\
\sqrt{3.00\,{}_{\wedge}00\,{}_{\wedge}00} \\
\phantom{\sqrt{}}\underline{1} \\
\phantom{\sqrt{}}200 \\
\phantom{\sqrt{}}\underline{189} \\
\phantom{\sqrt{}}1100 \\
\phantom{\sqrt{}}\underline{1029} \\
\phantom{\sqrt{}}7100 \\
\phantom{\sqrt{}}\underline{6924} \\
\phantom{\sqrt{}}176
\end{array}
$$

Check

$$
\begin{array}{r}
1.732 \\
\underline{\times\ 1.732} \\
3464 \\
5196 \\
12124 \\
\underline{1732} \\
2.999824
\end{array}
$$

Close!

Real Numbers

   Real numbers include all rational and irrational numbers. Every real number has a place on the number line.

Figure 1

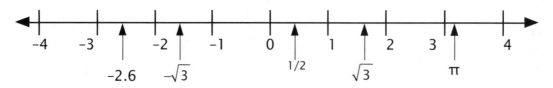

Student Solutions

Lesson Practice 1A

1. done
2. done
3. owe
4. owe
5. have
6. have
7. owe
8. owe
9. $(-5)+(-3)=-8$
10. $(-14)+(-69)=-83$
11. $(+93)+(-48)=+45$
12. $(-27)+(+56)=+29$
13. $(-8)+(-23)=-31$
14. $(-8)+(+10)=+2$
15. $(+6)+(-9)=-3$
16. $(+3)+(+8)=+11$
17. $(-15)+(+6)=-9$
18. $(\$+20)+(\$-9)=\$+11$
19. $(\$-11)+(\$-13)=\$-24$
20. $(\$+8)+(\$+15)=\$+23$

Lesson Practice 1B

1. owe
2. have
3. have
4. owe
5. owe
6. have
7. owe
8. owe
9. $(+5)+(-16)=-11$
10. $(+4)+(-2)=+2$
11. $(-7)+(-25)=-32$
12. $(+5)+(+10)=+15$
13. $(+6)+(-9)=-3$
14. $(-7)+(+20)=+13$

15. $(+14)+(-3)=+11$
16. $(+15)+(-6)=+9$
17. $(-6)+(+11)=+5$
18. $(+55)+(92)=+147$ points
19. $(\$-88)+(\$+24)=\$-64$
20. $(-36)+(-10)=-46$ gallons

Lesson Practice 1C

1. have
2. have
3. owe
4. owe
5. owe
6. have
7. owe
8. owe
9. $(+17)+(-5)=+12$
10. $(-63)+(-50)=-113$
11. $(+24)+(-36)=-12$
12. $(-32)+(-43)=-75$
13. $(+98)+(-44)=+54$
14. $(-76)+(+84)=+8$
15. $(-19)+(-35)=-54$
16. $(+48)+(-23)=+25$
17. $(-92)+(+29)=-63$
18. $(+33)+(+44)=+77$ coins
19. $(\$-91)+(\$+120)=\$+29$
20. $(\$-63)+(\$-54)=\$-117$

Systematic Review 1D

1. owe
2. have
3. owe
4. $(+24)+(+10)=+34$

5. $(+8)+(-15)=-7$
6. $(-9)+(-6)=-15$
7. $(+42)+(+54)=+96$
8. $(-17)+(-47)=-64$
9. $(-19)+(+8)=-11$
10. $(+78)+(-51)=+27$
11. $(-13)+(-12)=-25$
12. $(+14)+(-19)=-5$
13. done
14. $6\div3=2$
    $2\times1=2$
15. $20\div5=4$
    $4\times4=16$
16. $24\div6=4$
    $4\times5=20$
17. $18\div3=6$
    $6\times2=12$ brought their sons.
18. $100\div5=20$
    $20\times2=40$ jelly beans
19. $(-3)+(+5)=+2$
20. $(\$-500)+(\$-650)=\$-1,150$

Systematic Review 1E

1. owe
2. owe
3. have
4. $(-9)+(+4)=-5$
5. $(-8)+(+19)=+11$
6. $(+10)+(+23)=+33$
7. $(+4)+(-1)=+3$
8. $(-13)+(-6)=-19$
9. $(-9)+(+87)=+78$
10. $(-68)+(-41)=-109$
11. $(+17)+(+35)=+52$
12. $(-54)+(+16)=-38$
13. $12\div3=4$
    $4\times1=4$
14. $10\div5=2$
    $2\times2=4$

15. $18\div9=2$
    $2\times7=14$
16. $21\div3=7$
    $7\times2=14$
17. $24\div6=4$
    $4\times1=4$ hrs
18. $24\div8=3$
    $3\times1=3$ hrs
19. $4\times60=240$ min for Joshua
    $3\times60=180$ min for Jesse
    $240+180=420$ min total
20. $(\$+50)+(\$-60)=\$-10$

Systematic Review 1F

1. have
2. owe
3. owe
4. $(+6)+(-27)=-21$
5. $(-9)+(-3)=-12$
6. $(-8)+(-14)=-22$
7. $(+5)+(-8)=-3$
8. $(+23)+(+69)=+92$
9. $(+6)+(-15)=-9$
10. $(-3)+(-16)=-19$
11. $(-123)+(-341)=-464$
12. $(-45)+(+55)=+10$
13. $8\div4=2$
    $2\times1=2$
14. $10\div5=2$
    $2\times4=8$
15. $21\div7=3$
    $3\times3=9$
16. $20\div5=4$
    $4\times2=8$
17. $24\div3=8$
    $8\times2=16$ students
18. $24\div4=6$
    $6\times1=6$ students
19. $\$45+\$13=\$58$
20. $(+91)+(-46)=+45$ miles

Lesson Practice 2A

1. done
2. done
3. $(+5) - (-2) =$
   $(+5) + (+2) = +7$
4. $(-2) - (-3) =$
   $(-2) + (+3) = +1$
5. $(-6) - (+8) =$
   $(-6) + (-8) = -14$
6. $(+5) - (-7) =$
   $(+5) + (+7) = +12$
7. $(+10) - (+7) =$
   $(+10) + (-7) = +3$
8. $(-3) - (-3) =$
   $(-3) + (+3) = 0$
9. $(+9) - (+1) =$
   $(+9) + (-1) = +8$
10. $(-4) - (+2) =$
    $(-4) + (-2) = -6$
11. $(+30) - (-10) =$
    $(+30) + (+10) = +40$
12. $(+7) - (-8) =$
    $(+7) + (+8) = +15$
13. $(-81) - (-24) =$
    $(-81) + (+24) = -57$
14. $(-17) - (+11) =$
    $(-17) + (-11) = -28$
15. $(+63) - (-92) =$
    $(+63) + (+92) = +155$
16. $(+33) - (+8) =$
    $(+33) + (-8) = +25$
17. $(\$-10) + (\$-6) = \$-16$
18. $(+12) + (-6) = +6$ marshmallows
19. $(\$-20) + (\$+7) = \$-13$
20. $(-6) + \underline{(-8)} = -14$ cents

Lesson Practice 2B

1. $(-2) - (-5) =$
   $(-2) + (+5) = +3$

2. $(-3) - (+6) =$
   $(-3) + (-6) = -9$
3. $(+8) - (-6) =$
   $(+8) + (+6) = +14$
4. $(-4) - (-9) =$
   $(-4) + (+9) = +5$
5. $(-2) - (+7) =$
   $(-2) + (-7) = -9$
6. $(+2) - (-3) =$
   $(+2) + (+3) = +5$
7. $(+8) - (+5) =$
   $(+8) + (-5) = +3$
8. $(-6) - (-6) =$
   $(-6) + (+6) = 0$
9. $(+10) - (+11) =$
   $(+10) + (-11) = -1$
10. $(-5) - (+3) =$
    $(-5) + (-3) = -8$
11. $(+15) - (-7) =$
    $(+15) + (+7) = +22$
12. $(+25) - (-24) =$
    $(+25) + (+24) = +49$
13. $(-23) - (-8) =$
    $(-23) + (+8) = -15$
14. $(+10) - (-8) =$
    $(+10) + (+8) = +18$
15. $(-19) - (+6) =$
    $(-19) + (-6) = -25$
16. $(+81) - (-1) =$
    $(+81) + (+1) = +82$
17. $(\$+15) - (\$+17) =$
    $(\$+15) + (\$-17) = \$-2$
18. $(\$-35) + (\$-43) = \$-78$
19. $(+12) + (-3) = (+9)$ apples
20. $(-3) + (-5) = -8$ gallons

Lesson Practice 2C

1. $(+12) - (+13) =$
   $(+12) + (-13) = -1$

2. $(-27)-(+41)=$
$(-27)+(-41)=-68$

3. $(+17)-(-28)=$
$(+17)+(+28)=+45$

4. $(+27)-(-25)=$
$(+27)+(+25)=+52$

5. $(-8)-(-8)=$
$(-8)+(+8)=0$

6. $(-5)-(+12)=$
$(-5)+(-12)=-17$

7. $(+4)-(-12)=$
$(+4)+(+12)=+16$

8. $(-8)-(-3)=$
$(-8)+(+3)=-5$

9. $(+16)-(+45)=$
$(+16)+(-45)=-29$

10. $(-13)-(+14)=$
$(-13)+(-14)=-27$

11. $(+21)-(-43)=$
$(+21)+(+43)=+64$

12. $(+39)-(-12)=$
$(+39)+(+12)=+51$

13. $(-9)-(+15)=$
$(-9)+(-15)=-24$

14. $(-45)-(+11)=$
$(-45)+(-11)=-56$

15. $(-73)-(-24)=$
$(-73)+(+24)=-49$

16. $(+61)-(-13)=$
$(+61)+(+13)=+74$

17. $(\$-65)+(\$-149)=\$-214$

18. $(+15)-(+7)=+8$ plants

19. $(-6)+(-5)=-11$ gallons

20. $(-7)+(-12)=-19$ cents

Systematic Review 2D

1. $(-10)-(-10)=$
$(-10)+(+10)=0$

2. $(-6)-(+11)=$
$(-6)+(-11)=-17$

3. $(-6)-(-10)=$
$(-6)+(+10)=+4$

4. $(-5)-(+7)=$
$(-5)+(-7)=-12$

5. $(+4)-(-4)=$
$(+4)+(+4)=+8$

6. $(-8)-(-7)=$
$(-8)+(+7)=-1$

7. $(-42)+(-56)=-98$

8. $(+19)+(+24)=+43$

9. $(+38)-(+95)=$
$(+38)+(-95)=-57$

10. $(-43)+(+98)=+55$

11. $(+63)-(-22)=$
$(+63)+(+22)=+85$

12. $(-54)-(-58)=$
$(-54)+(+58)=+4$

13. $\dfrac{1}{3}+\dfrac{1}{3}=\dfrac{2}{3}$

14. $\dfrac{5}{8}-\dfrac{3}{8}=\dfrac{2}{8}$

15. $\dfrac{1}{7}+\dfrac{5}{7}=\dfrac{6}{7}$

16. $\dfrac{4}{9}-\dfrac{2}{9}=\dfrac{2}{9}$

17. $\dfrac{1}{5}+\dfrac{3}{5}=\dfrac{4}{5}$ of the book

18. $24\div6=4$
$4\times2=8$ apples used
$24-8=16$ apples left over

19. $(\$-7)+(\$+15)=\$+8$

20. $(+14)+(-9)=+5$ ft

Systematic Review 2E

1. $(+12)-(+13)=$
$(+12)+(-13)=-1$

2. $(-7)-(+4)=$
$(-7)+(-4)=-11$

3. $(+27)-(-25)=$
$(+27)+(+25)=+52$

4. $(-6)-(+3)=$
   $(-6)+(-3)=-9$

5. $(-13)-(-14)=$
   $(-13)+(+14)=+1$

6. $(+39)-(-8)=$
   $(+39)+(+8)=+47$

7. $(+76)-(+26)=$
   $(+76)+(-26)=+50$

8. $(-24)-(+85)=$
   $(-24)+(-85)=-109$

9. $(-35)+(-42)=-77$

10. $(+50)+(-51)=-1$

11. $(+62)-(-12)=$
    $(+62)+(+12)=+74$

12. $(-23)-(-8)=$
    $(-23)+(+8)=-15$

13. $\dfrac{3}{10}+\dfrac{3}{10}=\dfrac{6}{10}$

14. $\dfrac{5}{6}-\dfrac{1}{6}=\dfrac{4}{6}$

15. $\dfrac{3}{8}+\dfrac{2}{8}=\dfrac{5}{8}$

16. $\dfrac{9}{11}-\dfrac{6}{11}=\dfrac{3}{11}$

17. $\dfrac{3}{7}+\dfrac{2}{7}=\dfrac{5}{7}$ of the candies

18. $35 \div 5 = 7$
    $7 \times 3 = 21$ birds

19. $(-4)+(-5)=-9$ hours

20. 9 hours assigned -10 hours
    worked $= -1$ hours left to work

Systematic Review 2F

1. $(+4)-(+10)=$
   $(+4)+(-10)=-6$

2. $(-3)-(-6)=$
   $(-3)+(+6)=+3$

3. $(-2)-(+6)=$
   $(-2)+(-6)=-8$

4. $(+7)-(-14)=$
   $(+7)+(+14)=+21$

5. $(+12)-(-48)=$
   $(+12)+(+48)=+60$

6. $(-8)-(+5)=$
   $(-8)+(-5)=-13$

7. $(-13)+(-11)=-24$

8. $(-8)-(+25)=$
   $(-8)+(-25)=-33$

9. $(+37)-(-40)=$
   $(+37)+(+40)=+77$

10. $(-51)+(+73)=+22$

11. $(-62)+(-65)=-127$

12. $(-16)-(-18)=$
    $(-16)+(+18)=+2$

13. $\dfrac{1}{4}+\dfrac{2}{4}=\dfrac{3}{4}$

14. $\dfrac{6}{7}-\dfrac{5}{7}=\dfrac{1}{7}$

15. $\dfrac{5}{9}-\dfrac{1}{9}=\dfrac{4}{9}$

16. $\dfrac{7}{20}+\dfrac{3}{20}=\dfrac{10}{20}$

17. $30 \div 3 = 10$
    $10 \times 2 = 20$ right
    $30 - 20 = 10$ wrong

18. $\dfrac{4}{10}-\dfrac{3}{10}=\dfrac{1}{10}$ of a pie

19. $(\$-100)+(\$+60)=\$-40$

20. -7 brownies

Lesson Practice 3A

1. $(+5)\times(-6)=-30$

2. $(-6)\times(-7)=+42$

3. $(-9)\times(-10)=+90$

4. $(-10)\times(+12)=-120$

5. $(-5)\times(-8)=+40$

6. $(-16)\times(-11)=+176$

7. $(+4)\times(-15)=-60$

8. $(-18)\times(-6)=+108$

9. $(-16)\times(+12)=-192$

10. $(-17)\times(+3)=-51$

11. $(-18)\times(-4)=+72$

12. $(-24)\times(-5)=+120$

13. $(-11)\times(+16)=-176$
14. $(+3)\times(-24)=-72$
15. $(+8)\times(-12)=-96$
16. $(-10)\times(-16)=+160$
17. $(-3)\times(+6)=-18$ games
18. $(\$-.25)\times(+10)=\$-2.50$
19. $(\$-30)\times(+12)=\$-360$
20. $(+10)\times(+12)=+120\ \text{ft}^2$

Lesson Practice 3B

1. $(+36)\times(-4)=-144$
2. $(-4)\times(-19)=+76$
3. $(-6)\times(-8)=+48$
4. $(-24)\times(-6)=+144$
5. $(-25)\times(-3)=+75$
6. $(-10)\times(+19)=-190$
7. $(-8)\times(+6)=-48$
8. $(-42)\times(+16)=-672$
9. $(-50)\times(-19)=+950$
10. $(+25)\times(-6)=-150$
11. $(+23)\times(-13)=-299$
12. $(-46)\times(-8)=+368$
13. $(-16)\times(-24)=+384$
14. $(-8)\times(-16)=+128$
15. $(-42)\times(-15)=+630$
16. $(-17)\times(+48)=-816$
17. $(\$-3)\times(+2)=\$-6$
18. $(-10)\times(+5)=-50$ years
19. $(\$-682)\times(+4)=\$-2,728$
20. $(-3)\times(+9)=-27$ runs

Lesson Practice 3C

1. $(+8)\times(-5)=-40$
2. $(-6)\times(+10)=-60$
3. $(-3)\times(-4)=+12$
4. $(-20)\times(+12)=-240$
5. $(+17)\times(+3)=+51$

6. $(-8)\times(-9)=+72$
7. $(-90)\times(+4)=-360$
8. $(+24)\times(-8)=-192$
9. $(+42)\times(-6)=-252$
10. $(-10)\times(-10)=+100$
11. $(+7)\times(-6)=-42$
12. $(-18)\times(-4)=+72$
13. $(-36)\times(+4)=-144$
14. $(+13)\times(-4)=-52$
15. $(-17)\times(-3)=+51$
16. $(+19)\times(-51)=-969$
17. $(\$-2)\times(+5)=\$-10$
18. $(-32)\times(+21)=-672$ hairs
19. $(-4)\times(+10)=-40$ losses
20. $(+7)\times(+14)=+98\ \text{ft}^2$

Systematic Review 3D

1. $(+17)\times(-6)=-102$
2. $(+22)\times(-11)=-242$
3. $(-5)\times(-9)=+45$
4. $(-10)\times(+5)=-50$
5. $(+6)\times(-7)=-42$
6. $(-16)\times(+9)=-144$
7. $(+5)-(+10)=$
   $(+5)+(-10)=-5$
8. $(-6)+(-9)=-15$
9. $(+14)+(-3)=+11$
10. $20\div2=10$
    $10\times1=10$
11. $15\div3=5$
    $5\times2=10$
12. $27\div9=3$
    $3\times4=12$
13. $\dfrac{1}{10}+\dfrac{7}{10}=\dfrac{8}{10}$
14. $\dfrac{5}{7}-\dfrac{1}{7}=\dfrac{4}{7}$
15. $\dfrac{4}{8}+\dfrac{1}{8}=\dfrac{5}{8}$
16. $\dfrac{7}{12}-\dfrac{3}{12}=\dfrac{4}{12}$

17. $\dfrac{1}{3} = \dfrac{2}{6} = \dfrac{3}{9} = \dfrac{4}{12}$

18. $\dfrac{2}{5} = \dfrac{4}{10} = \dfrac{6}{15} = \dfrac{8}{20}$

19. $(-2) \times (+13) = -26$ gallons

20. $(+9) + (-2) = +7$ miles

Systematic Review 3E

1. $(+16) \times (-10) = -160$

2. $(+17) \times (-10) = -170$

3. $(+23) \times (+11) = +253$

4. $(-8) \times (-4) = +32$

5. $(-7) \times (-8) = +56$

6. $(+10) \times (-11) = -110$

7. $(+8) - (+19) =$
   $(+8) + (-19) = -11$

8. $(+17) + (-5) = +12$

9. $(-63) - (-50) =$
   $(-63) + (+50) = -13$

10. $18 \div 3 = 6$
    $6 \times 1 = 6$

11. $49 \div 7 = 7$
    $7 \times 3 = 21$

12. $44 \div 11 = 4$
    $4 \times 2 = 8$

13. $\dfrac{4}{5} - \dfrac{2}{5} = \dfrac{2}{5}$

14. $\dfrac{5}{6} + \dfrac{1}{6} = \dfrac{6}{6}$

15. $\dfrac{4}{13} + \dfrac{5}{13} = \dfrac{9}{13}$

16. $\dfrac{1}{4} = \dfrac{2}{8} = \dfrac{3}{12} = \dfrac{4}{16}$

17. $\dfrac{5}{8} = \dfrac{10}{16} = \dfrac{15}{24} = \dfrac{20}{32}$

18. $\dfrac{1}{8} + \dfrac{2}{8} = \dfrac{3}{8}$ of the house

19. $(\$+25) + (\$-30) = \$-5$

20. $(+5) \times (+5) = +25$ mi^2

Systematic Review 3F

1. $(+14) \times (-5) = -70$

2. $(-18) \times (+11) = -198$

3. $(-9) \times (-12) = +108$

4. $(+14) \times (-6) = -84$

5. $(-19) \times (-23) = +437$

6. $(-19) \times (+17) = -323$

7. $(+32) + (-18) = +14$

8. $(-94) + (-7) = -101$

9. $(+58) - (+100) =$
   $(+58) + (-100) = -42$

10. $20 \div 5 = 4$
    $1 \times 4 = 4$

11. $21 \div 3 = 7$
    $7 \times 2 = 14$

12. $50 \div 10 = 5$
    $5 \times 3 = 15$

13. $\dfrac{2}{3} - \dfrac{1}{3} = \dfrac{1}{3}$

14. $\dfrac{4}{7} - \dfrac{2}{7} = \dfrac{2}{7}$

15. $\dfrac{1}{9} + \dfrac{5}{9} = \dfrac{6}{9}$

16. $\dfrac{1}{6} = \dfrac{2}{12} = \dfrac{3}{18} = \dfrac{4}{24}$

17. $\dfrac{3}{7} = \dfrac{6}{14} = \dfrac{9}{21} = \dfrac{12}{28}$

18. $\dfrac{5}{12} - \dfrac{3}{12} = \dfrac{2}{12}$ of a pizza

19. $(\$+15) \times (+4) = \$+60$

20. $(\$-20) \times (+4) = \$-80$
    $(\$-80) + (\$60) = \$-20$

Lesson Practice 4A

1. $\dfrac{-35}{5} = -7$

2. $\dfrac{-49}{-7} = +7$

3. $\dfrac{48}{-6} = -8$

4. $\dfrac{54}{9} = +6$

5. $(15) \div (3) = +5$

6. $(24) \div (-6) = -4$

7. $(-28) \div (-4) = +7$

8. $(-56) \div (7) = -8$

9. $(-30) \div (6) = -5$

10. $(+48) \div (-8) = -6$

11. $(-15) \div (-5) = +3$

12. $(+63) \div (-9) = -7$

13. done

14. done

15. see graph

16. see graph

17. true

18. $(-25) \div (5) = -5$
   losses per week

19. $(\$-144) \div (+12) = \$-12$

20. $(-30) \div (-5) = +6$ days

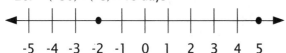

Lesson Practice 4B

1. $\dfrac{-16}{4} = -4$

2. $\dfrac{-81}{-9} = +9$

3. $\dfrac{42}{-6} = -7$

4. $\dfrac{40}{8} = +5$

5. $(-42) \div (-7) = +6$

6. $(+18) \div (-3) = -6$

7. $(-30) \div (-6) = +5$

8. $(-12) \div (2) = -6$

9. $(32) \div (-8) = -4$

10. $(+9) \div (+3) = +3$

11. $(-64) \div (-8) = +8$

12. $(-54) \div (+9) = -6$

13. see graph

14. see graph

15. see graph

16. see graph

17. true

18. $(-49) \div (7) = -7$
   cents per day

19. $(\$-72) \div (6) = \$-12$

20. $(\$-55) \div (\$-5) = +11$ months

Lesson Practice 4C

1. $\dfrac{-36}{-9} = +4$

2. $\dfrac{64}{8} = +8$

3. $\dfrac{-56}{8} = -7$

4. $\dfrac{35}{-7} = -5$

5. $(45) \div (-9) = -5$

6. $(36) \div (-6) = -6$

7. $(-5) \div (-1) = +5$

8. $(-42) \div (6) = -7$

9. $(-144) \div (+12) = -12$

10. $(56) \div (-7) = -8$

11. $(-20) \div (-5) = +4$

12. $(18) \div (-6) = -3$

13. done

14. $D = 1$; see graph

15. $D = 3$; see graph

16. $D = 8$; see graph

17. whole numbers

18. $(-36) \div (6) = -6$ qt

19. $(\$-56) \div (7) = \$-8$

20. $(-20) \div (-4) = +5$ weeks

Systematic Review 4D

1. $\dfrac{-25}{-5} = +5$

2. $(+24) \div (-3) = -8$
3. $(+14) \div (-7) = -2$
4. $(+13) \times (-21) = -273$
5. $(-10) \times (-16) = +160$
6. $(-99) \times (+11) = -1{,}089$
7. $(+9) + (-22) = -13$
8. $(-71) + (-41) = -112$
9. $(+68) - (+7) =$
   $(+68) + (-7) = +61$
10. $45 \div 5 = 9$
    $9 \times 3 = 27$
11. $64 \div 8 = 8$
    $8 \times 1 = 8$
12. $33 \div 11 = 3$
    $3 \times 7 = 21$
13. $\frac{3}{8} = \frac{6}{16} = \frac{9}{24} = \frac{12}{32}$
14. $\frac{1}{9} = \frac{2}{18} = \frac{3}{27} = \frac{4}{36}$
15. $\frac{8}{10} \div \frac{2}{2} = \frac{4}{5}$
16. $\frac{4}{24} \div \frac{4}{4} = \frac{1}{6}$
17. $\frac{6}{18} \div \frac{6}{6} = \frac{1}{3}$
18. $\frac{18}{30} \div \frac{6}{6} = \frac{3}{5}$
19. $1{,}500 \div 50 = 30$ minutes
20. $(10) \times (20) = 200$ fewer ft^2,
    or -200 ft^2

Systematic Review 4E

1. $\frac{42}{-7} = -6$
2. $(14) \div (-2) = -7$
3. $(-24) \div (8) = -3$
4. $(14) \times (-17) = -238$
5. $(-19) \times (-32) = +608$
6. $(-22) \times (+22) = -484$
7. $(+14) + (-16) = -2$
8. $(-83) + (-6) = -89$

9. $(+56) - (+8) =$
   $(+56) + (-8) = +48$
10. $50 \div 2 = 25$
    $25 \times 1 = 25$
11. $28 \div 7 = 4$
    $4 \times 3 = 12$
12. $40 \div 10 = 4$
    $4 \times 3 = 12$
13. $\frac{4}{5} - \frac{1}{5} = \frac{3}{5}$
14. $\frac{7}{8} + \frac{5}{8} = \frac{12}{8} \div \frac{4}{4} = \frac{3}{2}$
15. $\frac{2}{9} + \frac{1}{9} = \frac{3}{9} \div \frac{3}{3} = \frac{1}{3}$
16. $D = 1$
17. integers
18. $\frac{3}{7} + \frac{3}{7} = \frac{6}{7}$ of his route
19. $\frac{4}{6} \div \frac{2}{2} = \frac{2}{3}$ of the class
20. $(\$ - 48) + (\$ - 51) = \$ - 99$

Systematic Review 4F

1. $\frac{-36}{-6} = +6$
2. $(+18) \div (-3) = -6$
3. $(-63) \div (9) = -7$
4. $(+31) \times (-24) = -744$
5. $(-35) \times (-29) = +1{,}015$
6. $(-9) \times (46) = -414$
7. $(+9) + (-23) = -14$
8. $(-17) + (-8) = -25$
9. $(+51) - (+61) =$
   $(+51) + (-61) = -10$
10. $75 \div 3 = 25$
    $25 \times 2 = 50$
11. $18 \div 6 = 3$
    $3 \times 5 = 15$
12. $30 \div 15 = 2$
    $2 \times 2 = 4$
13. $\frac{1}{6} + \frac{4}{6} = \frac{5}{6}$

14. $\dfrac{5}{9} - \dfrac{2}{9} = \dfrac{3}{9} \div \dfrac{3}{3} = \dfrac{1}{3}$

15. $\dfrac{5}{13} + \dfrac{6}{13} = \dfrac{11}{13}$

16. $D = 4$

17. yes; yes

18. $\$35 \div 7 = \$5$
    $\$5 \times 3 = \$15$

19. $(\$10) - (\$10) = \$0$
    $(\$0) + (\$8) = \$8$

20. $(\$-50) + (3 \times \$5) =$
    $(\$-50) + (\$15) = \$-35$

Lesson Practice 5A

1. done
2. done
3. 4^2
4. 10^2
5. $(5)(5) = 25$
6. $(12)(12) = 144$
7. $(4)(4)(4) = 64$
8. $(6) = 6$
9. $\left(\dfrac{1}{3}\right)\left(\dfrac{1}{3}\right) = \dfrac{1}{9}$
10. $\left(\dfrac{3}{4}\right)\left(\dfrac{3}{4}\right) = \dfrac{9}{16}$
11. done
12. 3
13. 2
14. 5
15. 2
16. 3
17. done
18. $1^4 = 1 \times 1 \times 1 \times 1 = 1$
19. $12^2 = 12 \times 12 = 144$
20. $2^4 = 2 \times 2 \times 2 \times 2 = 16$

Lesson Practice 5B

1. 7^2
2. 12^2
3. $2 \times 2 \times 2 \times 2 = 16$
4. $1 \times 1 = 1$
5. $(10)(10) = 100$
6. $(8) = 8$
7. $(9)(9) = 81$
8. $(3)(3)(3) = 27$
9. $\left(\dfrac{1}{2}\right)\left(\dfrac{1}{2}\right)\left(\dfrac{1}{2}\right) = \dfrac{1}{8}$
10. $\left(\dfrac{2}{3}\right)\left(\dfrac{2}{3}\right) = \dfrac{4}{9}$
11. $\left(\dfrac{1}{9}\right)\left(\dfrac{1}{9}\right) = \dfrac{1}{81}$
12. 1
13. 2
14. 4
15. 3
16. 4
17. 2
18. $100^3 = (100)(100)(100) = 1{,}000{,}000$
19. $2^6 = (2)(2)(2)(2)(2)(2) = 64$
20. $6^2 = (6)(6) = 36$

Lesson Practice 5C

1. 1^2
2. 14^2
3. $(1)(1)(1) = 1$
4. $(11)(11) = 121$
5. $(25)(25) = 625$
6. $(3)(3)(3)(3) = 81$
7. $(8)(8) = 64$
8. $(16)(16) = 256$
9. $\left(\dfrac{1}{3}\right)\left(\dfrac{1}{3}\right)\left(\dfrac{1}{3}\right) = \dfrac{1}{27}$
10. $\left(\dfrac{2}{5}\right)\left(\dfrac{2}{5}\right) = \dfrac{4}{25}$

11. $\left(\dfrac{1}{6}\right)\left(\dfrac{1}{6}\right) = \dfrac{1}{36}$

12. 2

13. 3

14. 2

15. 3

16. 5

17. 3

18. $3^4 = (3)(3)(3)(3) = 81$

19. $100^2 = (100)(100) = 10,000$

20. $20^3 = (20)(20)(20) = 8,000$

Systematic Review 5D

1. $(6)(6) = 36$

2. $(13)(13) = 169$

3. $(1)(1)(1)(1)(1) = 1$

4. $\dfrac{1}{7} \cdot \dfrac{1}{7} = \dfrac{1}{49}$

5. 2

6. 3

7. 3

8. $(-9)(-5) = +45$

9. $(-25) \div (+5) = -5$

10. $(-81) + (-9) = -90$

11. $12 \div 3 = 4$
    $4 \times 1 = 4$

12. $12 \div 2 = 6$
    $6 \times 1 = 6$

13. $16 \div 8 = 2$
    $2 \times 7 = 14$

14. $\dfrac{2}{5} + \dfrac{3}{5} = \dfrac{5}{5} = 1$

15. $\dfrac{2}{6} + \dfrac{2}{6} = \dfrac{4}{6} \div \dfrac{2}{2} = \dfrac{2}{3}$

16. $\dfrac{3}{8} + \dfrac{1}{8} = \dfrac{4}{8} \div \dfrac{4}{4} = \dfrac{1}{2}$

17. $\dfrac{1}{4} = \dfrac{5}{20} ; \dfrac{3}{5} = \dfrac{12}{20} ; \dfrac{5}{20} < \dfrac{12}{20}$

18. $\dfrac{3}{4} = \dfrac{18}{24} ; \dfrac{1}{6} = \dfrac{4}{24} ; \dfrac{18}{24} > \dfrac{4}{24}$

19. $\dfrac{1}{3} = \dfrac{5}{15} ; \dfrac{2}{5} = \dfrac{6}{15} ; \dfrac{5}{15} < \dfrac{6}{15}$

20. $\dfrac{1}{5} = \dfrac{4}{20} ; \dfrac{3}{4} = \dfrac{15}{20} ; \dfrac{4}{20} < \dfrac{15}{20}$
    Thursday

Systematic Review 5E

1. $(2)(2)(2) = 8$

2. $(15)(15) = 225$

3. $(5)(5)(5) = 125$

4. $\left(\dfrac{3}{5}\right)\left(\dfrac{3}{5}\right) = \dfrac{9}{25}$

5. 4

6. 2

7. 2

8. $(21) - (29) =$
   $(21) + (-29) = -8$

9. $(-35) \div (-7) = +5$

10. $(-46) \times (-22) = +1,012$

11. $6 \div 2 = 3$
    $3 \times 1 = 3$

12. $10 \div 5 = 2$
    $2 \times 3 = 6$

13. $12 \div 3 = 4$
    $4 \times 2 = 8$

14. $\dfrac{1}{4} = \dfrac{6}{24} ; \dfrac{2}{6} = \dfrac{8}{24} ; \dfrac{6}{24} < \dfrac{8}{24}$

15. $\dfrac{1}{5} = \dfrac{6}{30} ; \dfrac{1}{6} = \dfrac{5}{30} ; \dfrac{6}{30} > \dfrac{5}{30}$

16. $\dfrac{3}{5} = \dfrac{9}{15} ; \dfrac{1}{3} = \dfrac{5}{15} ; \dfrac{9}{15} > \dfrac{5}{15}$

17. $5^3 = (5)(5)(5) = 125$

18. yes; yes

19. $\dfrac{1}{6} = \dfrac{3}{18} ; \dfrac{1}{3} = \dfrac{6}{18} ; \dfrac{3}{18} < \dfrac{6}{18}$
    Isaac ate more

20. $(-30) \div (-5) = +6$ months

Systematic Review 5F

1. $(13)(13) = 169$

2. $(1)(1)(1)(1)(1)(1)(1) = 1$

3. $(32)(32) = 1,024$

4. $\left(\dfrac{2}{3}\right)\left(\dfrac{2}{3}\right)\left(\dfrac{2}{3}\right) = \dfrac{8}{27}$

5. 2

6. 4

7. 2

8. $(-10)+(+44) = +34$

9. $(-15)-(-31) =$
   $(-15)+(+31) = 16$

10. $(11)\div(-11) = -1$

11. $\dfrac{1}{2} = \dfrac{6}{12}$

12. $\dfrac{1}{3} = \dfrac{12}{36}$

13. $\dfrac{2}{8} = \dfrac{4}{16}$

14. $\dfrac{1}{5} = \dfrac{6}{30}; \dfrac{2}{6} = \dfrac{10}{30}; \dfrac{6}{30} < \dfrac{10}{30}$

15. $\dfrac{1}{2} = \dfrac{5}{10}; \dfrac{2}{5} = \dfrac{4}{10}; \dfrac{5}{10} > \dfrac{4}{10}$

16. $\dfrac{3}{7} = \dfrac{24}{56}; \dfrac{1}{8} = \dfrac{7}{56}; \dfrac{24}{56} > \dfrac{7}{56}$

17. $\dfrac{1}{3} + \dfrac{1}{3} = \dfrac{2}{3}$ done

   $\dfrac{3}{3} - \dfrac{2}{3} = \dfrac{1}{3}$ left to do

18. $\dfrac{1}{3} = \dfrac{8}{24}; \dfrac{3}{8} = \dfrac{9}{24}; \dfrac{9}{24} > \dfrac{8}{24}$
   Monte ate more.

19. $1^6 = (1)(1)(1)(1)(1)(1) = 1$

20. $D = 5$

Lesson Practice 6A

1. $10^2; 10^1; 10^0$

2. $\overline{1000}\ \overline{100}\ \overline{10}\ \overline{1}\ \cdot\ \overline{\dfrac{1}{10}}\ \overline{\dfrac{1}{100}}\ \overline{\dfrac{1}{1000}}$

3. $(10)(10) = 100$

4. $(10) = 10$

5. $(10)(10)(10) = 1,000$

6. 4

7. 3

8. 1

9. done

10. done

11. $1\times1,000+4\times100+9\times1$

12. $1\times10^3+4\times10^2+9\times10^0$

13. 8,403

14. 70,060.04

15. 4,962

16. 3,530.3

17. done

18. 3; 2; 7
   $\$3.00+\$0.20+\$0.07 = \$3.27$

Lesson Practice 6B

1. $\overline{1000}\ \overline{100}\ \overline{10}\ \overline{1}\ \cdot\ \overline{\dfrac{1}{10}}\ \overline{\dfrac{1}{100}}\ \overline{\dfrac{1}{1000}}$

2. divide

3. multiply

4. $1\times10^0+9\times\dfrac{1}{10^1}+9\times\dfrac{1}{10^2}+9\times\dfrac{1}{10^3}$

5. $2\times10^1+3\times10^0+6\times\dfrac{1}{10^1}+5\times\dfrac{1}{10^2}$

6. $2\times10^2+3\times10^0+1\times\dfrac{1}{10^2}+6\times\dfrac{1}{10^3}$

7. $8\times10^3+4\times10^2+3\times10^1+9\times10^0+7\times\dfrac{1}{10^3}$

8. 2,401.613

9. 770.907

10. 5.8

11. dimes; pennies
   $\$0.40+\$0.05 = \$0.45$

12. 6; 9; 8
   $\$6.00+\$0.90+\$0.08 = \$6.98$

13. penny

14. dime

Lesson Practice 6C

1. $\overline{1000}\ \overline{100}\ \overline{10}\ \overline{1}\ \cdot\ \overline{\dfrac{1}{10}}\ \overline{\dfrac{1}{100}}\ \overline{\dfrac{1}{1000}}$

2. left

3. right

4. $3\times10^3+4\times10^0+5\times\dfrac{1}{10^3}$

5. $4 \times 10^0 + 6 \times \dfrac{1}{10^1} + 7 \times \dfrac{1}{10^2} + 9 \times \dfrac{1}{10^3}$

6. $6 \times 10^2 + 9 \times 10^1 + 1 \times 10^0 + 4 \times \dfrac{1}{10^1}$

7. $2 \times 10^1 + 5 \times 10^0 + 3 \times \dfrac{1}{10^2}$

8. 9,841.132

9. 3,006.084

10. 200.05

11. dollars; dimes; pennies
$7.00 + $0.40 + $0.05 = $7.45

12. 1; 1; 4
$1.00 + $0.10 + $0.04 = $1.14

13. dollar

14. dollar

Systematic Review 6D

1. $2 \times 10^3 + 3 \times 10^2 + 1 \times \dfrac{1}{10^1}$

2. $3 \times 10^1 + 8 \times 10^0 + 1 \times \dfrac{1}{10^1} + 2 \times \dfrac{1}{10^2} + 3 \times \dfrac{1}{10^3}$

3. $4 \times 10^3 + 3 \times 10^2 + 7 \times 10^1 + 6 \times 10^0$

4. 8×10^2

5. 8,715.546

6. 7,000.83

7. dollars; dime; pennies
$3.00 + $0.10 + $0.09 = $3.19

8. 9; 0; 4
$9.00 + $0.00 + $0.04 = $9.04

9. $(5)(5) = 25$

10. $(1)(1)(1)(1)(1) = 1$

11. 2

12. 3

13. $(-10) + (-16) = -26$

14. $(-6) \times (25) = -150$

15. $(45) \div (-9) = -5$

16. $(-21) - (+28) =$
$(-21) + (-28) = -49$

17. $\dfrac{1}{4} + \dfrac{3}{5} = \dfrac{5}{20} + \dfrac{12}{20} = \dfrac{17}{20}$

18. $\dfrac{3}{4} - \dfrac{1}{6} = \dfrac{18}{24} - \dfrac{4}{24} = \dfrac{14}{24} \div \dfrac{2}{2} = \dfrac{7}{12}$

19. $\dfrac{1}{3} + \dfrac{2}{5} = \dfrac{5}{15} + \dfrac{6}{15} = \dfrac{11}{15}$

20. $4.00 + $0.30 + $0.06 = $4.36

Systematic Review 6E

1. $6 \times \dfrac{1}{10^1} + 1 \times \dfrac{1}{10^2} + 5 \times \dfrac{1}{10^3}$

2. $1 \times 10^2 + 3 \times 10^1 + 5 \times 10^0 + 4 \times \dfrac{1}{10^3}$

3. $3 \times 10^2 + 1 \times 10^0$

4. $4 \times 10^3 + 2 \times 10^2 + 3 \times 10^1 + 8 \times 10^0$

5. 451.221

6. 10.607

7. dollar; dimes; pennies
$1.00 + $0.00 + $0.09 = $1.09

8. 6; 7; 5
$6.00 + $0.70 + $0.05 = $6.75

9. $(2)(2)(2)(2) = 16$

10. 1

11. $(+38) - (-11) =$
$(38) + (+11) = +49$

12. $(-31)(-15) = +465$

13. $(-36) \div (+6) = -6$

14. $(-76) + (+51) = -25$

15. $\dfrac{2}{3} + \dfrac{1}{5} = \dfrac{10}{15} + \dfrac{3}{15} = \dfrac{13}{15}$

16. $\dfrac{1}{2} - \dfrac{1}{7} = \dfrac{7}{14} - \dfrac{2}{14} = \dfrac{5}{14}$

17. $\dfrac{2}{9} + \dfrac{1}{3} = \dfrac{6}{27} + \dfrac{9}{27} = \dfrac{15}{27} \div \dfrac{3}{3} = \dfrac{5}{9}$

18. $\dfrac{1}{2} + \dfrac{1}{3} = \dfrac{3}{6} + \dfrac{2}{6} = \dfrac{5}{6}$ of the class

19. $10 \div 5 = 2$
$2 \times 4 = 8$ pizzas eaten

20. $1.00 + $0.40 + $0.07 = $1.47

Systematic Review 6F

1. $1 \times 10^0 + 1 \times \dfrac{1}{10^3}$

2. $4 \times 10^2 + 5 \times 10^0 + 1 \times \dfrac{1}{10^1} + 6 \times \dfrac{1}{10^3}$

3. $1 \times 10^3 + 1 \times 10^0$

4. $7 \times 10^2 + 3 \times 10^1 + 5 \times 10^0$

5. 6,528.05

6. 300.042

7. dollars; dimes; pennies
$\$9.00 + \$0.80 + \$0.07 = \$9.87$

8. 5; 0; 0
$\$5.00 + \$0.00 + \$0.00 = \$5.00$

9. 2

10. 1

11. $(-16) - (-16) =$
$(-16) + (+16) = 0$

12. $(+10) \times (-10) = -100$

13. $(-56) \div (-8) = +7$

14. $+(48) + (-24) = +24$

15. $\dfrac{1}{5} + \dfrac{2}{6} = \dfrac{6}{30} + \dfrac{10}{30} = \dfrac{16}{30} \div \dfrac{2}{2} = \dfrac{8}{15}$

16. $\dfrac{7}{8} - \dfrac{1}{2} = \dfrac{14}{16} - \dfrac{8}{16} = \dfrac{6}{16} \div \dfrac{2}{2} = \dfrac{3}{8}$

17. $\dfrac{1}{4} + \dfrac{4}{6} = \dfrac{6}{24} + \dfrac{16}{24} = \dfrac{22}{24} \div \dfrac{2}{2} = \dfrac{11}{12}$

18. $\dfrac{1}{4} + \dfrac{2}{5} = \dfrac{5}{20} + \dfrac{8}{20} = \dfrac{13}{20}$ of the
weeding finished

19. $20 \div 5 = 4$
$4 \times 3 = 12$ people who won prizes

20. $20 \div 5 = 4$
$4 \times 2 = 8$ who were good sports
or
$20 - 12 = 8$ who were good sports

Lesson Practice 7A

1. done

2. done

3. $(-2)^3 = (-2)(-2)(-2) = -8$

4. $-4^1 = -4$

5. $-(-6)^2 = -(-6)(-6) = -36$

6. $-(4)^2 = -(4)(4) = -16$

7. $1^3 = (1)(1)(1) = 1$

8. $-3^3 = -(3)(3)(3) = -27$

9. $-(8)^2 = -(8)(8) = -64$

10. $(-10)^2 = (-10)(-10) = 100$

11. $-9^2 = -(9)(9) = -81$

12. $-(2)^4 = -(2)(2)(2)(2) = -16$

13. $\left(-\dfrac{1}{2}\right)^5 = \left(-\dfrac{1}{2}\right)\left(-\dfrac{1}{2}\right)\left(-\dfrac{1}{2}\right)\left(-\dfrac{1}{2}\right)\left(-\dfrac{1}{2}\right)$
$= -\dfrac{1}{32}$

14. $-\left(\dfrac{1}{11}\right)^2 = -\left(\dfrac{1}{11}\right)\left(\dfrac{1}{11}\right) = -\dfrac{1}{121}$

15. $-\left(-\dfrac{1}{3}\right)^2 = -\left(-\dfrac{1}{3}\right)\left(-\dfrac{1}{3}\right) = -\dfrac{1}{9}$

16. $-5^2 = -(5)(5) = -25$

17. $6^2 = (6)(6) = 36$

18. $(-2)^4 = (-2)(-2)(-2)(-2) = 16$

19. $10^2 \times -1 = -(10)(10) = -100$

20. $(-15)^2 \times -1 = -(-15)(-15) = -225$

Lesson Practice 7B

1. $-(-5)^2 = -(-5)(-5) = -25$

2. $-(6)^2 = -(6)(6) = -36$

3. $-(+7)^2 = -(+7)(+7) = -49$

4. $-(3)^2 = -(3)(3) = -9$

5. $(-8)^2 = (-8)(-8) = 64$

6. $-11^2 = -(11)(11) = -121$

7. $-4^3 = -(4)(4)(4) = -64$

8. $-(-1)^4 = -(-1)(-1)(-1)(-1) = -1$

9. $(-14)^2 = (-14)(-14) = 196$

10. $-4^2 = -(4)(4) = -16$

11. $-(5)^2 = -(5)(5) = -25$

12. $-2^5 = -(2)(2)(2)(2)(2) = -32$

13. $(-2)^5 = (-2)(-2)(-2)(-2)(-2) = -32$

14. $-\left(\dfrac{2}{3}\right)^2 = -\left(\dfrac{2}{3}\right)\left(\dfrac{2}{3}\right) = -\dfrac{4}{9}$

15. $-\left(-\dfrac{1}{4}\right)^2 = -\left(-\dfrac{1}{4}\right)\left(-\dfrac{1}{4}\right) = -\dfrac{1}{16}$

16. $\left(-\dfrac{1}{10}\right)^2 = \left(-\dfrac{1}{10}\right)\left(-\dfrac{1}{10}\right) = \dfrac{1}{100}$

17. $\left(-\dfrac{1}{2}\right)^2 = \left(-\dfrac{1}{2}\right)\left(-\dfrac{1}{2}\right) = \dfrac{1}{4}$

18. $(-2)^4 = (-2)(-2)(-2)(-2) = 16$

19. $(-4)^2 \times -1 = -(-4)(-4) = -16$

20. $(-1)^5 = (-1)(-1)(-1)(-1)(-1) = -1$

Lesson Practice 7C

1. $-6^2 = -(6)(6) = -36$

2. $-(7)^2 = -(7)(7) = -49$

3. $-(12)^2 = -(12)(12) = -144$

4. $-(3)^1 = -3$

5. $-(-2)^3 = -(-2)(-2)(-2) = -(-8) = 8$

6. $-8^2 = -(8)(8) = -64$

7. $9^2 = (9)(9) = 81$

8. $-(-6)^2 = -(-6)(-6) = -36$

9. $-(5)^3 = -(5)(5)(5) = -125$

10. $(-10)^2 = (-10)(-10) = 100$

11. $\left(-\dfrac{1}{3}\right)^3 = \left(-\dfrac{1}{3}\right)\left(-\dfrac{1}{3}\right)\left(-\dfrac{1}{3}\right) = -\dfrac{1}{27}$

12. $-12^2 = -(12)(12) = -144$

13. $-(4)^5 = -(4)(4)(4)(4)(4) = -1{,}024$

14. $-10^2 = -(10)(10) = -100$

15. $-\left(-\dfrac{2}{5}\right)^2 = -\left(-\dfrac{2}{5}\right)\left(-\dfrac{2}{5}\right) = -\dfrac{4}{25}$

16. $-\left(\dfrac{4}{9}\right)^2 = -\left(\dfrac{4}{9}\right)\left(\dfrac{4}{9}\right) = -\dfrac{16}{81}$

17. $(-6)^2 = (-6)(-6) = 36$

18. $(-1)^3 = (-1)(-1)(-1) = -1$

19. $5^2 \times -1 = -(5)(5) = -25$

20. $(-11)^2 \times -1 = -(-11)(-11) = -121$

Systematic Review 7D

1. $-10^2 = -(10)(10) = -100$

2. $-(15)^2 = -(15)(15) = -225$

3. $(-4)^2 = (-4)(-4) = 16$

4. $-\left(\dfrac{2}{3}\right)^3 = -\left(\dfrac{2}{3}\right)\left(\dfrac{2}{3}\right)\left(\dfrac{2}{3}\right) = -\dfrac{8}{27}$

5. $4 \times \dfrac{1}{10^1} + 8 \times \dfrac{1}{10^2} + 5 \times \dfrac{1}{10^3}$

6. $2 \times 10^1 + 3 \times 10^0 + 1 \times \dfrac{1}{10^1} + 6 \times \dfrac{1}{10^2}$

7. 302.46

8. $3{,}000.815$

9. $(+16) + (-14) = +2$

10. $(-23) - (+9) =$
$(-23) + (-9) = -32$

11. $(-15) - (-35) =$
$(-15) + (+35) = +20$

12. $\dfrac{1}{6} + \dfrac{3}{4} = \dfrac{4}{24} + \dfrac{18}{24} = \dfrac{22}{24} \div \dfrac{2}{2} = \dfrac{11}{12}$

13. $\dfrac{3}{5} - \dfrac{1}{4} = \dfrac{12}{20} - \dfrac{5}{20} = \dfrac{7}{20}$

14. $\dfrac{3}{6} + \dfrac{2}{5} = \dfrac{15}{30} + \dfrac{12}{30} = \dfrac{27}{30} \div \dfrac{3}{3} = \dfrac{9}{10}$

15. $\dfrac{1}{\cancel{2}} \times \dfrac{\cancel{2}}{3} = \dfrac{1}{3}$

16. $\dfrac{\cancel{3}}{4} \times \dfrac{1}{\cancel{9}_3} = \dfrac{1}{12}$

17. $\dfrac{\cancel{6}^3}{8} \times \dfrac{1}{\cancel{2}} = \dfrac{3}{8}$

18. $(-10)^3 = (-10)(-10)(-10) = -1{,}000$

19. $\dfrac{1}{5} \times \dfrac{1}{2} = \dfrac{1}{10}$ of a pie

20. water and fat:
$\dfrac{1}{8} + \dfrac{1}{10} = \dfrac{10}{80} + \dfrac{8}{80} = \dfrac{18}{80} \div \dfrac{2}{2} = \dfrac{9}{40}$

$\dfrac{40}{40} - \dfrac{9}{40} = \dfrac{31}{40}$ pound of meat

Systematic Review 7E

1. $(-9)^2 = (-9)(-9) = 81$

2. $-(5)^2 = -(5)(5) = -25$

3. $-2^3 = -(2)(2)(2) = -8$

4. $-\left(\dfrac{3}{4}\right)^2 = -\left(\dfrac{3}{4}\right)\left(\dfrac{3}{4}\right) = -\dfrac{9}{16}$

5. $6 \times \dfrac{1}{10^1} + 3 \times \dfrac{1}{10^3}$

6. $1 \times 10^2 + 2 \times 10^1 + 5 \times 10^0 + 9 \times \dfrac{1}{10^2}$

7. $1{,}402.001$

8. $5{,}820.387$

9. $(10) \times (-2) = -20$

10. $(-14) + (-18) = -32$

11. $(-42) \div (-7) = +6$

12. $\dfrac{1}{3} + \dfrac{1}{4} = \dfrac{4}{12} + \dfrac{3}{12} = \dfrac{7}{12}$

13. $\dfrac{2}{5} + \dfrac{1}{8} = \dfrac{16}{40} + \dfrac{5}{40} = \dfrac{21}{40}$

14. $\dfrac{5}{6} - \dfrac{1}{4} = \dfrac{20}{24} - \dfrac{6}{24} = \dfrac{14}{24} \div \dfrac{2}{2} = \dfrac{7}{12}$

15. $\dfrac{2}{5} \times \dfrac{2}{3} = \dfrac{4}{15}$

16. $\dfrac{\overset{2}{\cancel{8}}}{9} \times \dfrac{1}{\cancel{4}} = \dfrac{2}{9}$

17. $\dfrac{1}{\cancel{5}} \times \dfrac{\cancel{5}}{6} = \dfrac{1}{6}$

18. $(-6)^2 \times -1 = -(-6)(-6) = -36$

19. $\dfrac{1}{2} \times \dfrac{1}{3} = \dfrac{1}{6}$ of his collection

20. $\dfrac{3}{4} - \dfrac{1}{4} = \dfrac{2}{4} \div \dfrac{2}{2} = \dfrac{1}{2}$ pizza

Systematic Review 7F

1. $-(13)^2 = -(13)(13) = -169$

2. $-8^2 = -(8)(8) = -64$

3. $(-4)^3 = (-4)(-4)(-4) = -64$

4. $\left(-\dfrac{2}{7}\right)^2 = \left(-\dfrac{2}{7}\right)\left(-\dfrac{2}{7}\right) = \dfrac{4}{49}$

5. $3 \times 10^0 + 8 \times \dfrac{1}{10^3}$

6. $4 \times 10^1 + 5 \times 10^0 + 1 \times \dfrac{1}{10^1} + 6 \times \dfrac{1}{10^2} + 3 \times \dfrac{1}{10^3}$

7. $5{,}367.02$

8. $7{,}070.215$

9. $(-77) + (15) = -62$

10. $(52) \times (-36) = -1{,}872$

11. $(-72) \div (9) = -8$

12. $\dfrac{1}{3} + \dfrac{1}{6} = \dfrac{6}{18} + \dfrac{3}{18} = \dfrac{9}{18} \div \dfrac{9}{9} = \dfrac{1}{2}$

13. $\dfrac{2}{4} + \dfrac{2}{5} = \dfrac{10}{20} + \dfrac{8}{20} = \dfrac{18}{20} \div \dfrac{2}{2} = \dfrac{9}{10}$

14. $\dfrac{4}{9} - \dfrac{1}{3} = \dfrac{12}{27} - \dfrac{9}{27} = \dfrac{3}{27} \div \dfrac{3}{3} = \dfrac{1}{9}$

15. $\dfrac{2}{5} \times \dfrac{2}{5} = \dfrac{4}{25}$

16. $\dfrac{\overset{2}{\cancel{4}}}{\underset{3}{\cancel{6}}} \times \dfrac{1}{\underset{5}{\cancel{10}}} = \dfrac{1}{15}$

17. $\dfrac{\cancel{7}}{8} \times \dfrac{1}{\underset{2}{\cancel{14}}} = \dfrac{1}{16}$

18. $\dfrac{1}{3} + \dfrac{2}{5} = \dfrac{5}{15} + \dfrac{6}{15} = \dfrac{11}{15}$ of the work

19. $\dfrac{\overset{2}{\cancel{4}}}{5} \times \dfrac{1}{\cancel{2}} = \dfrac{2}{5}$ of the total pay

20. $D = 2$

Lesson Practice 8A

1. done

2. $8 \times 8 = 64$ units^2

3. $12 \times 12 = 144$ units^2

4. done

5. $\sqrt{64} = 8$

6. $\sqrt{144} = 12$

7. $\sqrt{9} = 3$

8. $\sqrt{4} = 2$

9. $\sqrt{25} = 5$

10. $\sqrt{81} = 9$

11. $\sqrt{16} = 4$

12. $\sqrt{36} = 6$

13. $\sqrt{121} = 11$

14. $\sqrt{1} = 1$

15. $\sqrt{49} = 7$

16. $\sqrt{10^2} = 10$

17. $\sqrt{9^2} = 9$

18. $\sqrt{6^2} = 6$

19. $\sqrt{X^2} = X$

20. $\sqrt{144} = 12$ ft

Lesson Practice 8B

1. $7 \times 7 = 49$ units^2
2. $5 \times 5 = 25$ units^2
3. $9 \times 9 = 81$ units^2
4. $\sqrt{49} = 7$
5. $\sqrt{25} = 5$
6. $\sqrt{81} = 9$
7. $\sqrt{100} = 10$
8. $\sqrt{144} = 12$
9. $\sqrt{4} = 2$
10. $\sqrt{16} = 4$
11. $\sqrt{121} = 11$
12. $\sqrt{1} = 1$
13. $\sqrt{9} = 3$
14. $\sqrt{36} = 6$
15. $\sqrt{64} = 8$
16. $\sqrt{3^2} = 3$
17. $\sqrt{8^2} = 8$
18. $\sqrt{Y^2} = Y$
19. $\sqrt{A^2} = A$
20. $\sqrt{121} = 11$ soldiers

Lesson Practice 8C

1. $11 \times 11 = 121$ units^2
2. $6 \times 6 = 36$ units^2
3. $4 \times 4 = 16$ units^2
4. $\sqrt{121} = 11$
5. $\sqrt{36} = 6$
6. $\sqrt{16} = 4$
7. $\sqrt{144} = 12$
8. $\sqrt{49} = 7$
9. $\sqrt{64} = 8$
10. $\sqrt{1} = 1$
11. $\sqrt{81} = 9$
12. $\sqrt{4} = 2$
13. $\sqrt{100} = 10$
14. $\sqrt{9} = 3$
15. $\sqrt{25} = 5$

16. $\sqrt{4^2} = 4$
17. $\sqrt{11^2} = 11$
18. $\sqrt{X^2} = X$
19. $\sqrt{B^2} = B$
20. $\sqrt{36} = 6$ ft

Systematic Review 8D

1. $\sqrt{144} = 12$
2. $\sqrt{16} = 4$
3. $\sqrt{81} = 9$
4. $\sqrt{Z^2} = Z$
5. $-6^2 = -(6)(6) = -36$
6. $-(11)^2 = -(11)(11) = -121$
7. $(-4)^2 = (-4)(-4) = 16$
8. $-\left(\frac{1}{2}\right)^4 = -\left(\frac{1}{2}\right)\left(\frac{1}{2}\right)\left(\frac{1}{2}\right)\left(\frac{1}{2}\right) = -\frac{1}{16}$
9. 130.29
10. 27.004
11. $\frac{5}{12} = \frac{35}{84}; \frac{4}{7} = \frac{48}{84}; \frac{35}{84} < \frac{48}{84}$
12. $\frac{4}{5} = \frac{40}{50}; \frac{8}{10} = \frac{40}{50}; \frac{40}{50} = \frac{40}{50}$
13. $\frac{1}{2} \times \frac{1}{2} = \frac{1}{4}$
14. $\frac{3}{7} \times \frac{2}{3} = \frac{2}{7}$
15. $\frac{1}{5} \times \frac{4}{5} = \frac{4}{25}$
16. $\frac{3}{4} \div \frac{1}{2} = \frac{6}{8} \div \frac{4}{8} = \frac{6 \div 4}{8 \div 8} = \frac{6 \div 4}{1} = \frac{6}{4}$
17. $\frac{4}{5} \div \frac{1}{3} = \frac{12}{15} \div \frac{5}{15} = \frac{12 \div 5}{15 \div 15} =$
$\frac{12 \div 5}{1} = \frac{12}{5}$
18. $\frac{2}{3} \div \frac{1}{4} = \frac{8}{12} \div \frac{3}{12} = \frac{8 \div 3}{12 \div 12} =$
$\frac{8 \div 3}{1} = \frac{8}{3}$
19. $\frac{3}{4} \div \frac{1}{8} = \frac{24}{32} \div \frac{4}{32} = \frac{24 \div 4}{32 \div 32} =$
$\frac{6}{1} = 6$ times

20. $\dfrac{2}{3} \div \dfrac{1}{6} = \dfrac{12}{18} \div \dfrac{3}{18} = \dfrac{12 \div 3}{18 \div 18} =$

$\dfrac{4}{1} = 4$ pieces

Systematic Review 8E

1. $\sqrt{144} = 12$
2. $\sqrt{16} = 4$
3. $\sqrt{81} = 9$
4. $\sqrt{Z^2} = Z$
5. $-6^2 = -(6)(6) = -36$
6. $-(11)^2 = -(11)(11) = -121$
7. $(-4)^2 = (-4)(-4) = 16$
8. $-\left(\dfrac{1}{2}\right)^4 = -\left(\dfrac{1}{2}\right)\left(\dfrac{1}{2}\right)\left(\dfrac{1}{2}\right)\left(\dfrac{1}{2}\right) = -\dfrac{1}{16}$
9. 130.29
10. 27.004
11. $\dfrac{5}{12} = \dfrac{35}{84}; \dfrac{4}{7} = \dfrac{48}{84}; \dfrac{35}{84} < \dfrac{48}{84}$
12. $\dfrac{4}{5} = \dfrac{40}{50}; \dfrac{8}{10} = \dfrac{40}{50}; \dfrac{40}{50} = \dfrac{40}{50}$
13. $\dfrac{1}{2} \times \dfrac{1}{2} = \dfrac{1}{4}$
14. $\dfrac{3}{7} \times \dfrac{2}{3} = \dfrac{2}{7}$
15. $\dfrac{1}{5} \times \dfrac{4}{5} = \dfrac{4}{25}$
16. $\dfrac{3}{4} \div \dfrac{1}{2} = \dfrac{6}{8} \div \dfrac{4}{8} = \dfrac{6 \div 4}{8 \div 8} = \dfrac{6 \div 4}{1} = \dfrac{6}{4}$
17. $\dfrac{4}{5} \div \dfrac{1}{3} = \dfrac{12}{15} \div \dfrac{5}{15} = \dfrac{12 \div 5}{15 \div 15} =$

$\dfrac{12 \div 5}{1} = \dfrac{12}{5}$
18. $\dfrac{2}{3} \div \dfrac{1}{4} = \dfrac{8}{12} \div \dfrac{3}{12} = \dfrac{8 \div 3}{12 \div 12} =$

$\dfrac{8 \div 3}{1} = \dfrac{8}{3}$
19. $\dfrac{3}{4} \div \dfrac{1}{8} = \dfrac{24}{32} \div \dfrac{4}{32} = \dfrac{24 \div 4}{32 \div 32} =$

Systematic Review 8F

1. $\sqrt{49} = 7$
2. $\sqrt{36} = 6$
3. $\sqrt{100} = 10$
4. $\sqrt{8^2} = 8$
5. $(-4)^2 = (-4)(-4) = 16$
6. $-(12)^2 = -(12)(12) = -144$
7. $(6)^2 = (6)(6) = 36$
8. $\left(\dfrac{2}{3}\right)^3 = \left(\dfrac{2}{3}\right)\left(\dfrac{2}{3}\right)\left(\dfrac{2}{3}\right) = \dfrac{8}{27}$
9. $1 \times 10^1 + 2 \times 10^0 + 6 \times \dfrac{1}{10^1} + 7 \times \dfrac{1}{10^2} + 4 \times \dfrac{1}{10^3}$
10. $1 \times 10^3 + 2 \times \dfrac{1}{10^2}$
11. $\dfrac{1}{2} = \dfrac{2}{4} = \dfrac{3}{6} = \dfrac{4}{8}$
12. $\dfrac{3}{5} = \dfrac{6}{10} = \dfrac{9}{15} = \dfrac{12}{20}$
13. $\dfrac{2}{3} \times \dfrac{5}{8}_4 = \dfrac{5}{12}$
14. $\dfrac{1}{3}_6 \times \dfrac{^2 4}{5} = \dfrac{2}{15}$
15. $\dfrac{^3 6}{7} \times \dfrac{1}{2} = \dfrac{3}{7}$
16. $\dfrac{1}{3} \div \dfrac{1}{5} = \dfrac{5}{15} \div \dfrac{3}{15} = \dfrac{5 \div 3}{15 \div 15} = \dfrac{5 \div 3}{1} = \dfrac{5}{3}$
17. $\dfrac{4}{5} \div \dfrac{2}{5} = \dfrac{4 \div 2}{5 \div 5} = \dfrac{2}{1} = 2$

Lesson Practice 9A

1. done
2. done
3. done
4. done

These problems may be solved using either of the methods shown in the text.

5. $X - 9 = 63$
    $X - 9 + 9 = 63 + 9$
          $X = 72$

6. $X - 9 = 63 \Rightarrow (72) - 9 = 63$
                        $63 = 63$

7.
$$X + 5 = 35$$
$$X + 5 - 5 = 35 - 5$$
$$X = 30$$

8.
$$X + 5 = 35 \Rightarrow (30) + 5 = 35$$
$$35 = 35$$

9.
$$6X = 36 + 5X$$
$$6X - 5X = 36 + 5X - 5X$$
$$X = 36$$

10.
$$6X = 36 + 5X \Rightarrow 6(36) = 36 + 5(36)$$
$$216 = 36 + 180$$
$$216 = 216$$

11.
$$5X - 5 = 4X + 5$$
$$5X - 5 + 5 = 4X + 5 + 5$$
$$5X = 4X + 10$$
$$5X - 4X = 4X - 4X + 10$$
$$X = 10$$

12.
$$5X - 5 = 4X + 5 \Rightarrow 5(10) - 5 = 4(10) + 5$$
$$50 - 5 = 40 + 5$$
$$45 = 45$$

13.
$$-X + 4 = -2X - 6$$
$$-X + 2X + 4 = -2X + 2X - 6$$
$$X + 4 = -6$$
$$X + 4 - 4 = -6 - 4$$
$$X = -10$$

14.
$$-X + 4 = -2X - 6 \Rightarrow -(-10) + 4 = -2(-10) - 6$$
$$10 + 4 = 20 - 6$$
$$14 = 14$$

15. done
16. done

17.
$$R + 6 = 9$$
$$R + 6 - 6 = 9 - 6$$
$$R = 3 \text{ rabbits}$$

18.
$$2P - 10 = P + 20$$
$$2P - 10 + 10 = P + 20 + 10$$
$$2P = P + 30$$
$$2P - P = P - P + 30$$
$$P = 30 \text{ pennies}$$

Lesson Practice 9B

1.
$$X + 18 = 24$$
$$X + 18 - 18 = 24 - 18$$
$$X = 6$$

2.
$$X + 18 = 24 \Rightarrow (6) + 18 = 24$$
$$24 = 24$$

3.
$$Y - 16 = 36$$
$$Y - 16 + 16 = 36 + 16$$
$$Y = 52$$

4.
$$Y - 16 = 36 \Rightarrow (52) - 16 = 36$$
$$36 = 36$$

5.
$$X + 4 = 9$$
$$X + 4 - 4 = 9 - 4$$
$$X = 5$$

6.
$$X + 4 = 9 \Rightarrow (5) + 4 = 9$$
$$9 = 9$$

7.
$$X - 10 = -10$$
$$X - 10 + 10 = -10 + 10$$
$$X = 0$$

8.
$$X - 10 = -10 \Rightarrow (0) - 10 = -10$$
$$-10 = -10$$

9.
$$3B + 4 = 2B - 4$$
$$3B + 4 - 4 = 2B - 4 - 4$$
$$3B = 2B - 8$$
$$3B - 2B = 2B - 2B - 8$$
$$B = -8$$

10.
$$3B + 4 = 2B - 4 \Rightarrow 3(-8) + 4 = 2(-8) - 4$$
$$-24 + 4 = -16 - 4$$
$$-20 = -20$$

11.
$$-X + 11 = -2X + 16$$
$$-X + 11 - 11 = -2X + 16 - 11$$
$$-X = -2X + 5$$
$$-X + 2X = -2X + 2X + 5$$
$$X = 5$$

12.
$$-X + 11 = -2X + 16 \Rightarrow -(5) + 11 = -2(5) + 16$$
$$6 = -10 + 16$$
$$6 = 6$$

13.
$$18 = 6 + J$$
$$18 - 6 = 6 - 6 + J$$
$$12 = J$$
$$J = 12$$

14.
$$5N + 9 = 4N + 12$$
$$5N + 9 - 9 = 4N + 12 - 9$$
$$5N = 4N + 3$$
$$5N - 4N = 4N + 3 - 4N$$
$$N = 3$$

15.
$$D + \$200 = \$500$$
$$D + \$200 - \$200 = \$500 - \$200$$
$$D = \$300$$

16.
$$2D = 12 + D$$
$$2D - D = 12 + D - D$$
$$D = \$12$$

17.
$$X + 8 = 15$$
$$X + 8 - 8 = 15 - 8$$
$$X = 7 \text{ chores}$$

18.
$$2D = D + 30$$
$$2D - D = D + 30 - D$$
$$D = 30 \text{ dimes}$$

Lesson Practice 9C

1.
$$B - 24 = 52$$
$$B - 24 + 24 = 52 + 24$$
$$B = 76$$

2. $B - 24 = 52 \Rightarrow (76) - 24 = 52$
$$52 = 52$$

3.
$$X - 7 = -3$$
$$X - 7 + 7 = -3 + 7$$
$$X = 4$$

4. $X - 7 = -3 \Rightarrow (4) - 7 = -3$
$$-3 = -3$$

5.
$$A + 52 = 100$$
$$A + 52 - 52 = 100 - 52$$
$$A = 48$$

6. $A + 52 = 100 \Rightarrow (48) + 52 = 100$
$$100 = 100$$

7.
$$6Y - 2 = 5Y - 2$$
$$6Y - 2 + 2 = 5Y - 2 + 2$$
$$6Y = 5Y$$
$$6Y - 5Y = 5Y - 5Y$$
$$Y = 0$$

8. $6Y - 2 = 5Y - 2 \Rightarrow 6(0) - 2 = 5(0) - 2$
$$0 - 2 = 0 - 2$$
$$-2 = -2$$

9.
$$-X - 7 = -2X - 5$$
$$-X - 7 + 7 = -2X - 5 + 7$$
$$-X = -2X + 2$$
$$-X + 2X = -2X + 2 + 2X$$
$$X = 2$$

10. $-X - 7 = -2X - 5 \Rightarrow -(2) - 7 = -2(2) - 5$
$$-9 = -4 - 5$$
$$-9 = -9$$

11.
$$23D + 13 = 22D + 25$$
$$23D + 13 - 13 = 22D + 25 - 13$$
$$23D = 22D + 12$$
$$23D - 22D = 22D + 12 - 22D$$
$$D = 12$$

12. $23D + 13 = 22D + 25 \Rightarrow$
$$23(12) + 13 = 22(12) + 25$$
$$276 + 13 = 264 + 25$$
$$289 = 289$$

13.
$$F + 20 = 52$$
$$F + 20 - 20 = 52 - 20$$
$$F = 32$$

14.
$$7N - 8 = 6N - 5$$
$$7N - 8 + 8 = 6N - 5 + 8$$
$$7N = 6N + 3$$
$$7N - 6N = 6N + 3 - 6N$$
$$N = 3$$

15.
$$X + 6 = 10$$
$$X + 6 - 6 = 10 - 6$$
$$X = 4 \text{ pies}$$

16.
$$2R - 1 = R + 3$$
$$2R - 1 + 1 = R + 3 + 1$$
$$2R = R + 4$$
$$2R - R = R + 4 - R$$
$$R = 4 \text{ runs}$$

17.
$$T + 3 = 8$$
$$T + 3 - 3 = 8 - 3$$
$$T = 5 \text{ miles}$$

18.
$$3V + 1 = 2V + 6$$
$$3V + 1 - 1 = 2V + 6 - 1$$
$$3V = 2V + 5$$
$$3V - 2V = 2V + 5 - 2V$$
$$V = 5 \text{ ft}$$

Systematic Review 9D

1. $$X + 36 = 45$$
$$X + 36 - 36 = 45 - 36$$
$$X = 9$$

2. $X + 36 = 45 \Rightarrow (9) + 36 = 45$
$$45 = 45$$

3. $$3B - 6 = 2B + 7$$
$$3B - 6 + 6 = 2B + 7 + 6$$
$$3B = 2B + 13$$
$$3B - 2B = 2B + 13 - 2B$$
$$B = 13$$

4. $3B - 6 = 2B + 7 \Rightarrow 3(13) - 6 = 2(13) + 7$
$$39 - 6 = 26 + 7$$
$$33 = 33$$

5. $\sqrt{49} = 7$

6. $\sqrt{A^2} = A$

7. $(-6)^2 = (-6)(-6) = 36$

8. done

9. $\dfrac{3}{4} + \dfrac{5}{9} = \dfrac{27}{36} + \dfrac{20}{36} = \dfrac{47}{36}$ or $1\dfrac{11}{36}$

10. $\dfrac{1}{3} + \dfrac{7}{8} = \dfrac{8}{24} + \dfrac{21}{24} = \dfrac{29}{24}$ or $1\dfrac{5}{24}$

11. $\dfrac{3}{4} - \dfrac{1}{5} = \dfrac{15}{20} - \dfrac{4}{20} = \dfrac{11}{20}$

12. done

13. $\dfrac{7}{1} \times \dfrac{1}{7} = \dfrac{7}{7} = 1$

14. $\dfrac{9}{8} \times \dfrac{8}{9} = \dfrac{72}{72} = 1$

15. $\dfrac{1}{2} \times \dfrac{2}{1} = \dfrac{2}{2} = 1$

16. $4,000.312$

17. $2 \times 10^0 + 3 \times \dfrac{1}{10^3}$

18. $\dfrac{1}{3} \div \dfrac{1}{6} = \dfrac{6}{18} \div \dfrac{3}{18}$
$$= \dfrac{6 \div 3}{18 \div 18} = \dfrac{2}{1} = 2 \text{ boards}$$

19. $$B + 100 = 250$$
$$B + 100 - 100 = 250 - 100$$
$$B = 150 \text{ books}$$

20. $\dfrac{5}{6} = \dfrac{15}{18}; \dfrac{2}{3} = \dfrac{12}{18}; \dfrac{15}{18} > \dfrac{12}{18}$
Chris ate more.

Systematic Review 9E

1. $$7X + 10 = 6X + 4$$
$$7X + 10 - 10 = 6X + 4 - 10$$
$$7X = 6X - 6$$
$$7X - 6X = 6X - 6X - 6$$
$$X = -6$$

2. $7X + 10 = 6X + 4 \Rightarrow 7(-6) + 10 = 6(-6) + 4$
$$-42 + 10 = -36 + 4$$
$$-32 = -32$$

3. $$-F - 6 = -2F + 1$$
$$-F - 6 + 6 = -2F + 1 + 6$$
$$-F = -2F + 7$$
$$-F + 2F = -2F + 2F + 7$$
$$F = 7$$

4. $-F - 6 = -2F + 1 \Rightarrow -(7) - 6 = -2(7) + 1$
$$-13 = -14 + 1$$
$$-13 = -13$$

5. $\sqrt{144} = 12$

6. $\sqrt{X^2} = X$

7. $-2^4 = -(2)(2)(2)(2) = -16$

8. $\dfrac{2}{\sqrt{81}} = \dfrac{2}{9}$

9. $\dfrac{1}{5} + \dfrac{3}{4} = \dfrac{4}{20} + \dfrac{15}{20} = \dfrac{19}{20}$

10. $\dfrac{3}{5} - \dfrac{1}{3} = \dfrac{9}{15} - \dfrac{5}{15} = \dfrac{4}{15}$

11. $\dfrac{7}{8} - \dfrac{1}{6} = \dfrac{42}{48} - \dfrac{8}{48} = \dfrac{34}{48} \div \dfrac{2}{2} = \dfrac{17}{24}$

12. $\dfrac{3}{4} \div \dfrac{5}{8} = \dfrac{24}{32} \div \dfrac{20}{32} = \dfrac{24 \div 20}{32 \div 32} =$
$$\dfrac{24 \div 20}{1} = \dfrac{24}{20} \div \dfrac{4}{4} = \dfrac{6}{5}$$

13. $\dfrac{2}{3} \div \dfrac{5}{7} = \dfrac{14}{21} \div \dfrac{15}{21} = \dfrac{14 \div 15}{21 \div 21}$
$$= \dfrac{14 \div 15}{1} = \dfrac{14}{15}$$

14. $\dfrac{5}{9} \div \dfrac{2}{3} = \dfrac{15}{27} \div \dfrac{18}{27} = \dfrac{15 \div 18}{27 \div 27} =$
$$\dfrac{15 \div 18}{1} = \dfrac{15}{18} \div \dfrac{3}{3} = \dfrac{5}{6}$$

15. $\dfrac{1}{3} \times \dfrac{3}{1} = \dfrac{3}{3} = 1$

16. $\dfrac{9}{1} \times \dfrac{1}{9} = \dfrac{9}{9} = 1$

17. $\dfrac{4}{5} \times \dfrac{5}{4} = \dfrac{20}{20} = 1$

18. $\dfrac{9}{10} \div \dfrac{1}{10} = \dfrac{90}{100} \div \dfrac{10}{100} =$

$\qquad = \dfrac{90 \div 10}{100 \div 100} = \dfrac{9}{1}$

$\qquad = 9 \text{ volunteers}$

19. $6X + 8 = 5X - 5$

$\quad 6X + 8 - 8 = 5X - 5 - 8$

$\quad 6X = 5X - 13$

$\quad 6X - 5X = 5X - 5X - 13$

$\quad X = -13$

20. $24 \div 6 = 4; \ 4 \times 1 = 4 \text{ hours}$

Systematic Review 9F

1. $13R + 1 = 12R + 7$

$\quad 13R + 1 - 1 = 12R + 7 - 1$

$\quad 13R = 12R + 6$

$\quad 13R - 12R = 12R - 12R + 6$

$\quad R = 6$

2. $13R + 1 = 12R + 7 \Rightarrow 13(6) + 1 = 12(6) + 7$

$\qquad\qquad 78 + 1 = 72 + 7$

$\qquad\qquad 79 = 79$

3. $-6X - 3 = -7X + 5$

$\quad -6X - 3 + 3 = -7X + 5 + 3$

$\quad -6X = -7X + 8$

$\quad -6X + 7X = -7X + 7X + 8$

$\quad X = 8$

4. $-6X - 3 = -7X + 5 \Rightarrow -6(8) - 3 = -7(8) + 5$

$\qquad\qquad -48 - 3 = -56 + 5$

$\qquad\qquad -51 = -51$

5. $\sqrt{121} = 11$

6. $\sqrt{B^2} = B$

7. $-6^2 = -(6)(6) = -36$

8. $\left(\dfrac{2}{5}\right)^2 = \left(\dfrac{2}{5}\right)\left(\dfrac{2}{5}\right) = \dfrac{4}{25}$

9. $\dfrac{5}{6} + \dfrac{7}{10} = \dfrac{50}{60} + \dfrac{42}{60} = \dfrac{92}{60} \div \dfrac{4}{4} = \dfrac{23}{15}$

10. $\dfrac{8}{11} + \dfrac{3}{4} = \dfrac{32}{44} + \dfrac{33}{44} = \dfrac{65}{44}$

11. $\dfrac{5}{8} - \dfrac{1}{3} = \dfrac{15}{24} - \dfrac{8}{24} = \dfrac{7}{24}$

12. $\dfrac{1}{2} \div \dfrac{1}{3} = \dfrac{3}{6} \div \dfrac{2}{6} = \dfrac{3 \div 2}{6 \div 6} = \dfrac{3 \div 2}{1} = \dfrac{3}{2}$

13. $\dfrac{1}{3} \div \dfrac{5}{18} = \dfrac{18}{54} \div \dfrac{15}{54} = \dfrac{18 \div 15}{54 \div 54} =$

$\dfrac{18 \div 15}{1} = \dfrac{18}{15} \div \dfrac{3}{3} = \dfrac{6}{5}$

14. $\dfrac{7}{10} \div \dfrac{7}{12} = \dfrac{84}{120} \div \dfrac{70}{120} = \dfrac{84 \div 70}{120 \div 120} =$

$\dfrac{84 \div 70}{1} = \dfrac{84}{70} \div \dfrac{14}{14} = \dfrac{6}{5}$

15. $\dfrac{7}{8} \times \dfrac{8}{7} = \dfrac{56}{56} = 1$

16. $\dfrac{54}{1} \times \dfrac{1}{54} = \dfrac{54}{54} = 1$

17. $\dfrac{1}{6} \times \dfrac{6}{1} = \dfrac{6}{6} = 1$

18. $\dfrac{2}{3} \div \dfrac{1}{6} = \dfrac{12}{18} \div \dfrac{3}{18} = \dfrac{12 \div 3}{1} = 4 \text{ times}$

19. $2S - 10 = S + 10$

$\quad 2S - 10 + 10 = S + 10 + 10$

$\quad 2S = S + 20$

$\quad 2S - S = S - S + 20$

$\quad S = 20 \text{ points}$

20. Each received 20 points last week.

Gary: $2(20) - 10 = 30$ points this week

Jenna: $20 + 10 = 30$ points this week

Lesson Practice 10A

1. done

2. done

3. $6^2 + 8^2 = H^2$

$\quad 36 + 64 = H^2$

$\quad 100 = H^2$

$\quad H = 10 \text{ ft}$

4. $16^2 + L^2 = 20^2$

$\quad 256 + L^2 = 400$

$\quad 256 - 256 + L^2 = 400 - 256$

$\quad L^2 = 144$

$\quad L = 12 \text{ mi}$

5. done

6. $8^2 + 15^2 = 17^2$

$\quad 64 + 225 = 289$

$\quad 289 = 289; \text{ yes}$

7. $10^2 + 24^2 = 26^2$
$100 + 576 = 676$
$676 = 676$; yes

8. $4^2 + 5^2 = 6^2$
$16 + 25 = 36$
$41 \neq 36$; no

9. $12^2 + 9^2 = H^2$
$144 + 81 = H^2$
$225 = H^2$
$H = 15$ ft

10. $9^2 + 12^2 = 16^2$
$81 + 144 = 256$
$225 \neq 256$; no

7. $16^2 + 30^2 = 34^2$
$256 + 900 = 1,156$
$1,156 = 1,156$; yes

8. $1^2 + 3^2 = 4^2$
$1 + 9 = 16$
$10 \neq 16$; no

9. $3^2 + 4^2 = H^2$
$9 + 16 = H^2$
$25 = H^2$
$H = 5$ ft

10. $4^2 + 5^2 = 6^2$
$16 + 25 = 36$
$41 \neq 36$; no

Lesson Practice 10B

1. $9^2 + 12^2 = H^2$
$81 + 144 = H^2$
$225 = H^2$
$H = 15$ in

2. $L^2 + 6^2 = 10^2$
$L^2 + 36 = 100$
$L^2 + 36 - 36 = 100 - 36$
$L^2 = 64$
$L = 8$ ft

3. $15^2 + 36^2 = 39^2$
$225 + 1,296 = 1,521$
$1,521 = 1,521$; yes

4. $15^2 + 20^2 = 25^2$
$225 + 400 = 625$
$625 = 625$; yes

5. $5^2 + 10^2 = 12^2$
$25 + 100 = 144$
$125 \neq 144$; no

6. $3^2 + 4^2 = 8^2$
$9 + 16 = 64$
$25 \neq 64$; no

Lesson Practice 10C

1. $5^2 + 12^2 = H^2$
$25 + 144 = H^2$
$169 = H^2$
$H = 13$ ft

2. $16^2 + 12^2 = H^2$
$256 + 144 = H^2$
$400 = H^2$
$H = 20$ yd

3. $20^2 + 48^2 = 52^2$
$400 + 2,304 = 2,704$
$2,704 = 2,704$; yes

4. $4^2 + 8^2 = 16^2$
$16 + 64 = 256$
$80 \neq 256$; no

5. $7^2 + 24^2 = 25^2$
$49 + 576 = 625$
$625 = 625$; yes

6. $2^2 + 3^2 = 6^2$
$4 + 9 = 36$
$13 \neq 36$; no

7. $10^2 + 12^2 = 15^2$
$100 + 144 = 225$
$244 \neq 225$; no

8. $18^2 + 24^2 = 30^2$
    $324 + 576 = 900$
    $900 = 900$; yes

9. $6^2 + 8^2 = H^2$
    $36 + 64 = H^2$
    $100 = H^2$
    $H = 10$ in

10. $5^2 + 5^2 = 5^2$
    $25 + 25 = 25$
    $50 \neq 25$; no

Systematic Review 10D

1. $3^2 + 4^2 = H^2$
    $9 + 16 = H^2$
    $25 = H^2$
    $H = 5$ in

2. $3^2 + 4^2 = 10^2$
    $9 + 16 = 100$
    $25 \neq 100$; no

3. $X - 15 = 12$
    $X - 15 + 15 = 12 + 15$
    $X = 27$

4. $X - 15 = 12 \Rightarrow (27) - 15 = 12$
    $12 = 12$

5. $4X + 3 = 3X - 12$
    $4X + 3 - 3 = 3X - 12 - 3$
    $4X = 3X - 15$
    $4X - 3X = 3X - 3X - 15$
    $X = -15$

6. $4X + 3 = 3X - 12 \Rightarrow 4(-15) + 3 = 3(-15) - 12$
    $-60 + 3 = -45 - 12$
    $-57 = -57$

7. $\sqrt{16} = 4$

8. $\sqrt{X^2} = X$

9. $(-2)^3 = (-2)(-2)(-2) = -8$

10. $\left(\frac{1}{2}\right)^4 = \left(\frac{1}{2}\right)\left(\frac{1}{2}\right)\left(\frac{1}{2}\right)\left(\frac{1}{2}\right) = \frac{1}{16}$

11. $6 \times \frac{1}{10^1} + 5 \times \frac{1}{10^2} + 8 \times \frac{1}{10^3}$

12. $1 \times 10^3 + 4 \times 10^2 + 8 \times 10^1 + 2 \times 10^0$

13. $\frac{1}{\cancel{2}} \times \frac{\cancel{4}^2}{5} = \frac{2}{5}$

14. $\frac{\cancel{2}}{3} \times \frac{1}{\cancel{8}_4} = \frac{1}{12}$

15. $\frac{\cancel{5}}{\cancel{6}_3} \times \frac{\cancel{4}^2}{\cancel{10}_2} = \frac{1}{3}$

16. $\frac{3}{6} \div \frac{1}{2} = \frac{\cancel{3}}{\cancel{6}_3} \times \frac{\cancel{2}}{1} = 1$

17. $\frac{5}{8} \div \frac{2}{3} = \frac{5}{8} \times \frac{3}{2} = \frac{15}{16}$

18. $\frac{1}{2} \div \frac{1}{4} = \frac{1}{\cancel{2}} \times \frac{\cancel{4}^2}{1} = 2$

19. $\frac{2}{3} \div \frac{1}{3} = \frac{2}{\cancel{3}} \times \frac{\cancel{3}}{1} = \frac{2}{1} = 2$

20. $9^2 + 12^2 = 15^2$
    $81 + 144 = 225$
    $225 = 225$; yes

Systematic Review 10E

1. $5^2 + L^2 = 13^2$
    $25 + L^2 = 169$
    $25 - 25 + L^2 = 169 - 25$
    $L^2 = 144$
    $L = 12$ ft

2. $21^2 + 28^2 = 35^2$
    $441 + 784 = 1,225$
    $1,225 = 1,225$; yes

3. $X + 5 = 17$
    $X + 5 - 5 = 17 - 5$
    $X = 12$

4. $X + 5 = 17 \Rightarrow (12) + 5 = 17$
    $17 = 17$

5. $-B + 9 = -2B + 1$
    $-B + 9 - 9 = -2B + 1 - 9$
    $-B = -2B - 8$
    $-B + 2B = -2B - 8 + 2B$
    $B = -8$

6. $-B + 9 = -2B + 1 \Rightarrow -(-8) + 9 = -2(-8) + 1$
    $17 = 16 + 1$
    $17 = 17$

7. $\sqrt{100} = 10$

8. $-6^2 = -(6)(6) = -36$

9. $(-5)^2 = (-5)(-5) = 25$

10. $\sqrt{\dfrac{4}{9}} = \dfrac{2}{3}$

11. $2,700.051$

12. 0.963

13. $\dfrac{4}{5} \times \dfrac{\overset{2}{10}}{\underset{3}{12}} = \dfrac{2}{3}$

14. $\dfrac{1}{3} \times \dfrac{\overset{2}{6}}{7} = \dfrac{2}{7}$

15. $\dfrac{3}{\underset{2}{4}} \times \dfrac{2}{3} = \dfrac{1}{2}$

16. $\dfrac{1}{2} \div \dfrac{1}{4} = \dfrac{1}{2} \times \dfrac{\overset{2}{4}}{1} = 2$

17. $\dfrac{6}{8} \div \dfrac{1}{4} = \dfrac{\overset{3}{6}}{\underset{2}{8}} \times \dfrac{4}{1} = 3$

18. $\dfrac{5}{7} \div \dfrac{1}{7} = \dfrac{5}{7} \times \dfrac{7}{1} = 5$

19. $\begin{aligned} X + \$100 &= \$500 \\ X + \$100 - \$100 &= \$500 - \$100 \\ X &= \$400 \end{aligned}$

20. $\begin{aligned} 7^2 + 8^2 &= 9^2 \\ 49 + 64 &= 81 \\ 113 &\neq 81; \text{ no} \end{aligned}$

Systematic Review 10F

1. $\begin{aligned} 8^2 + 6^2 &= H^2 \\ 64 + 36 &= H^2 \\ 100 &= H^2 \\ H &= 10 \text{ ft} \end{aligned}$

2. $\begin{aligned} 5^2 + 10^2 &= 100^2 \\ 25 + 100 &= 10,000 \\ 125 &\neq 10,000; \text{ no} \end{aligned}$

3. $\begin{aligned} A - 3 &= 19 \\ A - 3 + 3 &= 19 + 3 \\ A &= 22 \end{aligned}$

4. $A - 3 = 19 \Rightarrow (22) - 3 = 19$
$ 19 = 19$

5. $\begin{aligned} 7X - 5 &= 6X - 10 \\ 7X - 5 + 5 &= 6X - 10 + 5 \\ 7X &= 6X - 5 \\ 7X - 6X &= 6X - 6X - 5 \\ X &= -5 \end{aligned}$

6. $7X - 5 = 6X - 10 \Rightarrow 7(-5) - 5 = 6(-5) - 10$
$ -35 - 5 = -30 - 10$
$ -40 = -40$

7. $\sqrt{49} = 7$

8. $(-9)^2 = (-9)(-9) = 81$

9. $4^2 = (4)(4) = 16$

10. $\sqrt{\dfrac{9}{25}} = \dfrac{3}{5}$

11. $\dfrac{2}{7} + \dfrac{4}{8} = \dfrac{16}{56} + \dfrac{28}{56} = \dfrac{44}{56} \div \dfrac{4}{4} = \dfrac{11}{14}$

12. $\dfrac{1}{4} + \dfrac{5}{9} = \dfrac{9}{36} + \dfrac{20}{36} = \dfrac{29}{36}$

13. $\dfrac{4}{5} - \dfrac{1}{2} = \dfrac{8}{10} - \dfrac{5}{10} = \dfrac{3}{10}$

14. $\dfrac{3}{4} \div \dfrac{1}{7} = \dfrac{3}{4} \times \dfrac{7}{1} = \dfrac{21}{4}$

15. $\dfrac{3}{5} \div \dfrac{1}{8} = \dfrac{3}{5} \times \dfrac{8}{1} = \dfrac{24}{5}$

16. $\dfrac{7}{10} \div \dfrac{1}{9} = \dfrac{7}{10} \times \dfrac{9}{1} = \dfrac{63}{10}$

17. $(-16) \times (-30) = 480$

18. $\begin{aligned} 6N - 8 &= 5N + 2 \\ 6N - 8 + 8 &= 5N + 2 + 8 \\ 6N &= 5N + 10 \\ 6N - 5N &= 5N - 5N + 10 \\ N &= 10 \end{aligned}$

19. $\dfrac{3}{4} \div \dfrac{1}{4} = \dfrac{3}{4} \times \dfrac{4}{1} = \dfrac{12}{4} = 3 \text{ times}$

20. $\begin{aligned} 12^2 + 16^2 &= H^2 \\ 144 + 256 &= H^2 \\ 400 &= H^2 \\ H &= 20 \text{ ft} \end{aligned}$

Lesson Practice 11A

1. done

2. done

3. $7 - 3 = 3 - 7$
$ 4 = -4 \text{ false}$

4. $7 \times 9 = 9 \times 7$
$63 = 63$ true

5. $(2+3)+8 = 2+(3+8)$
$5+8 = 2+11$
$13 = 13$ true

6. $(4 \times 5) \times 6 = 4 \times (5 \times 6)$
$20 \times 6 = 4 \times 30$
$120 = 120$ true

7. $(10-4)-1 = 10-(4-1)$
$6-1 = 10-3$
$5 = 7$ false

8. $(9 \div 3) \div 3 = 9 \div (3 \div 3)$
$3 \div 3 = 9 \div 1$
$1 = 9$ false

9. done
10. $7+5 = 5+7$
11. $X+2Y = 2Y+X$
12. $20 \times 13 = 13 \times 20$
13. done
14. $9 \times (5 \times 2) = (9 \times 5) \times 2$
15. $3B + (6B+4B) = (3B+6B) + 4B$
16. $(3 \times R) \times 5 = 3 \times (R \times 5)$
17. addition; multiplication
18. subtraction; division
19. addition; multiplication
20. subtraction; division

Lesson Practice 11B

1. $10+13 = 13+10$
$23 = 23$ true

2. $16 \div 2 = 2 \div 16$
$8 = \frac{2}{16}$ false

3. $8 \times 4 = 4 \times 8$
$32 = 32$ true

4. $12-5 = 5-12$
$7 = -7$ false

5. $(3 \times 4) \times 5 = 3 \times (4 \times 5)$
$12 \times 5 = 3 \times 20$
$60 = 60$ true

6. $(12 \div 6) \div 2 = 12 \div (6 \div 2)$
$2 \div 2 = 12 \div 3$
$1 = 4$ false

7. done
8. done
9. $5Q + 3C - C + Q + 4Q - 5C =$
$5Q + 3C + (-C) + Q + 4Q + (-5C) =$
$5Q + Q + 4Q + 3C + (-C) + (-5C) =$
$10Q - 3C$

10. $2X + 2 - X + 2X =$
$2X + 2 + (-X) + 2X =$
$2X + (-X) + 2X + 2 =$
$3X + 2$

11. $3Y - 1 + 2Y - 1 - 4Y =$
$3Y + (-1) + 2Y + (-1) + (-4Y) =$
$3Y + 2Y + (-4Y) + (-1) + (-1) =$
$Y - 2$

12. $5A - 6B - 3B + 10A - 8 =$
$5A + (-6B) + (-3B) + 10A + (-8) =$
$5A + 10A + (-6B) + (-3B) + (-8) =$
$15A - 9B - 8$

13. false
14. true
15. false
16. true
17. $5 \times 3 = 15$ calves
$3 \times 5 = 15$ calves
$15 = 15$; They had the same number of calves.

18. $(X)(Y) = XY$ guests
$(Y)(X) = XY$ guests
$XY = XY$; They had the same number of guests.

Lesson Practice 11C

1. $6 \times 7 = 7 \times 6$
$42 = 42$ true

2. $25 - 15 = 15 - 25$
$10 = -10$ false

3. $(-1) + (-8) = (-8) + (-1)$
$-9 = -9$ true

4. $45 \div 5 = 5 \div 45$

   $9 = \dfrac{5}{45}$ false

5. $(2+3)+4 = 2+(3+4)$

   $5+4 = 2+7$

   $9 = 9$ true

6. $(7-4)-1 = 7-(4-1)$

   $3-1 = 7-3$

   $2 = 4$ false

7. $4Q+2C-2C-2Q-3C =$

   $4Q+2C+(-2C)+(-2Q)+(-3C) =$

   $4Q-2Q+2C+(-2C)+(-3C) =$

   $2Q+(-3C) =$

   $2Q-3C$

8. $-5M-7+3M-4+5 =$

   $(-5M)+(-7)+3M+(-4)+5 =$

   $(-5M)+3M+(-7)+(-4)+5 =$

   $(-2M)+(-6) =$

   $-2M-6$

9. $4A-5-2A+7-1 =$

   $4A+(-5)+(-2A)+7+(-1) =$

   $4A+(-2A)+(-5)+7+(-1) =$

   $2A+1$

10. $15X+6-4Y-5Y-14X+10 =$

    $15X+6+(-4Y)+(-5Y)+(-14X)+10 =$

    $15X+(-14X)+(-4Y)+(-5Y)+6+10 =$

    $X+(-9Y)+16 =$

    $X-9Y+16$

11. $3A+2+7B+6A+8 =$

    $3A+6A+7B+2+8 =$

    $9A+7B+10$

12. $-5X+3+8X-4 =$

    $(-5X)+3+8X+(-4) =$

    $(-5X)+8X+3+(-4) =$

    $3X+(-1) =$

    $3X-1$

13. true

14. false

15. $\$25+\$10 = \$35$

    $\$10+\$25 = \$35$

    $\$35 = \$35$

16. $6 \div 3 = 2$ pies the first day

    $3 \div 6 = \dfrac{3}{6} \div \dfrac{3}{3} = \dfrac{1}{2}$ pie the next day

17. $\$100 - \$30 = \$70$

    $\$30 - \$100 = \$ -70$

    $\$70 > \$ -70$; David has more.

18. $5 \times \$8 = \$40$

    $8 \times \$5 = \$40$

    $\$40 = \$40$; their incomes are the same.

Systematic Review 11D

1. $8K-6+3K-2+3 =$

   $8K+(-6)+3K+(-2)+3 =$

   $8K+3K+(-6)+(-2)+3 =$

   $11K+(-5) =$

   $11K-5$

2. $13A-8Z-2A-12Z =$

   $13A+(-8Z)+(-2A)+(-12Z) =$

   $13A+(-2A)+(-8Z)+(-12Z) =$

   $11A+(-20Z) =$

   $11A-20Z$

3. $5^2+12^2 = H^2$

   $25+144 = H^2$

   $169 = H^2$

   $H = 13$ ft

4. $3^2+4^2 = 5^2$

   $9+16 = 25$

   $25 = 25$; yes

5. $-X+11 = -2X-7$

   $-X+11-11 = -2X-7-11$

   $-X = -2X-18$

   $-X+2X = -2X+2X-18$

   $X = -18$

6. $-X+11 = +2X-7 \Rightarrow -(-18)+11 = -2(-18)-7$

   $18+11 = 36-7$

   $29 = 29$

7. $A+10 = 23$

   $A+10-10 = 23-10$

   $A = 13$

8. $A + 10 = 23 \Rightarrow (13) + 10 = 23$
$$23 = 23$$

9. $1\frac{5}{8} = \frac{8}{8} + \frac{5}{8} = \frac{13}{8}$

10. $4\frac{1}{2} = \frac{8}{2} + \frac{1}{2} = \frac{9}{2}$

11. $\frac{11}{3} = \frac{9}{3} + \frac{2}{3} = 3\frac{2}{3}$

12. $\frac{8}{5} = \frac{5}{5} + \frac{3}{5} = 1\frac{3}{5}$

13. $\frac{3}{5} \times \frac{5}{9} = \frac{1}{3}$

14. $\frac{1}{3} \times \frac{7}{8} = \frac{7}{24}$

15. $\frac{2}{5} \div \frac{3}{10} = \frac{2}{5} \times \frac{10}{3} = \frac{4}{3} = 1\frac{1}{3}$

16. $\frac{3}{4} \div \frac{1}{6} = \frac{3}{4} \times \frac{6}{1} = \frac{9}{2} = 4\frac{1}{2}$

17. $3\frac{2}{3} = \frac{9}{3} + \frac{2}{3} = \frac{11}{3}$ of a mile

18. $\frac{14}{5} = \frac{10}{5} + \frac{4}{5} = 2\frac{4}{5}$ hours

19. $\frac{1}{3} \times \frac{2}{3} = \frac{2}{9}$ cup

20. yes

Systematic Review 11E

1. $10C - 3C - 9D + 3D - C =$
$10C + (-3C) + (-9D) + 3D + (-C) =$
$10C + (-3C) + (-C) + (-9D) + 3D =$
$6C + (-6D) =$
$6C - 6D$

2. $3 + 6X - 1 + X =$
$3 + 6X + (-1) + X =$
$6X + X + 3 + (-1) =$
$7X + 2$

3. $6^2 + L^2 = 10^2$
$36 + L^2 = 100$
$L^2 = 100 - 36$
$L^2 = 64$
$L = 8$ ft

4. $12^2 + L^2 = 15^2$
$144 + L^2 = 225$
$L^2 = 225 - 144$
$L^2 = 81$
$L = 9$ ft

5. done

6. done

7. $4B + 3B - 6B = 1 + 7$
$4B + 3B + (-6B) = 1 + 7$
$B = 8$

8. $4B + 3B - 6B = 1 + 7 \Rightarrow$
$4(8) + 3(8) - 6(8) = 1 + 7$
$32 + 24 - 48 = 8$
$8 = 8$

9. $\frac{3}{8} \times \frac{4}{5} = \frac{12}{40} \div \frac{4}{4} = \frac{3}{10}$

10. $\frac{1}{2} \times \frac{5}{9} = \frac{5}{18}$

11. $\frac{5}{9} \div \frac{3}{7} = \frac{5}{9} \times \frac{7}{3} = \frac{35}{27}$
$= \frac{27}{27} + \frac{8}{27} = 1\frac{8}{27}$

12. $\frac{4}{7} \div \frac{2}{3} = \frac{4}{7} \times \frac{3}{2} = \frac{6}{7}$

13. $(-24) \div (6) = -4$

14. $(25) \div (-5) = -5$

15. $(-54) \div (-9) = 6$

16. $(56) \div (8) = 7$

17. no

18. $3Q + 4Q - 5 = 6Q + 7$
$7Q - 5 = 6Q + 7$
$7Q - 5 + 5 = 6Q + 7 + 5$
$7Q = 6Q + 12$
$7Q - 6Q = 6Q - 6Q + 12$
$Q = 12$

19. $\frac{2}{3} \times \frac{1}{4} = \frac{1}{6}$ of the people

20. One sixth of 36:
$36 \div 6 = 6$
$6 \times 1 = 6$ brown-haired girls

Systematic Review 11F

1. $A + B + 2A - 3B =$
    $A + B + 2A + (-3B) =$
    $A + 2A + B + (-3B) =$
    $3A + (-2B) =$
    $3A - 2B$

2. $2F + 3 - F + 1 =$
    $2F + 3 + (-F) + 1 =$
    $2F + (-F) + 3 + 1 =$
    $F + 4$

3. $16^2 + L^2 = 20^2$
    $256 + L^2 = 400$
    $L^2 = 400 - 256$
    $L^2 = 144$
    $L = 12$ in

4. $24^2 + L^2 = 25^2$
    $576 + L^2 = 625$
    $L^2 = 625 - 576$
    $L^2 = 49$
    $L = 7$ in

5. $6D - 10 + 5 = 5D - 34$
    $6D - 5 = 5D - 34$
    $6D - 5 + 5 = 5D - 34 + 5$
    $6D = 5D - 29$
    $6D - 5D = 5D - 5D - 29$
    $D = -29$

6. $6D - 10 + 5 = 5D - 34 \Rightarrow$
    $6(-29) - 10 + 5 = 5(-29) - 34$
    $-174 - 10 + 5 = -145 - 34$
    $-179 = -179$

7. $4A - 3A = 7 + 9$
    $A = 16$

8. $4A - 3A = 7 + 9 \Rightarrow 4(16) - 3(16) = 7 + 9$
    $64 - 48 = 16$
    $16 = 16$

9. $\frac{1}{\cancel{2}} \times \frac{\cancel{6}^{\,3}}{8} = \frac{3}{8}$

10. $\frac{7}{\cancel{9}_{\,3}} \times \frac{\cancel{3}}{4} = \frac{7}{12}$

11. $\frac{11}{12} \div \frac{1}{3} = \frac{11}{\cancel{12}_{\,4}} \times \frac{\cancel{3}}{1} = \frac{11}{4} = \frac{8}{4} + \frac{3}{4} = 2\frac{3}{4}$

12. $\frac{7}{7} \div \frac{1}{8} = \frac{7}{7} \times \frac{8}{1} = \frac{56}{7} = 8$

13. $\sqrt{64} = 8$

14. $\sqrt{X^2} = X$

15. $(9)^2 = (9)(9) = 81$

16. $\left(\frac{1}{3}\right)^3 = \left(\frac{1}{3}\right)\left(\frac{1}{3}\right)\left(\frac{1}{3}\right) = \frac{1}{27}$

17. no

18. $4A + 6 - 2A = A + 12$
    $4A + 6 + (-2A) = A + 12$
    $4A + (-2A) + 6 = A + 12$
    $2A + 6 = A + 12$
    $2A + 6 - 6 = A + 12 - 6$
    $2A = A + 6$
    $2A - A = A - A + 6$
    $A = 6$

19. $\frac{4}{6} \div \frac{2}{6} = \frac{\cancel{4}^{\,2}}{\cancel{6}} \times \frac{\cancel{6}}{\cancel{2}} = \frac{2}{1} = 2$ times

20. $\frac{9}{10} \div \frac{1}{5} = \frac{9}{{}_{2}\cancel{10}} \times \frac{\cancel{5}}{1} =$
    $\frac{9}{2} = \frac{8}{2} + \frac{1}{2} = 4\frac{1}{2}$ hours

Lesson Practice 12A

1. done

2. $7(A - B) =$
    $7A - 7B$

3. $3(4D + 3A) =$
    $(3)(4D) + (3)(3A) =$
    $12D + 9A$

4. $5(2X + 3Y) =$
    $(5)(2X) + (5)(3Y) =$
    $10X + 15Y$

5. $6(2 - B) =$
    $(6)(2) - (6)(B) =$
    $12 - 6B$

6. $8(A + 3B) =$
    $(8)(A) + (8)(3B) =$
    $8A + 24B$

7. done

8. done

9. $6D + 18E =$
$(6)(D) + (6)(3E) =$
$6(D + 3E)$

10. $15X + 10A =$
$(5)(3X) + (5)(2A) =$
$5(3X + 2A)$

11. $7X + 14Y =$
$(7)(X) + (7)(2Y) =$
$7(X + 2Y)$

12. $2M - 6N =$
$(2)(M) - 2(3N) =$
$2(M - 3N)$

13. done

14. $6(2 + 3) = 6(2) + 6(3)$
$6(5) = 12 + 18$
$30 = 30$

15. $8(1 + 9) = 8(1) + 8(9)$
$8(10) = 8 + 72$
$80 = 80$

16. $3(4 - 2) = 3(4) - 3(2)$
$3(2) = 12 - 6$
$6 = 6$

17. $10(4 + 5) = 10(4) + 10(5)$
$10(9) = 40 + 50$
$90 = 90$

18. $-2(3 + 3) = -2(3) + -2(3)$
$-2(6) = -6 + (-6)$
$-12 = -12$

Lesson Practice 12B

1. $9(C + D) =$
$9C + 9D$

2. $4(2A + 4B) =$
$4(2A) + 4(4B) =$
$8A + 16B$

3. $3(2X + Y) =$
$3(2X) + 3(Y) =$
$6X + 3Y$

4. $2(3X - 2Y) =$
$2(3X) - 2(2Y) =$
$6X - 4Y$

5. $4(8 - X) =$
$4(8) - 4(X) =$
$32 - 4X$

6. $A(X + Y) =$
$AX + AY$

7. $8X + 12Y =$
$4(2X) + 4(3Y) =$
$4(2X + 3Y)$

8. $18A + 24B =$
$6(3A) + 6(4B) =$
$6(3A + 4B)$

9. $6B + 10 =$
$2(3B) + 2(5) =$
$2(3B + 5)$

10. $6A + 3 =$
$3(2A) + 3(1) =$
$3(2A + 1)$

11. $2X + 2Y =$
$2(X + Y)$

12. $4A - 12B =$
$4(A) - 4(3B) =$
$4(A - 3B)$

13. $8(5 + 2) = 8(5) + 8(2)$
$8(7) = 40 + 16$
$56 = 56$

14. $5(4 - 3) = 5(4) - 5(3)$
$5(1) = 20 - 15$
$5 = 5$

15. $2(1 + 1) = 2(1) + 2(1)$
$2(2) = 2 + 2$
$4 = 4$

16. $11(7 - 4) = 11(7) - 11(4)$
$11(3) = 77 - 44$
$33 = 33$

17. $4(3 + 6) = 4(3) + 4(6)$
$4(9) = 12 + 24$
$36 = 36$

18.
$$-5(10+11)=-5(10)+-5(11)$$
$$-5(21)=-50+(-55)$$
$$-105=-105$$

Lesson Practice 12C

1. $4(A+B)=$
$4A+4B$

2. $5(X-Y)=$
$5X-5Y$

3. $3(2Q-4)=$
$3(2Q)-3(4)=$
$6Q-12$

4. $3(4X+3Y)=$
$3(4X)+3(3Y)=$
$12X+9Y$

5. $6(2-B)=$
$6(2)-6(B)=$
$12-6B$

6. $B(3+4)=$
$B3+B4$ or
$3B+4B$
It is customary to put the number first.

7. $15Y+30X=$
$15(Y)+15(2X)=$
$15(Y+2X)$

8. $12Q+6Y=$
$6(2Q)+6(Y)=$
$6(2Q+Y)$

9. $24A+18B=$
$(6)4A+(6)3B=$
$6(4A+3B)$

10. $5X-5=$
$5(X)-5(1)=$
$5(X-1)$

11. $6X+7X=$
$X(6)+X(7)=$
$X(6+7)$ or $13X$

12. $2A+2AB=$
$2A(1)+2A(B)=$
$2A(1+B)$

13.
$$6(1+8)=6(1)+6(8)$$
$$6(9)=6+48$$
$$54=54$$

14.
$$3(5-2)=3(5)-3(2)$$
$$3(3)=15-6$$
$$9=9$$

15. done

16.
$$3(X+X)=3(X)+3(X)$$
$$3(2X)=3X+3X$$
$$6X=6X$$

17.
$$9(7-2)=9(7)-9(2)$$
$$9(5)=63-18$$
$$45=45$$

18.
$$-1(14+21)=-1(14)+-1(21)$$
$$-1(35)=-14+(-21)$$

Systematic Review 12D

1. $3(A-B)=$
$3A-3B$

2. $5(3A+9)=$
$5(3A)+5(9)=$
$15A+45$

3. $Q(X+3)=$
$QX+3Q$

4. $10X-25Y=$
$5(2X)-5(5Y)=$
$5(2X-5Y)$

5. $24A+12B=$
$12(2A)+12(B)=$
$12(2A+B)$

6. $3XY+4XZ=$
$X(3Y)+X(4Z)=$
$X(3Y+4Z)$

7. $3A-2B+5A+6=$
$3A+(-2B)+5A+6=$
$3A+5A+(-2B)+6=$
$8A+-2B+6=$
$8A-2B+6$

8. $4X + 5Y + 8X - 3Y =$
$4X + 5Y + 8X + (-3Y) =$
$4X + 8X + 5Y + (-3Y) =$
$12X + 2Y$

9. $11X - 7 = 10X + 6$
$11X - 7 + 7 = 10X + 6 + 7$
$11X = 10X + 13$
$11X - 10X = 10X - 10X + 13$
$X = 13$

10. $11X - 7 = 10X + 6 \Rightarrow 11(13) - 7 = 10(13) + 6$
$143 - 7 = 130 + 6$
$136 = 136$

11. $X + 15 = 45$
$X + 15 - 15 = 45 - 15$
$X = 30$

12. $X + 15 = 45 \Rightarrow (30) + 15 = 45$
$45 = 45$

13. done

14. $\dfrac{6}{16} \div \dfrac{2}{2} = \dfrac{3}{8}$ in

15. $\dfrac{12}{16} \div \dfrac{4}{4} = \dfrac{3}{4}$ in

16. $\dfrac{4}{16} \div \dfrac{4}{4} = \dfrac{1}{4}$ in

17. $\dfrac{9}{16}$ in

18. $\dfrac{14}{16} \div \dfrac{2}{2} = \dfrac{7}{8}$ in

19. $12^2 + 15^2 = 20^2$
$144 + 225 = 400$
$369 \neq 400$; no

20. $1 \times 10^2 + 2 \times 10^1 + 3 \times 10^0 + 4 \times \dfrac{1}{10^1} + 2 \times \dfrac{1}{10^2}$

Systematic Review 12E

1. $-2(Q + 2R) =$
$-2Q + (-2)(2R) =$
$-2Q - 4R$

2. $X(Y + 2) =$
$XY + 2X$

3. $5(3B + B) =$
$5(3B) + 5B =$
$15B + 5B$ or $20B$

4. $4A - 16B =$
$4(A) - 4(4B) =$
$4(A - 4B)$

5. $20A - 40D =$
$20(A) - 20(2D) =$
$20(A - 2D)$

6. $6Q + 12G =$
$6(Q) + 6(2G) =$
$6(Q + 2G)$

7. $2 + 6X - 5 + 2X =$
$2 + 6X + (-5) + 2X =$
$2 + (-5) + 6X + 2X =$
$-3 + 8X$ or $8X - 3$

8. $3B + 6B - 7Y + 9Y =$
$3B + 6B + (-7Y) + 9Y =$
$9B + 2Y$

9. $4X + 8 = 3X - 4$
$4X + 8 - 8 = 3X - 4 - 8$
$4X = 3X - 12$
$4X - 3X = 3X - 3X - 12$
$X = -12$

10. $4X + 8 = 3X - 4 \Rightarrow 4(-12) + 8 = 3(-12) - 4$
$-48 + 8 = -36 - 4$
$-40 = -40$

11. $-B + 5 = -2B + 16$
$-B + 5 - 5 = -2B + 16 - 5$
$-B = -2B + 11$
$-B + 2B = -2B + 2B + 11$
$B = 11$

12. $-B + 5 = -2B + 16 \Rightarrow -(11) + 5 = -2(11) + 16$
$-6 = -22 + 16$
$-6 = -6$

13. $\dfrac{3}{16}$ in

14. $\dfrac{10}{16} \div \dfrac{2}{2} = \dfrac{5}{8}$ in

15. $\dfrac{13}{16}$ in

16. $L^2 + 3^2 = 5^2$
$L^2 + 9 = 25$
$L^2 = 25 - 9$
$L^2 = 16$
$L = 4$ ft

17. $L^2 + 5^2 = 13^2$
$L^2 + 25 = 169$
$L^2 = 169 - 25$
$L^2 = 144$
$L = 12$ ft

18. $\dfrac{12}{8} \div \dfrac{4}{4} = \dfrac{3}{2} = \dfrac{2}{2} + \dfrac{1}{2} = 1\dfrac{1}{2}$ yd

19. $5N + 3N = 7N + 4$
$8N = 7N + 4$
$8N - 7N = 7N - 7N + 4$
$N = 4$

20. 834.179

Systematic Review 12F

1. $6(2X + 4Y) =$
$6(2X) + 6(4Y) =$
$12X + 24Y$

2. $A(3 + 4) =$
$3A + 4A$

3. $5X(3A + 2B) =$
$5X(3A) + 5X(2B) =$
$15AX + 10BX$

4. $6X + 30Y =$
$6(X) + 6(5Y) =$
$6(X + 5Y)$

5. $6A + 7A =$
$(A)6 + (A)7 =$
$A(6 + 7)$ or $13A$

6. $3AB + 6AD =$
$3A(B) + 3A(2D) =$
$3A(B + 2D)$

7. $4F + 3G + 5F - G =$
$4F + 3G + 5F + (-G) =$
$4F + 5F + 3G + (-G) =$
$9F + 2G$

8. $4 - 6Y - 10 =$
$4 + (-6Y) + (-10) =$
$4 + (-10) + (-6Y) =$
$-6 + (-6Y) =$
$-6 - 6Y$ or $-6Y - 6$

9. $10X + 3 = 9X + 3$
$10X + 3 - 3 = 9X + 3 - 3$
$10X = 9X$
$10X - 9X = 9X - 9X$
$X = 0$

10. $10X + 3 = 9X + 3 \Rightarrow 10(0) + 3 = 9(0) + 3$
$0 + 3 = 0 + 3$
$3 = 3$

11. $B - 18 = 21$
$B - 18 + 18 = 21 + 18$
$B = 39$

12. $B - 18 = 21 \Rightarrow (39) - 18 = 21$
$21 = 21$

13. $\dfrac{2}{16} \div \dfrac{2}{2} = \dfrac{1}{8}$ in

14. $\dfrac{5}{16}$ in

15. $\dfrac{3}{16}$ in

16. $9^2 + 12^2 = H^2$
$81 + 144 = H^2$
$225 = H^2$
$H = 15$ ft

17. $21^2 + 28^2 = 35^2$
$441 + 784 = 1{,}225$
$1{,}225 = 1{,}225$ yes

18. $\dfrac{5}{\cancel{6}_{3}} \times \dfrac{\cancel{2}}{3} = \dfrac{5}{9}$ of the dogs

19. $\dfrac{2}{3} \div \dfrac{1}{6} = \dfrac{2}{\cancel{3}} \times \dfrac{\cancel{6}^{2}}{1} = \dfrac{4}{1} = 4$ pieces

20. $\dfrac{2}{5} + \dfrac{1}{2} = \dfrac{4}{10} + \dfrac{5}{10} = \dfrac{9}{10}$ of the apples used

$\dfrac{9}{10}$ of 20:

$20 \div 10 = 2$

$2 \times 9 = 18$ apples used

$20 - 18 = 2$ apples left

alternate method:

Find fractions of 20 first.

$\dfrac{2}{5}$ of 20 :

$20 \div 5 = 4$

$4 \times 2 = 8$ apples used by Rachel

$\dfrac{1}{2}$ of 20 :

$20 \div 2 = 10$

$10 \times 1 = 10$ apples used by Mother

$8 + 10 = 18$ apples used

$20 - 18 = 2$ apples left

Lesson Practice 13A

1. done

2. done

3. $3D = 9$

$\dfrac{1}{3} \times 3D = \dfrac{1}{3} \times 9$

$D = 3$

4. $3D = 9 \Rightarrow 3(3) = 9$

$\qquad\qquad\qquad 9 = 9$

5. done

6. done

7. $5Y + 8Y + 2 = 9Y + 22$

$13Y + 2 = 9Y + 22$

$13Y + 2 - 2 = 9Y + 22 - 2$

$13Y = 9Y + 20$

$13Y - 9Y = 9Y - 9Y + 20$

$4Y = 20$

$\dfrac{1}{4} \times 4Y = \dfrac{1}{4} \times 20$

$Y = 5$

8. $5Y + 8Y + 2 = 9Y + 22 \Rightarrow$

$5(5) + 8(5) + 2 = 9(5) + 22$

$25 + 40 + 2 = 45 + 22$

$67 = 67$

9. $5D - 3 + D = 2D + 7 - D$

$6D - 3 = D + 7$

$6D - 3 + 3 = D + 7 + 3$

$6D = D + 10$

$6D - D = D - D + 10$

$5D = 10$

$\dfrac{1}{5} \times D = \dfrac{1}{5} \times 10$

$D = 2$

10. $5D - 3 + D = 2D + 7 - D \Rightarrow$

$5(2) - 3 + (2) = 2(2) + 7 - (2)$

$10 - 3 + 2 = 4 + 7 - 2$

$9 = 9$

11. done

12. done

13. $2(5X - 8) + 2 = 2(X + 5)$

$2(5X) - 2(8) + 2 = 2X + 2(5)$

$10X - 16 + 2 = 2X + 10$

$10X - 14 = 2X + 10$

$10X - 14 + 14 = 2X + 10 + 14$

$10X = 2X + 24$

$10X - 2X = 2X - 2X + 24$

$8X = 24$

$\dfrac{1}{8} \times 8X = \dfrac{1}{8} \times 24$

$X = 3$

14. $2(5X - 8) + 2 = 2(X + 5) \Rightarrow$

$2(5(3) - 8) + 2 = 2((3) + 5)$

$2(15 - 8) + 2 = 2(8)$

$2(7) + 2 = 16$

$14 + 2 = 16$

$16 = 16$

15.
$$-3(2H-3)=29-H$$
$$-3(2H)-3(-3)=29-H$$
$$-6H+9=29-H$$
$$-6H+9-9=29-9-H$$
$$-6H=20-H$$
$$-6H+H=20-H+H$$
$$-5H=20$$
$$-\frac{1}{5}\times-5H=-\frac{1}{5}\times20$$
$$H=-4$$

16.
$$-3(2H-3)=29-H\Rightarrow$$
$$-3(2(-4)-3)=29-(-4)$$
$$-3(-8-3)=29+4$$
$$-3(-11)=33$$
$$33=33$$

17.
$$2X=16$$
$$\frac{1}{2}\times2X=\frac{1}{2}\times16$$
$$X=8 \text{ years old}$$

18.
$$3R+5=23$$
$$3R+5-5=23-5$$
$$3R=18$$
$$\frac{1}{3}\times3R=\frac{1}{3}\times18$$
$$R=6 \text{ things}$$

Lesson Practice 13B

1.
$$6X=36$$
$$\frac{1}{6}\times6X=\frac{1}{6}\times36$$
$$X=6$$

2.
$$6X=36\Rightarrow 6(6)=36$$
$$36=36$$

3.
$$11F=121$$
$$\frac{1}{11}\times11F=\frac{1}{11}\times121$$
$$F=11$$

4.
$$11F=121\Rightarrow 11(11)=121$$
$$121=121$$

5.
$$2B+7+5B=42$$
$$7B+7=42$$
$$7B+7-7=42-7$$
$$7B=35$$
$$\frac{1}{7}\times7B=\frac{1}{7}\times35$$
$$B=5$$

6.
$$2B+7+5B=42\Rightarrow 2(5)+7+5(5)=42$$
$$10+7+25=42$$
$$42=42$$

7.
$$5Q-6+5Q=2+3+9$$
$$10Q-6=14$$
$$10Q-6+6=14+6$$
$$10Q=20$$
$$\frac{1}{10}\times10Q=\frac{1}{10}\times20$$
$$Q=2$$

8.
$$5Q-6+5Q=2+3+9\Rightarrow$$
$$5(2)-6+5(2)=2+3+9$$
$$10-6+10=2+3+9$$
$$14=14$$

9.
$$3C-8+7=12+1+C$$
$$3C-1=13+C$$
$$3C-1+1=13+1+C$$
$$3C=14+C$$
$$3C-C=14+C-C$$
$$2C=14$$
$$\frac{1}{2}\times2C=\frac{1}{2}\times14$$
$$C=7$$

10.
$$3C-8+7=12+1+C\Rightarrow$$
$$3(7)-8+7=12+1+(7)$$
$$21-8+7=20$$
$$20=20$$

11.
$$2(Q+2)=Q+1$$
$$2Q+2(2)=Q+1$$
$$2Q+4=Q+1$$
$$2Q+4-4=Q+1-4$$
$$2Q=Q-3$$
$$2Q-Q=Q-Q-3$$
$$Q=-3$$

12.
$$2(Q+2)=Q+1\Rightarrow 2((-3)+2)=(-3)+1$$
$$2(-1)=-3+1$$
$$-2=-2$$

13.
$$4(C+5) = -1(C+20)$$
$$4C + 4(5) = -1C + -1(20)$$
$$4C + 20 = -C - 20$$
$$4C + 20 - 20 = -C - 20 - 20$$
$$4C = -C - 40$$
$$4C + C = -C + C - 40$$
$$5C = -40$$
$$\frac{1}{5} \times 5C = \frac{1}{5} \times -40$$
$$C = -8$$

14.
$$4(C+5) = -1(C+20) \Rightarrow$$
$$4((-8)+5) = -1((-8)+20)$$
$$4(-3) = -1(12)$$
$$-12 = -12$$

15.
$$2(B+2) + B = 2(6+B) + 1$$
$$2B + 2(2) + B = 2(6) + 2B + 1$$
$$2B + 4 + B = 12 + 2B + 1$$
$$3B + 4 = 13 + 2B$$
$$3B + 4 - 4 = 13 - 4 + 2B$$
$$3B = 9 + 2B$$
$$3B - 2B = 9 + 2B - 2B$$
$$B = 9$$

16.
$$2(B+2) + B = 2(6+B) + 1 \Rightarrow$$
$$2((9)+2) + (9) = 2(6+(9)) + 1$$
$$2(11) + 9 = 2(15) + 1$$
$$22 + 9 = 30 + 1$$
$$31 = 31$$

17.
$$P + 5P = P + 3P + 20$$
$$6P = 4P + 20$$
$$6P - 4P = 4P - 4P + 20$$
$$2P = 20$$
$$\frac{1}{2} \times 2P = \frac{1}{2} \times 20$$
$$P = 10 \text{ points}$$

18.
$$3(M+2) = 21$$
$$3M + 3(2) = 21$$
$$3M + 6 = 21$$
$$3M + 6 - 6 = 21 - 6$$
$$3M = 15$$
$$\frac{1}{3} \times 3M = \frac{1}{3} \times 15$$
$$M = \$5$$

Lesson Practice 13C

1.
$$8X - 6 = 50$$
$$8X - 6 + 6 = 50 + 6$$
$$8X = 56$$
$$\frac{1}{8} \times 8X = \frac{1}{8} \times 56$$
$$X = 7$$

2.
$$8X - 6 = 50 \Rightarrow 8(7) - 6 = 50$$
$$56 - 6 = 50$$
$$50 = 50$$

3.
$$10R = 110$$
$$\frac{1}{10} \times 10R = \frac{1}{10} \times 110$$
$$R = 11$$

4.
$$10R = 110 \Rightarrow 10(11) = 110$$
$$110 = 110$$

5.
$$10B - 2B + 3 = 5B + 21$$
$$8B + 3 = 5B + 21$$
$$8B + 3 - 3 = 5B + 21 - 3$$
$$8B = 5B + 18$$
$$8B - 5B = 5B - 5B + 18$$
$$3B = 18$$
$$\frac{1}{3} \times 3B = \frac{1}{3} \times 18$$
$$B = 6$$

6.
$$10B - 2B + 3 = 5B + 21 \Rightarrow$$
$$10(6) - 2(6) + 3 = 5(6) + 21$$
$$60 - 12 + 3 = 30 + 21$$
$$51 = 51$$

7.
$$6D - 10 = -2D - 34$$
$$6D - 10 + 10 = -2D - 34 + 10$$
$$6D = -2D - 24$$
$$6D + 2D = -2D + 2D - 24$$
$$8D = -24$$
$$\frac{1}{8} \times 8D = \frac{1}{8} \times -24$$
$$D = -3$$

8.
$$6D - 10 = -2D - 34 \Rightarrow$$
$$6(-3) - 10 = -2(-3) - 34$$
$$-18 - 10 = 6 - 34$$
$$-28 = -28$$

9.
$$-6B + 4 + 10B - 7 = 77$$
$$-6B + 10B + 4 - 7 = 77$$
$$4B - 3 = 77$$
$$4B - 3 + 3 = 77 + 3$$
$$4B = 80$$
$$\frac{1}{4} \times 4B = \frac{1}{4} \times 80$$
$$B = 20$$

10. $-6B + 4 + 10B - 7 = 77 \Rightarrow$
$$-6(20) + 4 + 10(20) - 7 = 77$$
$$-120 + 4 + 200 - 7 = 77$$
$$77 = 77$$

11.
$$2(W + 5) + 4W = 4(3 + W)$$
$$2W + 2(5) + 4W = 4(3) + 4W$$
$$2W + 10 + 4W = 12 + 4W$$
$$6W + 10 = 12 + 4W$$
$$6W + 10 - 10 = 12 - 10 + 4W$$
$$6W = 2 + 4W$$
$$6W - 4W = 2 + 4W - 4W$$
$$2W = 2$$
$$\frac{1}{2} \times 2W = \frac{1}{2} \times 2$$
$$W = 1$$

12. $2(W + 5) + 4W = 4(3 + W) \Rightarrow$
$$2((1) + 5) + 4(1) = 4(3 + (1))$$
$$2(6) + 4 = 4(4)$$
$$12 + 4 = 16$$
$$16 = 16$$

13.
$$(F - 1) - 2F = -2(3 + F)$$
$$F - 1 + (-2F) = -2(3) + -2F$$
$$-F - 1 = -6 - 2F$$
$$-F - 1 + 1 = -6 + 1 - 2F$$
$$-F = -5 - 2F$$
$$-F + 2F = -5 - 2F + 2F$$
$$F = -5$$

14. $(F - 1) - 2F = -2(3 + F) \Rightarrow$
$$((-5) - 1) - 2(-5) = -2(3 + (-5))$$
$$(-6) - (-10) = -2(-2)$$
$$(-6) + (10) = 4$$
$$4 = 4$$

15.
$$2(X - 7) + 5X = 5(X - 2)$$
$$2X - 2(7) + 5X = 5X - 5(2)$$
$$2X - 14 + 5X = 5X - 10$$
$$7X - 14 = 5X - 10$$
$$7X - 14 + 14 = 5X - 10 + 14$$
$$7X = 5X + 4$$
$$7X - 5X = 5X - 5X + 4$$
$$2X = 4$$
$$\frac{1}{2} \times 2X = \frac{1}{2} \times 4$$
$$X = 2$$

16. $2(X - 7) + 5X = 5(X - 2) \Rightarrow$
$$2((2) - 7) + 5(2) = 5((2) - 2)$$
$$2(-5) + 10 = 5(0)$$
$$-10 + 10 = 0$$
$$0 = 0$$

17.
$$8W = 32$$
$$\frac{1}{8} \times 8W = \frac{1}{8} \times 32$$
$$W = 4 \text{ miles}$$

18.
$$2B + 7 = B + 17$$
$$2B + 7 - 7 = B + 17 - 7$$
$$2B = B + 10$$
$$2B - B = B - B + 10$$
$$B = 10 \text{ pages}$$

Systematic Review 13D

1.
$$7X = 77$$
$$\frac{1}{7} \times 7X = \frac{1}{7} \times 77$$
$$X = 11$$

2. $7X = 77 \Rightarrow 7(11) = 77$
$$77 = 77$$

3.
$$6X - 10 = -46$$
$$6X - 10 + 10 = -46 + 10$$
$$6X = -36$$
$$\frac{1}{6} \times 6X = \frac{1}{6} \times -36$$
$$X = -6$$

4. $6X - 10 = -46 \Rightarrow 6(-6) - 10 = -46$
$$-36 - 10 = -46$$
$$-46 = -46$$

5.
$$-3A - 5 + 4A - 6 + 2A = 19$$
$$(-3A) + (-5) + 4A + (-6) + 2A = 19$$
$$(-3A) + 4A + 2A + (-5) + (-6) = 19$$
$$3A - 11 = 19$$
$$3A - 11 + 11 = 19 + 11$$
$$3A = 30$$
$$\frac{1}{3} \times 3A = \frac{1}{3} \times 30$$
$$A = 10$$

6. $-3A - 5 + 4A - 6 + 2A = 19 \Rightarrow$
$$-3(10) - 5 + 4(10) - 6 + 2(10) = 19$$
$$-30 - 5 + 40 - 6 + 20 = 19$$
$$19 = 19$$

7.
$$4(V + 4) = 3(V + 3)$$
$$4V + 4(4) = 3V + 3(3)$$
$$4V + 16 = 3V + 9$$
$$4V + 16 - 16 = 3V + 9 - 16$$
$$4V = 3V - 7$$
$$4V - 3V = 3V - 3V - 7$$
$$V = -7$$

8. $4(V + 4) = 3(V + 3) \Rightarrow$
$$4((-7) + 4) = 3((-7) + 3)$$
$$4(-3) = 3(-4)$$
$$-12 = -12$$

9.
$$7^2 + 9^2 = 15^2$$
$$49 + 81 = 225$$
$$130 \neq 225; \text{ no}$$

10. $\frac{7}{16}$ in

11. $\frac{8}{16} \div \frac{8}{8} = \frac{1}{2}$ in

12. $\frac{14}{16} \div \frac{2}{2} = \frac{7}{8}$ in

13. $\frac{\cancel{2}}{\cancel{8}_4} \times \frac{1}{3} = \frac{1}{12}$

14. $\frac{7}{8} \div \frac{3}{4} = \frac{7}{\cancel{8}_2} \times \frac{\cancel{4}}{3} = \frac{7}{6} = \frac{6}{6} + \frac{1}{6} = 1\frac{1}{6}$

15. $\frac{1}{2} \div \frac{1}{5} = \frac{1}{2} \times \frac{5}{1} = \frac{5}{2} = \frac{4}{2} + \frac{1}{2} = 2\frac{1}{2}$

16. done

17. $3\frac{3}{8} + 1\frac{4}{5} = 3\frac{15}{40} + 1\frac{32}{40} = 4\frac{47}{40} =$
$$4 + \frac{40}{40} + \frac{7}{40} = 4 + 1 + \frac{7}{40} = 5\frac{7}{40}$$

18. $2\frac{1}{10} + 3\frac{5}{8} = 2\frac{8}{80} + 3\frac{50}{80} = 5\frac{58}{80} = 5\frac{29}{40}$

19. $4\frac{1}{2} + 3\frac{3}{8} = 4\frac{8}{16} + 3\frac{6}{16} = 7\frac{14}{16} = 7\frac{7}{8}$ tons

20.
$$2X + 2 = 62$$
$$2X + 2 - 2 = 62 - 2$$
$$2X = 60$$
$$\frac{1}{2} \times 2X = \frac{1}{2} \times 60$$
$$X = \$30$$

Systematic Review 13E

1.
$$8X = -88$$
$$\frac{1}{8} \times 8X = \frac{1}{8} \times -88$$
$$X = -11$$

2. $8X = -88 \Rightarrow 8(-11) = -88$
$$-88 = -88$$

3.
$$7X + 5 = -23$$
$$7X + 5 - 5 = -23 - 5$$
$$7X = -28$$
$$\frac{1}{7} \times 7X = \frac{1}{7} \times -28$$
$$X = -4$$

4. $7X + 5 = -23 \Rightarrow 7(-4) + 5 = -23$
$$-28 + 5 = -23$$
$$-23 = -23$$

5.
$$-5Y + 4 - 6Y = 13 - 2Y$$
$$(-5Y) + 4 + (-6Y) = 13 - 2Y$$
$$(-5Y) + (-6Y) + 4 = 13 - 2Y$$
$$-11Y + 4 = 13 - 2Y$$
$$-11Y + 4 - 4 = 13 - 4 - 2Y$$
$$-11Y = 9 - 2Y$$
$$-11Y + 2Y = 9 - 2Y + 2Y$$
$$-9Y = 9$$
$$-\frac{1}{9} \times -9Y = -\frac{1}{9} \times 9$$
$$Y = -1$$

6. $-5Y + 4 - 6Y = 13 - 2Y \Rightarrow$
$$-5(-1) + 4 - 6(-1) = 13 - 2(-1)$$
$$5 + 4 - (-6) = 13 - (-2)$$
$$5 + 4 + 6 = 13 + 2$$
$$15 = 15$$

7. $4(L + 1) = 3(L + 4)$
$$4L + 4(1) = 3L + 3(4)$$
$$4L + 4 = 3L + 12$$
$$4L + 4 - 4 = 3L + 12 - 4$$
$$4L = 3L + 8$$
$$4L - 3L = 3L - 3L + 8$$
$$L = 8$$

8. $4(L + 1) = 3(L + 4) \Rightarrow 4((8) + 1) = 3((8) + 4)$
$$4(9) = 3(12)$$
$$36 = 36$$

9. $L^2 + 6^2 = 10^2$
$$L^2 + 36 = 100$$
$$L^2 = 100 - 36$$
$$L^2 = 64$$
$$L = 8 \text{ ft}$$

10. $\frac{11}{16}$ in

11. $\frac{12}{16} \div \frac{4}{4} = \frac{3}{4}$ in

12. $\frac{2}{16} \div \frac{2}{2} = \frac{1}{8}$ in

13. $\frac{6}{10} \div \frac{1}{5} = \frac{\cancel{6}^{3}}{\cancel{10}_{2}} \times \frac{\cancel{5}^{1}}{1} = \frac{3}{1} = 3$

14. $\frac{5}{8} \div \frac{1}{3} = \frac{5}{8} \times \frac{3}{1} = \frac{15}{8} = \frac{8}{8} + \frac{7}{8} = 1\frac{7}{8}$

15. $\frac{\cancel{5}^{1}}{\cancel{9}_{3}} \times \frac{\cancel{3}}{\cancel{5}} = \frac{1}{3}$

16. $1\frac{3}{5} + 4\frac{5}{6} = 1\frac{18}{30} + 4\frac{25}{30} = 5\frac{43}{30} =$
$$5 + \frac{30}{30} + \frac{13}{30} = 5 + 1 + \frac{13}{30} = 6\frac{13}{30}$$

17. $3\frac{1}{3} + 5\frac{1}{4} = 3\frac{4}{12} + 5\frac{3}{12} = 8\frac{7}{12}$

18. $4\frac{2}{3} + 1\frac{2}{5} = 4\frac{10}{15} + 1\frac{6}{15} = 5\frac{16}{15} =$
$$5 + \frac{15}{15} + \frac{1}{15} = 5 + 1 + \frac{1}{15} = 6\frac{1}{15}$$

19. $\frac{2}{3} \times \frac{1}{5} = \frac{2}{15}$ of the horses

20. $3N + 23 = 29$
$$3N + 23 - 23 = 29 - 23$$
$$3N = 6$$
$$\frac{1}{3} \times 3N = \frac{1}{3} \times 6$$
$$N = 2 \text{ years old}$$

Systematic Review 13F

1. $3X + 7 = 43$
$$3X + 7 - 7 = 43 - 7$$
$$3X = 36$$
$$\frac{1}{3} \times 3X = \frac{1}{3} \times 36$$
$$X = 12$$

2. $3X + 7 = 43 \Rightarrow 3(12) + 7 = 43$
$$36 + 7 = 43$$
$$43 = 43$$

3. $7Q + 3 = 3Q + 7 + 40$
$$7Q + 3 = 3Q + 47$$
$$7Q + 3 - 3 = 3Q + 47 - 3$$
$$7Q = 3Q + 44$$
$$7Q - 3Q = 3Q - 3Q + 44$$
$$4Q = 44$$
$$\frac{1}{4} \times 4Q = \frac{1}{4} \times 44$$
$$Q = 11$$

4. $7Q + 3 = 3Q + 7 + 40 \Rightarrow$
$$7(11) + 3 = 3(11) + 7 + 40$$
$$77 + 3 = 33 + 7 + 40$$
$$80 = 80$$

5.
$$11A - 4A - 18 = 2A + A + 10$$
$$7A - 18 = 3A + 10$$
$$7A - 18 + 18 = 3A + 10 + 18$$
$$7A = 3A + 28$$
$$7A - 3A = 3A - 3A + 28$$
$$4A = 28$$
$$\frac{1}{4} \times 4A = \frac{1}{4} \times 28$$
$$A = 7$$

6.
$$11A - 4A - 18 = 2A + A + 10 \Rightarrow$$
$$11(7) - 4(7) - 18 = 2(7) + (7) + 10$$
$$77 - 28 - 18 = 14 + 7 + 10$$
$$31 = 31$$

7.
$$-2(X + 7) + 4X = 4(X - 9)$$
$$-2X + (-2)(7) + 4X = 4X - (4)(9)$$
$$-2X - 14 + 4X = 4X - 36$$
$$-2X + 4X - 14 = 4X - 36$$
$$2X - 14 = 4X - 36$$
$$2X - 14 + 14 = 4X - 36 + 14$$
$$2X = 4X - 22$$
$$2X - 4X = 4X - 4X - 22$$
$$-2X = -22$$
$$-\frac{1}{2} \times -2X = -\frac{1}{2} \times -22$$
$$X = 11$$

8.
$$-2(X + 7) + 4X = 4(X - 9) \Rightarrow$$
$$-2((11) + 7) + 4(11) = 4((11) - 9)$$
$$-2(18) + 44 = 4(2)$$
$$-36 + 44 = 8$$
$$8 = 8$$

9.
$$5^2 + 12^2 = H^2$$
$$25 + 144 = H^2$$
$$169 = H^2$$
$$H = 13 \text{ miles}$$

10. $\frac{10}{16} \div \frac{2}{2} = \frac{5}{8}$ in

11. $\frac{13}{16}$ in

12. $\frac{4}{16} \div \frac{4}{4} = \frac{1}{4}$ in

13. $\frac{3}{8} \times \frac{1}{7} = \frac{3}{56}$

14. $\frac{3}{4} \div \frac{1}{8} = \frac{3}{\cancel{4}} \times \frac{\cancel{8}^2}{1} = \frac{6}{1} = 6$

15.
$$\frac{6}{7} \div \frac{5}{14} = \frac{6}{\cancel{7}} \times \frac{\cancel{14}^2}{5} = \frac{12}{5}$$
$$= \frac{10}{5} + \frac{2}{5} = 2\frac{2}{5}$$

16.
$$3\frac{3}{5} + 2\frac{7}{10} = 3\frac{30}{50} + 2\frac{35}{50} = 5\frac{65}{50} =$$
$$5 + \frac{50}{50} + \frac{15}{50} = 5 + 1 + \frac{15}{50} = 6 + \frac{3}{10} = 6\frac{3}{10}$$

17.
$$2\frac{9}{10} + 4\frac{5}{8} = 2\frac{72}{80} + 4\frac{50}{80} = 6\frac{122}{80} =$$
$$6 + \frac{80}{80} + \frac{42}{80} = 6 + 1 + \frac{21}{40} = 7\frac{21}{40}$$

18.
$$3\frac{1}{2} + 1\frac{3}{4} = 3\frac{4}{8} + 1\frac{6}{8} = 4\frac{10}{8} =$$
$$4 + \frac{8}{8} + \frac{2}{8} = 4 + 1 + \frac{2}{8} = 5 + \frac{1}{4} = 5\frac{1}{4}$$

19. $\frac{6}{8} \div \frac{2}{8} = \frac{\cancel{6}^3}{\cancel{8}} \times \frac{\cancel{8}}{\cancel{2}} = \frac{3}{1} = 3$ people

20.
$$4N + 8N = 24$$
$$12N = 24$$
$$\frac{1}{12} \times 12N = \frac{1}{12} \times 24$$
$$N = 2 \text{ pies}$$

Lesson Practice 14A

1. done

2.
$$20 \times 9 - 14 \div 7 =$$
$$180 - 2 = 178$$

3.
$$120 \div 10 - 2 \times 3 + 5 =$$
$$12 - 6 + 5 = 11$$

4.
$$8^2 \div 2 - 6 \times 2^2 =$$
$$64 \div 2 - 6 \times 4 =$$
$$32 - 24 = 8$$

5.
$$5 \times 3 + 6^3 - 10 \times 3 =$$
$$5 \times 3 + 216 - 10 \times 3 =$$
$$15 + 216 - 30 = 201$$

6.
$$11^2 \times 5 - 7 \times 8 - 16 =$$
$$121 \times 5 - 7 \times 8 - 16 =$$
$$605 - 56 - 16 = 533$$

7.
$$(16 + 9) + (8 + 3) \times 8 =$$
$$(25) + (11) \times 8 =$$
$$25 + 88 = 113$$

8. $(4+6)\times 9^2 - 6\times 5 =$
$(10)\times 9^2 - 6\times 5 =$
$10\times 81 - 6\times 5 =$
$810 - 30 = 780$

9. $112 - \left(2^2 + 3\right)\times 5 =$
$112 - (4+3)\times 5 =$
$112 - (7)\times 5 =$
$112 - 35 = 77$

10. $\left(18\times 9 + 3\right)^2 - 10^2 =$
$(162+3)^2 - 10^2 =$
$(165)^2 - 10^2 =$
$27,225 - 100 = 27,125$

11. $15^2 + (12\div 3)^3 - (6\div 3)^2 =$
$15^2 + (4)^3 - (2)^2 =$
$225 + 64 - 4 = 285$

12. $(8+1)^2 \div 3 + 9\times 3^2 =$
$(9)^2 \div 3 + 9\times 3^2 =$
$81\div 3 + 9\times 9 =$
$27 + 81 = 108$

13. done

14. done

15. done

16. $9(2A-6) = 2(7) + 2^2 \Rightarrow$
$9(2(4)-6) = 2(7) + 2^2$
$9(8-6) = 2(7) + 2^2$
$9(2) = 2(7) + 2^2$
$9(2) = 2(7) + 4$
$18 = 14 + 4$
$18 = 18$

17. $4^2 - 2^2 + 5\times 3 = A$
$16 - 4 + 5\times 3 = A$
$16 - 4 + 15 = A$
$27 = A$

18. $4^2 - 2^2 + 5\times 3 = A \Rightarrow 4^2 - 2^2 + 5\times 3 = (27)$
$16 - 4 + 5\times 3 = 27$
$16 - 4 + 15 = 27$
$27 = 27$

Lesson Practice 14B

1. $8\times 6 + 9\times 3 =$
$48 + 27 = 75$

2. $48\div 6 - 2\times 3 =$
$8 - 6 = 2$

3. $69\div 3 + 5\times 8 - 6 =$
$23 + 40 - 6 = 57$

4. $9^2 + 2 - 6\times 3^2 =$
$81 + 2 - 6\times 9 =$
$81 + 2 - 54 = 29$

5. $10^2 \div 5 - 4^3 - 12 =$
$100\div 5 - 64 - 12 =$
$20 - 64 - 12 = -56$

6. $150 - 11^2 + 6^3 \div 12 =$
$150 - 121 + 216\div 12 =$
$150 - 121 + 18 = 47$

7. $(17+9)\times 5 - (18\times 6) =$
$(26)\times 5 - (108) =$
$130 - 108 = 22$

8. $(5+3)\times\left(8^2 - 2\right)\times 20 =$
$(8)\times(64-2)\times 20 =$
$8\times(62)\times 20 = 9,920$

9. $148 - \left(4^2 + 3\right)\times 6 =$
$148 - (16+3)\times 6 =$
$148 - (19)\times 6 =$
$148 - 114 = 34$

10. $(72\div 12\times 5)^2 - 18\times 3^2 =$
$(30)^2 - 18\times 3^2 =$
$900 - 18\times 9 =$
$900 - 162 = 738$

11. $8^2 - 4^2 \div (3+1)^2 + 6 =$
$8^2 - 4^2 \div (4)^2 + 6 =$
$64 - 16\div 16 + 6 =$
$64 - 1 + 6 = 69$

12. $(3+9)^2 - (2+3)^3 + (90\div 10) =$
$(12)^2 - (5)^3 + (9) =$
$144 - 125 + 9 = 28$

13. $2(5-4+1)+F-2=5$
$2(2)+F-2=5$
$4+F-2=5$
$2+F=5$
$2-2+F=5-2$
$F=3$

14. $2(5-4+1)+F-2=5 \Rightarrow$
$2(5-4+1)+(3)-2=5$
$2(2)+3-2=5$
$4+3-2=5$
$5=5$

15. $-X(-2+1)=5^2+2^3$
$-X(-1)=25+8$
$X=33$

16. $-X(-2+1)=5^2+2^3 \Rightarrow -(33)(-2+1)=5^2+2^3$
$-33(-1)=25+8$
$33=33$

17. $(3+X)-(2\cdot2)+6=9$
$3+X-4+6=9$
$X+5=9$
$X+5-5=9-5$
$X=4$

18. $(3+X)-(2\cdot2)+6=9 \Rightarrow$
$(3+(4))-(2\cdot2)+6=9$
$(7)-(4)+6=9$
$9=9$

Lesson Practice 14C

1. $8\times9+18\div3=$
$72+6=78$

2. $14+18\times5-100\div5=$
$14+90-20=84$

3. $36+120\div12-6\times3=$
$36+10-18=28$

4. $5^2-3\times6+4^3-10=$
$25-3\times6+64-10=$
$25-18+64-10=61$

5. $6^2\times5-24+18\times2=$
$36\times5-24+18\times2=$
$180-24+36=192$

6. $9^2+7^2-5^3+3^3-2^5=$
$81+49-125+27-32=0$

7. $(18+3)\times6-(20\times2)=$
$21\times6-40=$
$126-40=86$

8. $(6+5)\times(4^3-10)\times5=$
$11\times(64-10)\times5=$
$11\times54\times5=2,970$

9. $498-(5^3+3)\times3=$
$498-(125+3)\times3=$
$498-128\times3=$
$498-384=114$

10. $(64\div16\times3)^3-20\times4^3=$
$(12)^3-20\times4^3=$
$1,728-20\times64=$
$1,728-1,280=448$

11. $12^2-5^2\div(2+3)^2+7^2=$
$144-25\div(5)^2+49=$
$144-25\div25+49=$
$144-1+49=192$

12. $(4+5)^3-(7+8)^2+(5\times2)^3=$
$9^3-15^2+10^3=$
$729-225+1,000=1,504$

13. $A+2^2-6\cdot8+2=-41$
$A+4-6\cdot8+2=-41$
$A+4-48+2=-41$
$A-42=-41$
$A=-41+42$
$A=1$

14. $A+2^2-6\cdot8+2=-41 \Rightarrow$
$(1)+2^2-6\cdot8+2=-41$
$1+4-48+2=-41$
$-41=-41$

15.
$$3(Q-6) = 2(2Q+7)+10$$
$$3Q-3(6) = 2(2Q)+2(7)+10$$
$$3Q-18 = 4Q+14+10$$
$$3Q-18 = 4Q+24$$
$$3Q = 4Q+24+18$$
$$3Q = 4Q+42$$
$$3Q-4Q = 42$$
$$-Q = 42$$
$$-1 \times -Q = -1 \times 42$$
$$Q = -42$$

16.
$$3(Q-6) = 2(2Q+7)+10 \Rightarrow$$
$$3((-42)-6) = 2(2(-42)+7)+10$$
$$3(-48) = 2(-84+7)+10$$
$$-144 = 2(-77)+10$$
$$-144 = -154+10$$
$$-144 = -144$$

17.
$$8^2(F-1) = 5(6F-6)$$
$$64(F-1) = 30F-30$$
$$64F-64 = 30F-30$$
$$64F-30F = -30+64$$
$$34F = 34$$
$$\frac{1}{34} \times 34F = \frac{1}{34} \times 34$$
$$F = 1$$

18.
$$8^2(F-1) = 5(6F-6) \Rightarrow$$
$$8^2((1)-1) = 5(6(1)-6)$$
$$8^2(0) = 5(0)$$
$$64(0) = 0$$
$$0 = 0$$

Systematic Review 14D

1.
$$6 \times 7+9 \times 3 =$$
$$42+27 = 69$$

2.
$$30 \times 6-35 \div 7 =$$
$$180-5 = 175$$

3.
$$140 \div 10-3 \times 4+16 =$$
$$14-12+16 = 18$$

4.
$$5^3 \div 5^2 \times 6-9 \times 1 =$$
$$125 \div 25 \times 6-9 \times 1 =$$
$$5 \times 6-9 =$$
$$30-9 = 21$$

5.
$$2X(-8-4)+3^2 = -5X(2^3)-7$$
$$2X(-12)+9 = -5X(8)-7$$
$$-24X+9 = -40X-7$$
$$-24X+40X = -7-9$$
$$16X = -16$$
$$\frac{1}{16} \times 16X = \frac{1}{16} \times -16$$
$$X = -1$$

6.
$$2X(-8-4)+3^2 = -5X(2^3)-7 \Rightarrow$$
$$2(-1)(-8-4)+3^2 = -5(-1)(2^3)-7$$
$$2(-1)(-12)+3^2 = -5(-1)(8)-7$$
$$2(-1)(-12)+9 = -5(-1)(8)-7$$
$$24+9 = 40-7$$
$$33 = 33$$

7.
$$5X-X+4-5 = 3X+2X-3$$
$$4X-1 = 5X-3$$
$$4X-5X = -3+1$$
$$-X = -2$$
$$-1 \times -X = -1 \times -2$$
$$X = 2$$

8.
$$5X-X+4-5 = 3X+2X-3 \Rightarrow$$
$$5(2)-(2)+4-5 = 3(2)+2(2)-3$$
$$10-2+4-5 = 6+4-3$$
$$7 = 7$$

9. $(-8)^2 = (-8)(-8) = 64$

10. $-(4)^2 = -(4)(4) = -16$

11. $-1^3 = -(1)(1)(1) = -1$

12. $-\left(\frac{1}{2}\right)^2 = -\left(\frac{1}{2}\right)\left(\frac{1}{2}\right) = -\frac{1}{4}$

13. done

14. done

15.
$$5\frac{1}{4} = 5\frac{3}{12} = 4\frac{15}{12}$$
$$-1\frac{2}{3} = -1\frac{8}{12} = -1\frac{8}{12}$$
$$\overline{}$$
$$3\frac{7}{12}$$

16.

$$7\frac{1}{2} = 7\frac{3}{6} = 6\frac{9}{6}$$
$$-3\frac{2}{3} = -3\frac{4}{6} = -3\frac{4}{6}$$
$$3\frac{5}{6} \text{ miles}$$

17. $3^2 + 6^2 = 8^2$
$9 + 36 = 64$
$45 \neq 64; \text{ no}$

18. $2M = 4 \times 5$
$2M = 20$
$\frac{1}{2} \times 2M = \frac{1}{2} \times 20$
$M = 10 \text{ years old}$

Systematic Review 14E

1. $8 \times 3 + 7^3 - 12 \times 5 =$
$8 \times 3 + 343 - 12 \times 5 =$
$24 + 343 - 60 = 307$

2. $12^2 \times 2^6 + 3^3 - 4^3 =$
$144 \times 64 + 27 - 64 =$
$9,216 + 27 - 64 = 9,179$

3. $(20 \times 11) + (12 - 8) \times 10 =$
$220 + 4 \times 10 =$
$220 + 40 = 260$

4. $(5 + 8) + (6^2 - 7) \times 8 =$
$13 + (36 - 7) \times 8 =$
$13 + 29 \times 8 =$
$13 + 232 = 245$

5. $4A = 3^2(10 - 6)$
$4A = 9(4)$
$\frac{1}{4} \times 4A = \frac{1}{4} \times 9(4)$
$A = 9$

6. $4A = 3^2(10 - 6) \Rightarrow 4(9) = 3^2(10 - 6)$
$36 = 3^2(4)$
$36 = 9(4)$
$36 = 36$

7. $1 \cdot 2 \cdot 3 \cdot 4 - A(5 \cdot 6) = -7(8A + 4)$
$1 \cdot 2 \cdot 3 \cdot 4 - A(30) = -7(8A) + -7(4)$
$24 - 30A = -56A - 28$
$-30A = -56A - 28 - 24$
$-30A = -56A - 52$
$-30A + 56A = -52$
$26A = -52$
$\frac{1}{26} \times 26A = \frac{1}{26} \times -52$
$A = -2$

8. $1 \cdot 2 \cdot 3 \cdot 4 - A(5 \cdot 6) = -7(8A + 4)$
$1 \cdot 2 \cdot 3 \cdot 4 - (-2)(5 \cdot 6) = -7(8(-2) + 4)$
$1 \cdot 2 \cdot 3 \cdot 4 - (-2)(30) = -7(-16 + 4)$
$24 - (-60) = -7(-12)$
$84 = 84$

9. $(-3)^2 = (-3)(-3) = 9$

10. $5^2 = (5)(5) = 25$

11. $-2^4 = -(2)(2)(2)(2) = -16$

12. $\left(\frac{3}{4}\right)^2 = \left(\frac{3}{4}\right)\left(\frac{3}{4}\right) = \frac{9}{16}$

13.

$$6\frac{1}{3} = 6\frac{5}{15} = 5\frac{20}{15}$$
$$-5\frac{4}{5} = -5\frac{12}{15} = -5\frac{12}{15}$$
$$\frac{8}{15}$$

14.

$$9 = 8\frac{8}{8}$$
$$-6\frac{3}{8} = -6\frac{3}{8}$$
$$2\frac{5}{8}$$

15.

$$8\frac{2}{3} = 8\frac{16}{24}$$
$$-7\frac{1}{8} = -7\frac{3}{24}$$
$$1\frac{13}{24}$$

16.

$$9\frac{3}{4} = 9\frac{6}{8}$$
$$-2\frac{1}{2} = -2\frac{4}{8}$$
$$7\frac{2}{8} = 7\frac{1}{4} \text{ ft}$$

17. $\frac{15}{1} \times \frac{1}{8} = \frac{15}{8} = \frac{8}{8} + \frac{7}{8} = 1\frac{7}{8}$ pies

18.
$$2M + 20 = 80$$
$$2M = 80 - 20$$
$$2M = 60$$
$$\frac{1}{2} \times 2M = \frac{1}{2} \times 60$$
$$M = 30 \text{ mph}$$

Systematic Review 14F

1.
$$\left(6^3 \times 5\right) - \left(5^2 + 54\right) \times 3 =$$
$$\left(216 \times 5\right) - \left(25 + 54\right) \times 3 =$$
$$1,080 - 79 \times 3 =$$
$$1,080 - 237 = 843$$

2.
$$\left(24 \times 11 + 6\right)^2 - \left(154 \div 11\right)^3 =$$
$$\left(264 + 6\right)^2 - \left(14\right)^3 =$$
$$270^2 - 14^3 =$$
$$72,900 - 2,744 = 70,156$$

3.
$$18^2 + \left(14 \div 7\right)^6 - \left(8 \div 2\right)^2 =$$
$$18^2 + 2^6 - 4^2 =$$
$$324 + 64 - 16 = 372$$

4.
$$\left(9 + 3\right)^2 \div 9 + 8 \times 4^4 =$$
$$12^2 \div 9 + 8 \times 4^4 =$$
$$144 \div 9 + 8 \times 256 =$$
$$16 + 2,048 = 2,064$$

5.
$$3(X + 2) + 5X = 6X + 14$$
$$3X + 3(2) + 5X = 6X + 14$$
$$8X + 6 = 6X + 14$$
$$8X = 6X + 14 - 6$$
$$8X = 6X + 8$$
$$8X - 6X = 8$$
$$2X = 8$$
$$\frac{1}{2} \times 2X = \frac{1}{2} \times 8$$
$$X = 4$$

6.
$$3(X + 2) + 5X = 6X + 14 \Rightarrow$$
$$3((4) + 2) + 5(4) = 6(4) + 14$$
$$3(6) + 5(4) = 6(4) + 14$$
$$18 + 20 = 24 + 14$$
$$38 = 38$$

7.
$$2(A + 3) = 2^5 + 6$$
$$2A + 2(3) = 32 + 6$$
$$2A + 6 = 38$$
$$2A = 38 - 6$$
$$2A = 32$$
$$\frac{1}{2} \times 2A = \frac{1}{2} \times 32$$
$$A = 16$$

8.
$$2(A + 3) = 2^5 + 6 \Rightarrow \quad 2((16) + 3) = 2^5 + 6$$
$$2(19) = 32 + 6$$
$$38 = 38$$

9. $\sqrt{100} = 10$

10. $\sqrt{A^2} = A$

11. $-3^2 = -(3)(3) = -9$

12. $\frac{1}{\sqrt{49}} = \frac{1}{7}$

13.
$$\begin{array}{r} 8\frac{3}{5} = \quad 8\frac{27}{45} = \quad 7\frac{72}{45} \\ - 2\frac{8}{9} = \quad - 2\frac{40}{45} = \quad - 2\frac{40}{45} \\ \hline 5\frac{32}{45} \end{array}$$

14.
$$\begin{array}{r} 5\frac{1}{4} = \quad 5\frac{6}{24} = \quad 4\frac{30}{24} \\ - 1\frac{5}{6} = \quad - 1\frac{20}{24} = \quad - 1\frac{20}{24} \\ \hline 3\frac{10}{24} = 3\frac{5}{12} \end{array}$$

15.
$$\begin{array}{r} 4\frac{3}{4} = \quad 4\frac{24}{32} = \quad 3\frac{56}{32} \\ - 1\frac{7}{8} = \quad - 1\frac{28}{32} = \quad - 1\frac{28}{32} \\ \hline 2\frac{28}{32} = 2\frac{7}{8} \end{array}$$

16. $2\frac{3}{4} + 1\frac{1}{10} = 2\frac{30}{40} + 1\frac{4}{40} = 3\frac{34}{40} = 3\frac{17}{20}$ in

17. $\frac{2}{3} \div \frac{1}{4} = \frac{2}{3} \times \frac{4}{1} = \frac{8}{3} = \frac{6}{3} + \frac{2}{3} = 2\frac{2}{3}$ times

18.

$$7N - 4 = 4N + 20$$
$$7N = 4N + 20 + 4$$
$$7N = 4N + 24$$
$$7N - 4N = 24$$
$$3N = 24$$
$$\frac{1}{3} \times 3N = \frac{1}{3} \times 24$$
$$N = 8$$

Lesson Practice 15A

1. done

2. $6(4 \times 4) = 96 \text{ ft}^2$

3. $2(10 \times 15) + 2(15 \times 6) + 2(6 \times 10) =$
   $\quad 2(150) + 2(90) + 2(60) =$
   $\qquad 300 + 180 + 120 = 600 \text{ in}^2$

4. $2(5 \times 16) + 2(16 \times 10) + 2(10 \times 5) =$
   $\quad 2(80) + 2(160) + 2(50) =$
   $\qquad 160 + 320 + 100 = 580 \text{ ft}^2$

5. done

6. $4\left(6 \times 8 \times \frac{1}{2}\right) + (6 \times 6) =$
   $\quad 4(24) + 36 =$
   $\qquad 96 + 36 = 132 \text{ in}^2$

7. $2(20 \times 30) =$
   $\quad 2(600) = 1{,}200 \text{ ft}^2$

8. $1{,}200 \div 100 = 12 \text{ squares}$

Lesson Practice 15B

1. $2(5 \times 6) + 2(6 \times 4) + 2(4 \times 5) =$
   $\quad 2(30) + 2(24) + 2(20) =$
   $\qquad 60 + 48 + 40 = 148 \text{ ft}^2$

2. $6(7 \times 7) =$
   $\quad 6(49) = 294 \text{ units}^2$

3. $2(8 \times 10) + 2(10 \times 3) + 2(3 \times 8) =$
   $\quad 2(80) + 2(30) + 2(24) =$
   $\qquad 160 + 60 + 48 = 268 \text{ in}^2$

4. $2(2 \times 7) + 2(7 \times 5) + 2(5 \times 2) =$
   $\quad 2(14) + 2(35) + 2(10) =$
   $\qquad 28 + 70 + 20 = 118 \text{ yd}^2$

5. $4\left(5 \times 8 \times \frac{1}{2}\right) + (5 \times 5) =$
   $\quad 4(20) + 25 =$
   $\qquad 80 + 25 = 105 \text{ units}^2$

6. $4\left(5 \times 4 \times \frac{1}{2}\right) + (4 \times 4) =$
   $\quad 4(10) + 16 =$
   $\qquad 40 + 16 = 56 \text{ in}^2$

7. rectangles:
   $2(24 \times 10) + 2(30 \times 10) =$
   $\quad 2(240) + 2(300) =$
   $\qquad 480 + 600 = 1{,}080 \text{ ft}^2$

   triangles:
   $2\left(24 \times 16 \times \frac{1}{2}\right) =$
   $\quad 2(192) = 384 \text{ ft}^2$

   total:
   $1{,}080 + 384 = 1{,}464 \text{ ft}^2$

8. $1{,}464 \div 100 = 14.64$
   Round up to next whole number.
   15 squares

Lesson Practice 15C

1. $2(4 \times 6) + 2(6 \times 3) + 2(3 \times 4) =$
   $\quad 2(24) + 2(18) + 2(12) =$
   $\qquad 48 + 36 + 24 = 108 \text{ ft}^2$

2. $6(9 \times 9) =$
   $\quad 6(81) = 486 \text{ units}^2$

3. $2(8 \times 12) + 2(12 \times 4) + 2(4 \times 8) =$
   $\quad 2(96) + 2(48) + 2(32) =$
   $\qquad 192 + 96 + 64 = 352 \text{ in}^2$

4. $2(7 \times 20) + 2(20 \times 10) + 2(10 \times 7) =$
   $\quad 2(140) + 2(200) + 2(70) =$
   $\qquad 280 + 400 + 140 = 820 \text{ ft}^2$

5. $4\left(9 \times 7 \times \frac{1}{2}\right) + (7 \times 7) =$
   $\quad 4(31.5) + (49) =$
   $\qquad 126 + 49 = 175 \text{ units}^2$

6. $4\left(10\times12\times\dfrac{1}{2}\right)+(10\times10)=$

$4(60)+100=$

$240+100=340$ in^2

7. $1{,}464$ ft$^2\div400=3.66$ gallons

rounding up:

4 gallons must be purchased

8. 4 gal $\times3=12$ gal

$12\times\$25=\$300$

If the student used the unrounded answer from problem 7 to figure this answer, it would result in an answer of 11 gallons for three coats of paint. The cost in that case would be $275.

Systematic Review 15D

1. $2(6\times8)+2(8\times2)+2(2\times6)=$

$2(48)+2(16)+2(12)=$

$96+32+24=152$ in^2

2. $4\left(8\times8\times\dfrac{1}{2}\right)+(8\times8)=$

$4(32)+64=$

$128+64=192$ in^2

3. $-2A+8A-18+5A=-A+14-5A+2$

$11A-18=-6A+16$

$11A=-6A+16+18$

$11A=-6A+34$

$11A+6A=34$

$17A=34$

$A=2$

4. $-2A+8A-18+5A=-A+14-5A+2\Rightarrow$

$-2(2)+8(2)-18+5(2)=-(2)+14-5(2)+2$

$-4+16-18+10=-2+14-10+2$

$4=4$

5. $D(3-7)-12=0$

$3D-7D-12=0$

$-4D-12=0$

$-4D=12$

$D=-3$

6. $D(3-7)-12=0\Rightarrow(-3)(3-7)-12=0$

$(-3)(-4)-12=0$

$12-12=0$

$0=0$

7. $\begin{array}{ccc}4\dfrac{5}{8}&=&4\dfrac{20}{32}\\[2mm]+\;1\dfrac{1}{4}&=&+\;1\dfrac{8}{32}\\[1mm]\hline&&5\dfrac{28}{32}=5\dfrac{7}{8}\end{array}$

8. $\begin{array}{ccc}7\dfrac{3}{4}&=&7\dfrac{15}{20}\\[2mm]+\;9\dfrac{3}{5}&=&+\;9\dfrac{12}{20}\\[1mm]\hline&&16\dfrac{27}{20}=17\dfrac{7}{20}\end{array}$

9. $\begin{array}{cccc}8\dfrac{3}{5}&=&8\dfrac{27}{45}&=&7\dfrac{72}{45}\\[2mm]-\;2\dfrac{8}{9}&=&-\;2\dfrac{40}{45}&=&-\;2\dfrac{40}{45}\\[1mm]\hline&&&&5\dfrac{32}{45}\end{array}$

10. $\begin{array}{cccc}5\dfrac{1}{4}&=&5\dfrac{6}{24}&=&4\dfrac{30}{24}\\[2mm]-\;1\dfrac{5}{6}&=&-\;1\dfrac{20}{24}&=&-\;1\dfrac{20}{24}\\[1mm]\hline&&&&3\dfrac{10}{24}=3\dfrac{5}{12}\end{array}$

11. done

12. $5\dfrac{5}{8}\times1\dfrac{2}{5}\times2\dfrac{1}{7}=$

$\dfrac{\cancel{45}^9}{8}\times\dfrac{\cancel{7}}{\cancel{5}}\times\dfrac{15}{\cancel{7}}=\dfrac{135}{8}=16\dfrac{7}{8}$

13. $1\dfrac{2}{3}\times2\dfrac{1}{3}\times\dfrac{9}{10}=$

$\dfrac{\cancel{5}}{\cancel{3}}\times\dfrac{7}{\cancel{3}}\times\dfrac{\cancel{9}^{3}}{\cancel{10}_{2}}=\dfrac{7}{2}=3\dfrac{1}{2}$

14. $5\dfrac{2}{8}\times3\dfrac{5}{6}\times1\dfrac{5}{7}=$

$\dfrac{\cancel{42}^{7}}{\cancel{8}_{2}}\times\dfrac{23}{\cancel{6}}\times\dfrac{\cancel{12}^{3}}{\cancel{7}}=\dfrac{69}{2}=34\dfrac{1}{2}$

15. $\dfrac{\cancel{3}}{4}\times\dfrac{1}{\cancel{2}}\times\dfrac{\cancel{2}}{\cancel{3}}=\dfrac{1}{4}$ of the jelly beans

16. $\dfrac{1}{4}$ of 48

$48\div4=12$

$12\times1=12$ jelly beans

17. $9^2 + 12^2 = H^2$

$81 + 144 = H^2$

$225 = H^2$

$H = 15 \text{ in}$

18. $6N - 7 = N + 33$

$6N = N + 33 + 7$

$6N = N + 40$

$6N - N = 40$

$5N = 40$

$N = 8$

Systematic Review 15E

1. $2(20 \times 30) + 2(30 \times 10) + 2(10 \times 20) =$

$2(600) + 2(300) + 2(200) =$

$1,200 + 600 + 400 = 2,200 \text{ units}^2$

2. $4\left(4 \times 8 \times \dfrac{1}{2}\right) + (4 \times 4) =$

$4(16) + 16 =$

$64 + 16 = 80 \text{ in}^2$

3. $X(2+2) + 5X - 12 = X(4+4) - 2$

$X(4) + 5X - 12 = X(8) - 2$

$4X + 5X - 12 = 8X - 2$

$9X - 12 = 8X - 2$

$9X = 8X - 2 + 12$

$9X = 8X + 10$

$9X - 8X = 10$

$X = 10$

4. $X(2+2) + 5X - 12 = X(4+4) - 2 \Rightarrow$

$(10)(2+2) + 5(10) - 12 = (10)(4+4) - 2$

$(10)(4) + 50 - 12 = (10)(8) - 2$

$40 + 50 - 12 = 80 - 2$

$78 = 78$

5. $17 - Y + 8 + 3Y + 2Y = Y(-7+3) + 1^2$

$4Y + 25 = -7Y + 3Y + 1$

$4Y + 25 = -4Y + 1$

$8Y = 1 - 25$

$8Y = -24$

$Y = -3$

6. $17 - Y + 8 + 3Y + 2Y = Y(-7+3) + 1^2 \Rightarrow$

$17 - (-3) + 8 + 3(-3) + 2(-3) = (-3)(-7+3) + 1^2$

$17 + 3 + 8 - 9 - 6 = -3(-4) + 1$

$13 = 12 + 1$

$13 = 13$

7. $X + 32 \div 4 - 2^2 = 3^2 + 3$

$X + 32 \div 4 - 4 = 9 + 3$

$X + 8 - 4 = 12$

$X + 4 = 12$

$X = 12 - 4$

$X = 8$

8. $X + 32 \div 4 - 2^2 = 3^2 + 3 \Rightarrow$

$(8) + 32 \div 4 - 2^2 = 3^2 + 3$

$(8) + 32 \div 4 - 4 = 9 + 3$

$(8) + 8 - 4 = 12$

$12 = 12$

9. $4\dfrac{3}{4} \quad = \quad 4\dfrac{6}{8} \quad = \quad 3\dfrac{14}{8}$

$-\ 1\dfrac{7}{8} \quad = \quad -\ 1\dfrac{7}{8} \quad = \quad -\ 1\dfrac{7}{8}$

$\overline{\qquad\qquad\qquad\qquad\qquad 2\dfrac{7}{8}}$

10. $7\dfrac{2}{5} \quad = \quad 7\dfrac{4}{10} \quad = \quad 6\dfrac{14}{10}$

$-\ 3\dfrac{7}{10} \quad = \quad -\ 3\dfrac{7}{10} \quad = \quad -\ 3\dfrac{7}{10}$

$\overline{\qquad\qquad\qquad\qquad\qquad 3\dfrac{7}{10}}$

11. $6\dfrac{3}{4} \quad = \quad 6\dfrac{18}{24}$

$+\ 6\dfrac{5}{6} \quad = \quad +\ 6\dfrac{20}{24}$

$\overline{\qquad\qquad 12\dfrac{38}{24} = 13\dfrac{14}{24} = 13\dfrac{7}{12}}$

12. $3\dfrac{1}{3} \times 1\dfrac{7}{8} \times \dfrac{4}{5} = \dfrac{10}{3} \times \dfrac{\overset{3}{\cancel{15}}}{\underset{2}{\cancel{8}}} \times \dfrac{\cancel{4}}{\cancel{5}} = \dfrac{30}{6} = 5$

13. $2\dfrac{2}{5} \times 1\dfrac{1}{4} \times 1\dfrac{1}{7} = \dfrac{12}{5} \times \dfrac{\cancel{5}}{\cancel{4}} \times \dfrac{\overset{2}{\cancel{8}}}{7} = \dfrac{24}{7} = 3\dfrac{3}{7}$

14. $\dfrac{4}{5} \times 2\dfrac{3}{4} \times 3\dfrac{1}{3} = \dfrac{\cancel{4}}{\cancel{5}} \times \dfrac{11}{\cancel{4}} \times \dfrac{\overset{2}{\cancel{10}}}{3} = \dfrac{22}{3} = 7\dfrac{1}{3}$

15. $\dfrac{\cancel{7}}{\underset{2}{\cancel{10}}} \times \dfrac{\cancel{5}}{6} \times \dfrac{1}{\cancel{7}} = \dfrac{1}{12}$ of the profits

16. $\$144 \div 12 = \$12$

$\$12 \times 1 = \$12$

17.
$$L^2 + 16^2 = 20^2$$
$$L^2 + 256 = 400$$
$$L^2 = 400 - 256$$
$$L^2 = 144$$
$$L = 12 \text{ ft}$$

18.
$$2M - 17 = 81$$
$$2M = 81 + 17$$
$$2M = 98$$
$$M = 49 \text{ years old}$$

19.
$$4\left(15 \times 14 \times \frac{1}{2}\right) = 420 \text{ ft}^2$$
$$420 \div 100 = 4.2: \text{ rounds up to 5 squares}$$

20.
$$4(15 \times 9) = 4(135) = 540 \text{ ft}^2$$
$$540 \div 100 = 5.4: \text{ rounds up to 6 squares}$$

Systematic Review 15F

1.
$$6(3 \times 3) =$$
$$6(9) = 54 \text{ in}^2$$

2.
$$4\left(7 \times 12 \times \frac{1}{2}\right) + (7 \times 7) =$$
$$4(42) + (49) =$$
$$168 + 49 = 217 \text{ in}^2$$

3.
$$-X + 7 + 2X - 5 - 3X = 3(3 - X) - 4(1 - X)$$
$$-2X + 2 = 9 - 3X - 4 + 4X$$
$$-2X + 2 = X + 5$$
$$-2X = X + 5 - 2$$
$$-2X - X = 3$$
$$-3X = 3$$
$$X = -1$$

4.
$$-X + 7 + 2X - 5 - 3X = 3(3 - X) - 4(1 - X) \Rightarrow$$
$$-(-1) + 7 + 2(-1) - 5 - 3(-1) = 3(3 - (-1)) - 4(1 - (-1))$$
$$1 + 7 - 2 - 5 + 3 = 3(4) - 4(2)$$
$$4 = 12 - 8$$
$$4 = 4$$

5.
$$3(8 + 4) + 4^2 - X = 4X + 3 \cdot 4$$
$$3(12) + 16 - X = 4X + 12$$
$$36 + 16 - X = 4X + 12$$
$$36 + 16 - 12 = 4X + X$$
$$40 = 5X$$
$$X = 8$$

6.
$$3(8 + 4) + 4^2 - X = 4X + 3 \cdot 4$$
$$3(8 + 4) + 4^2 - (8) = 4(8) + 3 \cdot 4$$
$$3(12) + 16 - 8 = 32 + 12$$
$$36 + 16 - 8 = 44$$
$$44 = 44$$

7.
$$-5R + (9^2 - 3^2) + 13 = 7R + 5R$$
$$-5R + (81 - 9) + 13 = 12R$$
$$81 - 9 + 13 = 12R + 5R$$
$$85 = 17R$$
$$R = 5$$

8.
$$-5R + (9^2 - 3^2) + 13 = 7R + 5R \Rightarrow$$
$$-5(5) + (9^2 - 3^2) + 13 = 7(5) + 5(5)$$
$$-25 + (81 - 9) + 13 = 35 + 25$$
$$60 = 60$$

9.
$$4\frac{1}{6} = 4\frac{1}{6} = 3\frac{7}{6}$$
$$-1\frac{1}{2} = -1\frac{3}{6} = -1\frac{3}{6}$$
$$\overline{}$$
$$2\frac{4}{6} = 2\frac{2}{3}$$

10.
$$6\frac{5}{9} = 6\frac{5}{9}$$
$$-1\frac{1}{3} = -1\frac{3}{9}$$
$$\overline{}$$
$$5\frac{2}{9}$$

11.
$$9\frac{7}{8} = 9\frac{70}{80}$$
$$+2\frac{5}{10} = +2\frac{40}{80}$$
$$\overline{}$$
$$11\frac{110}{80} = 12\frac{30}{80} = 12\frac{3}{8}$$

12.
$$1\frac{2}{8} \times 3\frac{1}{5} \times 3\frac{1}{5} =$$
$$\frac{\overset{2}{\cancel{10}}}{\cancel{8}} \times \frac{\overset{2}{\cancel{16}}}{\cancel{5}} \times \frac{16}{5} = \frac{64}{5} = 12\frac{4}{5}$$

13.
$$2\frac{1}{9} \times 2\frac{1}{4} \times 1\frac{1}{11} =$$
$$\frac{19}{\cancel{9}} \times \frac{\cancel{9}}{\cancel{4}} \times \frac{\overset{3}{\cancel{12}}}{11} = \frac{57}{11} = 5\frac{2}{11}$$

14.
$$\frac{1}{3} \times 2\frac{6}{7} \times 1\frac{1}{5} =$$
$$\frac{1}{\cancel{3}} \times \frac{\overset{4}{\cancel{20}}}{7} \times \frac{\overset{2}{\cancel{6}}}{\cancel{5}} = \frac{8}{7} = 1\frac{1}{7}$$

15. $2\frac{1}{2} \times \frac{3}{5} \times \frac{1}{3} = \frac{5}{2} \times \frac{3}{5} \times \frac{1}{3} =$

$\frac{1}{2}$ of a bushel

16. $\frac{3}{4} \div \frac{1}{6} = \frac{3}{4} \times \frac{6}{1} = \frac{9}{2} = 4\frac{1}{2}$ days

17. $10^2 + 12^2 = 15^2$
$100 + 144 = 225$
$244 \neq 225;$ no

18. $5P + 4P = 225$
$9P = 225$
$P = 25$ pennies

19. $2(24 \times 26) = 2(624) = 1{,}248 \text{ ft}^2$
$1{,}248 \div 100 = 12.48:$
round up to 13 squares

20. $2\left(15 \times 22 \times \frac{1}{2}\right) = 2(165) = 330 \text{ ft}^2$

Lesson Practice 16A

1. done

2. $60 \times \frac{9}{5} + 32 = F$
$108 + 32 = F$
$F = 140°$

3. $10 \times \frac{9}{5} + 32 = F$
$18 + 32 = F$
$F = 50°$

4. $65 \times \frac{9}{5} + 32 = F$
$117 + 32 = F$
$F = 149°$

5. $200 \times \frac{9}{5} + 32 = F$
$360 + 32 = F$
$F = 392°$

6. $85 \times \frac{9}{5} + 32 = F$
$153 + 32 = F$
$F = 185°$

7. $30 \times \frac{9}{5} + 32 = F$
$54 + 32 = F$
$F = 86°$

8. $110 \times \frac{9}{5} + 32 = F$
$198 + 32 = F$
$F = 230°$

9. $215 \times \frac{9}{5} + 32 = F$
$387 + 32 = F$
$F = 419°$

10. done

11. $2 \times \frac{9}{5} + 32 = F$
$\frac{18}{5} + 32 = F$
$3\frac{3}{5} + 32 = F$
$F = 35\frac{3}{5} = 35.6°$

OR
$2 \times 1.8 + 32 =$
$3.6 + 32 = 35.6°$

12. $24 \times \frac{9}{5} + 32 = F$
$43.2 + 32 = F$
$F = 75.2°$

13. $0°C;$ $32°F$
14. $100°C;$ $212°F$
15. $37°C;$ $98.6°F$

16. $15 \times \frac{9}{5} + 32 = F$
$27 + 32 = F$
$F = 59°;$ no

17. $45 \times \frac{9}{5} + 32 = F$
$81 + 32 = F$
$F = 113°;$ it is warmer.

18. $20 \times \frac{9}{5} + 32 = F$
$36 + 32 = F$
$F = 68°$

Lesson Practice 16B

1. $^{24}\cancel{120} \times \dfrac{9}{5} + 32 = F$

    $216 + 32 = F$

        $F = 248°$

2. $^{15}\cancel{75} \times \dfrac{9}{5} + 32 = F$

    $135 + 32 = F$

        $F = 167°$

3. $^{3}\cancel{15} \times \dfrac{9}{5} + 32 = F$

    $27 + 32 = F$

        $F = 59°$

4. $^{5}\cancel{25} \times \dfrac{9}{5} + 32 = F$

    $45 + 32 = F$

        $F = 77°$

5. $^{7}\cancel{35} \times \dfrac{9}{5} + 32 = F$

    $63 + 32 = F$

        $F = 95°$

6. $^{11}\cancel{55} \times \dfrac{9}{5} + 32 = F$

    $99 + 32 = F$

        $F = 131°$

7. $^{21}\cancel{105} \times \dfrac{9}{5} + 32 = F$

    $189 + 32 = F$

        $F = 221°$

8. $^{16}\cancel{80} \times \dfrac{9}{5} + 32 = F$

    $144 + 32 = F$

        $F = 176°$

9. $^{19}\cancel{95} \times \dfrac{9}{5} + 32 = F$

    $171 + 32 = F$

        $F = 203°$

10. $13 \times 1.8 + 32 = F$

    $23.4 + 32 = F$

        $F = 55.4°$

11. $42 \times 1.8 + 32 = F$

    $75.6 + 32 = F$

        $F = 107.6°$

12. $28 \times 1.8 + 32 = F$

    $50.4 = F$

        $F = 82.4°$

13. $32°F$

14. $98.6°F$

15. $212°F$

16. $^{10}\cancel{50} \times \dfrac{9}{5} + 32 = F$

    $90 + 32 = F$

        $F = 122°$

17. $^{1}\cancel{5} \times \dfrac{9}{5} + 32 = F$

    $9 + 32 = F$

        $F = 41°;$ yes

18. $^{16}\cancel{80} \times \dfrac{9}{5} + 32 = F$

    $144 + 32 = F$

        $F = 176°;$ no

Lesson Practice 16C

1. $32°$

2. $^{1}\cancel{5} \times \dfrac{9}{5} + 32 = F$

    $9 + 32 = F$

        $F = 41°$

3. $^{25}\cancel{125} \times \dfrac{9}{5} + 32 = F$

    $225 + 32 = F$

        $F = 257°$

4. $^{14}\cancel{70} \times \dfrac{9}{5} + 32 = F$

    $126 + 32 = F$

        $F = 158°$

5. $^{35}\cancel{175} \times \dfrac{9}{5} + 32 = F$

    $315 + 32 = F$

        $F = 347°$

6. $^{4}\cancel{20} \times \dfrac{9}{5} + 32 = F$

    $36 + 32 = F$

        $F = 68°$

7. $45 \times \dfrac{9}{5} + 32 = F$

$81 + 32 = F$

$F = 113°$

8. $50 \times \dfrac{9}{5} + 32 = F$

$90 + 32 = F$

$F = 122°$

9. $60 \times \dfrac{9}{5} + 32 = F$

$108 + 32 = F$

$F = 140°$

10. $17 \times 1.8 + 32 = F$

$30.6 + 32 = F$

$F = 62.6°$

11. $22 \times 1.8 + 32 = F$

$39.6 + 32 = F$

$F = 71.6°$

12. $9 \times 1.8 + 32 = F$

$16.2 + 32 = F$

$F = 48.2°$

13. $0°C;\ 32°F$

14. $100°C;\ 212°F$

15. $37°C;\ 98.6°F$

16. $35 \times \dfrac{9}{5} + 32 = F$

$63 + 32 = F$

$F = 95°;\ yes$

17. $85 \times \dfrac{9}{5} + 32 = F$

$153 + 32 = F$

$F = 185°$

18. $32°F$

Systematic Review 16D

1. $70 \times \dfrac{9}{5} + 32 = F$

$126 + 32 = F$

$F = 158°$

2. $80 \times \dfrac{9}{5} + 32 = F$

$144 + 32 = F$

$F = 176°$

3. $108 \times 1.8 + 32 = F$

$194.4 + 32 = F$

$F = 226.4°$

4. $2(4 \times 7) + 2(7 \times 2) + 2(2 \times 4) =$

$2(28) + 2(14) + 2(8) =$

$56 + 28 + 16 = 100 \text{ in}^2$

5. $4\left(5 \times 10 \times \dfrac{1}{2}\right) + (5 \times 5) =$

$4(25) + 25 =$

$100 + 25 = 125 \text{ in}^2$

6. $15 \div 3 \cdot 8 + 10 = B$

$40 + 10 = B$

$B = 50$

7. $15 \div 3 \cdot 8 + 10 = B \Rightarrow\ 15 \div 3 \cdot 8 + 10 = (50)$

$40 + 10 = 50$

$50 = 50$

8. $\left(4^2 - X\right) + (8 \div 4)^2 = X + 2$

$(16 - X) + (2)^2 = X + 2$

$16 - X + 4 = X + 2$

$-X - X = 2 - 16 - 4$

$-2X = -18$

$X = 9$

9. $\left(4^2 - X\right) + (8 \div 4)^2 = X + 2 \Rightarrow$

$\left(4^2 - (9)\right) + (8 \div 4)^2 = (9) + 2$

$(16 - 9) + (2)^2 = 11$

$7 + 4 = 11$

$11 = 11$

10. done

11. done

12. $8.84 + 3.09 = 11.93$

13. $4.9 - 1.36 = 3.54$

14. $9.2 - 3.5 = 5.7 \text{ mi}$

15. $12 + 10.6 = 22.6 \text{ gal}$

16. $75 \times \dfrac{9}{5} + 32 = F$

$135 + 32 = F$

$F = 167°$

17. $0°$

18. $\dfrac{\cancel{7}^{1}}{\cancel{8}_{2}} \times \dfrac{1}{\cancel{7}} \times \dfrac{\cancel{4}}{5} = \dfrac{1}{10}$ of the problems

Systematic Review 16E

1. $\overset{18}{\cancel{9}}\cancel{0} \times \dfrac{9}{\cancel{5}} + 32 = F$

   $162 + 32 = F$

   $F = 194°$

2. $212°$

3. $46 \times 1.8 + 32 = F$

   $82.8 + 32 = F$

   $F = 114.8°$

4. $2(2 \times 3) + 2(3 \times 1) + 2(1 \times 2) =$

   $2(6) + 2(3) + 2(2) =$

   $12 + 6 + 4 = 22 \text{ in}^2$

5. $4\left(3 \times 4 \times \dfrac{1}{2}\right) + (3 \times 3) =$

   $4(6) + (9) =$

   $24 + 9 = 33 \text{ in}^2$

6. $14 \div 2 - 6X = 2 \cdot 7 - 13$

   $7 - 6X = 14 - 13$

   $7 - 6X = 1$

   $-6X = 1 - 7$

   $-6X = -6$

   $X = 1$

7. $14 \div 2 - 6X = 2 \cdot 7 - 13 \Rightarrow$

   $14 \div 2 - 6(1) = 2 \cdot 7 - 13$

   $7 - 6 = 14 - 13$

   $1 = 1$

8. $7C - 4C + 5 - 8 + C = 5^2 + 4$

   $4C - 3 = 25 + 4$

   $4C = 25 + 4 + 3$

   $4C = 32$

   $C = 8$

9. $7C - 4C + 5 - 8 + C = 5^2 + 4 \Rightarrow$

   $7(8) - 4(8) + 5 - 8 + (8) = 5^2 + 4$

   $56 - 32 + 5 - 8 + 8 = 25 + 4$

   $29 = 29$

10. $7.4 + 1.0 = 8.4$

11. $4.17 - 2.03 = 2.14$

12. $3 + 1.19 = 4.19$

13. $5.6 - 2.9 = 2.7$

14. $1 \times 10^3 + 4 \times 10^2 + 5 \times 10^1 + 6 \times 10^0$

15. $5 \times \dfrac{1}{10^1} + 4 \times \dfrac{1}{10^2} + 1 \times \dfrac{1}{10^3}$

16. $4 \times 10^1 + 6 \times \dfrac{1}{10^2}$

17. $8.3 + 3.25 = 11.55 \text{ lb}$

18. $100°$

19. $E + 15 + 5 = 46$

    $E + 20 = 46$

    $E = 46 - 20$

    $E = 26 \text{ jelly beans}$

20. $4\left(10 \times 10 \times \dfrac{1}{2}\right) + (10 \times 10) =$

    $4(50) + (100) =$

    $200 + 100 = 300 \text{ ft}^2$

    $300 \div 100 = 3 \text{ qt}$

Systematic Review 16F

1. $\overset{26}{1}\cancel{3}\cancel{0} \times \dfrac{9}{\cancel{5}} + 32 = F$

   $234 + 32 = F$

   $F = 266°$

2. $\overset{3}{1}\cancel{5} \times \dfrac{9}{\cancel{5}} + 32 = F$

   $27 + 32 = F$

   $F = 59°$

3. $66 \times 1.8 + 32 = F$

   $118.8 + 32 = F$

   $F = 150.8°$

4. $2(6 \times 9) + 2(9 \times 3) + 2(3 \times 6) =$

   $2(54) + 2(27) + 2(18) =$

   $108 + 54 + 36 = 198 \text{ units}^2$

5. $4\left(8 \times 8 \times \dfrac{1}{2}\right) + (8 \times 8) =$

   $4(32) + (64) =$

   $128 + 64 = 192 \text{ ft}^2$

6.
$$4Q - 2 = 10$$
$$4Q = 10 + 2$$
$$4Q = 12$$
$$Q = 3$$

7.
$$4Q - 2 = 10 \Rightarrow 4(3) - 2 = 10$$
$$12 - 2 = 10$$
$$10 = 10$$

8.
$$12^2(X + 3) = 6^2(48 + 32)$$
$$144(X + 3) = 36(80)$$
$$144X + 144(3) = 2{,}880$$
$$144X + 432 = 2{,}880$$
$$144X = 2{,}880 - 432$$
$$144X = 2{,}448$$
$$\frac{1}{144} \times 144X = \frac{1}{144} \times 2{,}448$$
$$X = 17$$

9.
$$12^2(X + 3) = 6^2(48 + 32) \Rightarrow$$
$$12^2((17) + 3) = 6^2(48 + 32)$$
$$12^2(20) = 6^2(80)$$
$$144(20) = 36(80)$$
$$2{,}880 = 2{,}880$$

10. $8.90 + 0.61 = 9.51$

11. $10.3 - 4 = 6.3$

12. $5.11 + 4.4 = 9.51$

13. $23.7 - 3.8 = 19.9$

14. $1 \times 10^4 + 1 \times \dfrac{1}{10^1}$

15. $1 \times 10^0 + 6 \times \dfrac{1}{10^1} + 7 \times \dfrac{1}{10^2}$

16. $4 \times 10^1 + 5 \times 10^0 + 8 \times \dfrac{1}{10^1} + 9 \times \dfrac{1}{10^2} + 2 \times \dfrac{1}{10^3}$

17. $10 - 4.5 = 5.5$ in

18. 37°

19. $2\dfrac{3}{8} \div 1\dfrac{1}{4} = \dfrac{19}{8} \div \dfrac{5}{4} =$

$\dfrac{19}{\overset{}{8}} \times \dfrac{\overset{1}{\cancel{4}}}{5} = \dfrac{19}{10} = 1\dfrac{9}{10}$ days; no

20. $\dfrac{4}{20} = \dfrac{1}{5}$ of the problems

Lesson Practice 17A

1. done

2.
$$(86 - 32)\frac{5}{9} = C$$
$$\left(\overset{6}{5}\cancel{4}\right)\frac{5}{\cancel{9}} = C$$
$$C = 30^\circ$$

3.
$$(59 - 32)\frac{5}{9} = C$$
$$\left(\overset{3}{2}\cancel{7}\right)\frac{5}{\cancel{9}} = C$$
$$C = 15^\circ$$

4.
$$(68 - 32)\frac{5}{9} = C$$
$$\left(\overset{4}{3}\cancel{6}\right)\frac{5}{\cancel{9}} = C$$
$$C = 20^\circ$$

5.
$$(41 - 32)\frac{5}{9} = C$$
$$\left(\overset{1}{\cancel{9}}\right)\frac{5}{\cancel{9}} = C$$
$$C = 5^\circ$$

6.
$$(122 - 32)\frac{5}{9} = C$$
$$\left(\overset{10}{9}\cancel{0}\right)\frac{5}{\cancel{9}} = C$$
$$C = 50^\circ$$

7. done

8.
$$(80 - 32)\frac{5}{9} = C \qquad \text{or: } (80 - 32)(0.56) = C$$
$$(48)\frac{5}{9} = C \qquad\qquad\quad (48)(0.56) = C$$
$$\frac{240}{9} = C \qquad\qquad\qquad\quad C \approx 26.9^\circ$$
$$C \approx 26.7^\circ$$

9.
$$(54 - 32)\frac{5}{9} = C \qquad \text{or: } (54 - 32)(0.56) = C$$
$$(22)\frac{5}{9} = C \qquad\qquad\quad (22)(0.56) = C$$
$$\frac{110}{9} = C \qquad\qquad\qquad\quad C \approx 12.3^\circ$$
$$C \approx 12.2^\circ$$

10. $(72-32)\dfrac{5}{9}=C$ or: $(72-32)(0.56)=C$
$(40)\dfrac{5}{9}=C$ $(40)(0.56)=C$
$\dfrac{200}{9}=C$ $C=22.4°$
$C\approx22.2°$

11. $(40-32)\dfrac{5}{9}=C$ or: $(40-32)(0.56)=C$
$(8)\dfrac{5}{9}=C$ $(8)(0.56)=C$
$\dfrac{40}{9}=C$ $C\approx4.5°$
$C\approx4.4°$

12. $(62-32)\dfrac{5}{9}=C$ or: $(62-32)(0.56)=C$
$(30)\dfrac{5}{9}=C$ $(30)(0.56)=C$
$\dfrac{150}{9}=C$ $C=16.8°$
$C\approx16.7$

13. 98.6°F; 37°C

14. $(47-32)\dfrac{5}{9}=C$ or: $(47-32)(0.56)=C$
$\left(^5\!15\right)\dfrac{5}{9_3}=C$ $(15)(0.56)=C$
$\dfrac{25}{3}=C$ $C\approx8.4°$
$C\approx8.3°$

no

15. $(88-32)\dfrac{5}{9}=C$ or: $(88-32)(0.56)=C$
$(56)\dfrac{5}{9}=C$ $(56)(0.56)=C$
$\dfrac{280}{9}=C$ $C\approx31.4°$
$C\approx31.1°$

16. 32°F; 0°C

17. $(72-32)\dfrac{5}{9}=C$ or: $(72-32)(0.56)=C$
$(40)\dfrac{5}{9}=C$ $(40)(0.56)=C$
$\dfrac{200}{9}=C$ $C=22.4°$
$C\approx22.2°$

18. $(110-32)\dfrac{5}{9}=C$ or: $(110-32)(0.56)=C$
$(78)\dfrac{5}{9}=C$ $(78)(0.56)=C$
$\dfrac{390}{9}=C$ $C\approx43.7°$
$C\approx43.3°$

Lesson Practice 17B

1. $(23-32)\dfrac{5}{9}=C$
$\left(^{-1}\!9\right)\dfrac{5}{9}=C$
$C=-5°$

2. $(140-32)\dfrac{5}{9}=C$ or: $(140-32)(0.56)=C$
$\left(^{12}\!108\right)\dfrac{5}{9}=C$ $(108)(0.56)=C$
$C=60°$ $C\approx60.5°$

3. $(77-32)\dfrac{5}{9}=C$ or: $(77-32)(0.56)=C$
$\left(^5\!45\right)\dfrac{5}{9}=C$ $(45)(0.56)=C$
$C=25°$ $C\approx25.2°$

4. $(5-32)\dfrac{5}{9}=C$ or: $(5-32)(0.56)=C$
$\left(^{-3}\!\!-27\right)\dfrac{5}{9}=C$ $(-27)(0.56)=C$
$C=-15°$ $C\approx-15.1°$

5. $(176-32)\dfrac{5}{9}=C$ or: $(176-32)(0.56)=C$
$\left(^{16}\!144\right)\dfrac{5}{9}=C$ $(144)(0.56)=C$
$C=80°$ $C\approx80.6°$

6. $(95-32)\dfrac{5}{9}=C$ or: $(95-32)(0.56)=C$
$\left(^7\!63\right)\dfrac{5}{9}=C$ $(63)(0.56)=C$
$C=35°$ $C\approx35.3°$

7. $(33-32)\dfrac{5}{9} = C$ or: $(33-32)(0.56) = C$

     $(1)\dfrac{5}{9} = C$ $(1)(0.56) = C$

                      $C \approx 0.6°$

     $\dfrac{5}{9} = C$

     $C \approx 0.6°$

8. $(55-32)\dfrac{5}{9} = C$ or: $(55-32)(0.56) = C$

     $(23)\dfrac{5}{9} = C$ $(23)(0.56) = C$

                      $C \approx 12.9°$

     $\dfrac{115}{9} = C$

     $C \approx 12.8°$

9. $(78-32)\dfrac{5}{9} = C$ or: $(78-32)(0.56) = C$

     $(46)\dfrac{5}{9} = C$ $(46)(0.56) = C$

                      $C \approx 25.8°$

     $\dfrac{230}{9} = C$

     $C \approx 25.6$

10. $(2-32)\dfrac{5}{9} = C$ or: $(2-32)(0.56) = C$

     $(-30)\dfrac{5}{9} = C$ $(-30)(0.56) = C$

                      $C = -16.8°$

     $\dfrac{-150}{9} = C$

     $C \approx -16.7°$

11. $(91-32)\dfrac{5}{9} = C$ or: $(91-32)(0.56) = C$

     $(59)\dfrac{5}{9} = C$ $(59)(0.56) = C$

                      $C \approx 33.0°$

     $\dfrac{295}{9} = C$

     $C \approx 32.8°$

12. $(-10-32)\dfrac{5}{9} = C$ or: $(-10-32)(0.56) = C$

     $(-42)\dfrac{5}{9} = C$ $(-42)(0.56) = C$

                      $C \approx -23.5°$

     $\dfrac{-210}{9} = C$

     $C \approx -23.3°$

13. $212°F$; $100°C$

14. $(53-32)\dfrac{5}{9} = C$ or: $(53-32)(0.56) = C$

     $(21)\dfrac{5}{9} = C$ $(21)(0.56) = C$

                      $C \approx 11.8°$

     $\dfrac{105}{9} = C$

     $C \approx 11.7°$

15. $(101-32)\dfrac{5}{9} = C$ or: $(101-32)(0.56) = C$

     $(69)\dfrac{5}{9} = C$ $(69)(0.56) = C$

                      $C \approx 38.6°$

     $\dfrac{345}{9} = C$

     $C \approx 38.3°$

     she has a fever

16. $(104-32)\dfrac{5}{9} = C$ or: $(104-32)(0.56) = C$

     $(^872)\dfrac{5}{9} = C$ $(72)(0.56) = C$

                      $C \approx 40.3°$

     $C = 40°$

17. Fahrenheit

18. Fahrenheit

Lesson Practice 17C

1. $(-4-32)\dfrac{5}{9} = C$ or: $(-4-32)(0.56) = C$

     $(^{-4}\!-36)\dfrac{5}{9} = C$ $(-36)(0.56) = C$

                      $C \approx -20.2°$

     $C = -20°$

2. $(131-32)\dfrac{5}{9} = C$ or: $(131-32)(0.56) = C$

     $(^{11}99)\dfrac{5}{9} = C$ $(99)(0.56) = C$

                      $C \approx 55.4°$

     $C = 55°$

3. $(212-32)\dfrac{5}{9} = C$ or: $(212-32)(0.56) = C$

     $(^{20}180)\dfrac{5}{9} = C$ $(180)(0.56) = C$

                      $C \approx 100.8°$

     $C = 100°$

4. $(86-32)\dfrac{5}{9} = C$ or: $(86-32)(0.56) = C$

     $(^654)\dfrac{5}{9} = C$ $(54)(0.56) = C$

                      $C \approx 30.2°$

     $C = 30°$

5. $(14-32)\dfrac{5}{9}=C$ or: $(14-32)(0.56)=C$
$\qquad\qquad\qquad\qquad\qquad\quad(-18)(0.56)=C$
$(^{-2}\!\cancel{-18})\dfrac{5}{9}=C\qquad\qquad C\approx-10.1°$
$\qquad C=-10°$

6. $(167-32)\dfrac{5}{9}=C$ or: $(167-32)(0.56)=C$
$\qquad\qquad\qquad\qquad\qquad\quad(135)(0.56)=C$
$(^{15}\!\cancel{135})\dfrac{5}{9}=C\qquad\qquad C\approx75.6°$
$\qquad C=75°$

7. $(76-32)\dfrac{5}{9}=C$ or: $(76-32)(0.56)=C$
$\qquad\qquad\qquad\qquad\qquad\quad(44)(0.56)=C$
$(44)\dfrac{5}{9}=C\qquad\qquad\quad C\approx24.6°$
$\dfrac{220}{9}=C$
$\qquad C\approx24.4°$

8. $(64-32)\dfrac{5}{9}=C$ or: $(64-32)(0.56)=C$
$\qquad\qquad\qquad\qquad\qquad\quad(32)(0.56)=C$
$(32)\dfrac{5}{9}=C\qquad\qquad\quad C\approx17.9°$
$\dfrac{160}{9}=C$
$\qquad C\approx17.8°$

9. $(127-32)\dfrac{5}{9}=C$ or: $(127-32)(0.56)=C$
$\qquad\qquad\qquad\qquad\qquad\quad(95)(0.56)=C$
$(95)\dfrac{5}{9}=C\qquad\qquad\quad C\approx53.2°$
$\dfrac{475}{9}=C$
$\qquad C\approx52.8°$

10. $(98-32)\dfrac{5}{9}=C$ or: $(98-32)(0.56)=C$
$\qquad\qquad\qquad\qquad\qquad\quad(66)(0.56)=C$
$(66)\dfrac{5}{9}=C\qquad\qquad\quad C\approx37.0°$
$\dfrac{330}{9}=C$
$\qquad C\approx36.7°$

11. $(60-32)\dfrac{5}{9}=C$ or: $(60-32)(0.56)=C$
$\qquad\qquad\qquad\qquad\qquad\quad(28)(0.56)=C$
$(28)\dfrac{5}{9}=C\qquad\qquad\quad C\approx15.7°$
$\dfrac{140}{9}=C$
$\qquad C\approx15.6°$

12. $(-25-32)\dfrac{5}{9}=C$ or: $(-25-32)(0.56)=C$
$\qquad\qquad\qquad\qquad\qquad\quad(-57)(0.56)=C$
$(-57)\dfrac{5}{9}=C\qquad\qquad C\approx-31.9°$
$\dfrac{-285}{9}=C$
$\qquad C\approx-31.7°$

13. 32°F; 0°C

14. $(350-32)\dfrac{5}{9}=C$ or: $(350-32)(0.56)=C$
$\qquad\qquad\qquad\qquad\qquad\quad(318)(0.56)=C$
$(318)\dfrac{5}{9}=C\qquad\qquad C\approx178.1°$
$\dfrac{1,590}{9}=C$
$\qquad C\approx176.7°$

15. $(450-32)\dfrac{5}{9}=C$ or: $(450-32)(0.56)=C$
$\qquad\qquad\qquad\qquad\qquad\quad(418)(0.56)=C$
$(418)\dfrac{5}{9}=C\qquad\qquad C\approx234.1°$
$\dfrac{2,090}{9}=C$
$\qquad C\approx232.2°$

16. Fahrenheit

17. $(250-32)\dfrac{5}{9}=C$ or: $(250-32)(0.56)=C$
$\qquad\qquad\qquad\qquad\qquad\quad(218)(0.56)=C$
$(218)\dfrac{5}{9}=C\qquad\qquad C\approx122.1°$
$\dfrac{1,090}{9}=C$
$\qquad C\approx121.1°$

18. Fahrenheit

Systematic Review 17D

1. $(149-32)\dfrac{5}{9}=C$ or: $(149-32)(0.56)=C$
$\qquad\qquad\qquad\qquad\qquad\quad(117)(0.56)=C$
$(^{13}\!\cancel{117})\dfrac{5}{9}=C\qquad\qquad C\approx65.5°$
$\qquad C=65°$

2. $(120-32)\dfrac{5}{9}=C$ or: $(120-32)(0.56)=C$
$\qquad\qquad\qquad\qquad\qquad\quad(88)(0.56)=C$
$(88)\dfrac{5}{9}=C\qquad\qquad\quad C\approx49.3°$
$\dfrac{440}{9}=C$
$\qquad C\approx48.9°$

3. $(75-32)\dfrac{5}{9} = C$ or: $(75-32)(0.56) = C$

$(43)\dfrac{5}{9} = C$ $\qquad (43)(0.56) = C$

$\qquad\qquad\qquad\qquad C \approx 24.1°$

$\dfrac{215}{9} = C$

$\qquad C \approx 23.9°$

4. $^{6}\!\!\not\!30 \times \dfrac{9}{\not\!5} + 32 = F$

$54 + 32 = F$

$\qquad F = 86°$

5. $59 \times \dfrac{9}{5} + 32 = F$

$\dfrac{531}{5} + 32 = F$

$106.2 + 32 = F$

$\qquad F = 138.2°$

6. $32°F$

7. $4 \cdot 7 + 3^2 = X + 7$

$4 \cdot 7 + 9 = X + 7$

$28 + 9 = X + 7$

$37 = X + 7$

$37 - 7 = X$

$X = 30$

8. $4 \cdot 7 + 3^2 = X + 7 \Rightarrow 4 \cdot 7 + 3^2 = (30) + 7$

$4 \cdot 7 + 9 = 37$

$28 + 9 = 37$

$37 = 37$

9. $3C - 6 + 2C = 10C - 2C + 6$

$5C - 6 = 8C + 6$

$5C = 8C + 6 + 6$

$5C = 8C + 12$

$5C - 8C = 12$

$-3C = 12$

$C = -4$

10. $3C - 6 + 2C = 10C - 2C + 6 \Rightarrow$

$3(-4) - 6 + 2(-4) = 10(-4) - 2(-4) + 6$

$-12 - 6 - 8 = -40 + 8 + 6$

$-26 = -26$

11. done

12. $2.1 \times .6 = 1.26$

13. $0.22 \times 1.3 = 0.286$

14. $4.1 \text{ ft} \times 3.6 \text{ ft} = 14.76 \text{ ft}^2$

15. $29 \times \dfrac{9}{5} + 32 = F$

$\dfrac{261}{5} + 32 = F$

$52.2 + 32 = F$

$\qquad F = 84.2°; \text{ yes}$

16. $6 \times \$1.79 = \$10.74$

$\$20.00 - \$10.74 = \$9.26 \text{ in change}$

17. $\dfrac{3}{1} \div \dfrac{1}{8} = \dfrac{3}{1} \times \dfrac{8}{1} = \dfrac{24}{1} = 24 \text{ eighths}$

18. $2(6 \times 9) + 2(9 \times 3) + 2(3 \times 6) =$

$2(54) + 2(27) + 2(18) =$

$108 + 54 + 36 = 198 \text{ in}^2$

Systematic Review 17E

1. $(113-32)\dfrac{5}{9} = C$ or: $(113-32)(0.56) = C$

$(^9\!\not\!81)\dfrac{5}{\not\!9} = C$ $\qquad (81)(0.56) = C$

$\qquad\qquad\qquad\qquad C \approx 45.4°$

$\qquad C = 45°$

2. $(59-32)\dfrac{5}{9} = C$ or: $(59-32)(0.56) = C$

$(^3\!\not\!27)\dfrac{5}{\not\!9} = C$ $\qquad (27)(0.56) = C$

$\qquad\qquad\qquad\qquad C \approx 15.1°$

$\qquad C = 15°$

3. $(12-32)\dfrac{5}{9} = C$ or: $(12-32)(0.56) = C$

$(-20)\dfrac{5}{9} = C$ $\qquad (-20)(0.56) = C$

$\qquad\qquad\qquad\qquad C = -11.2°$

$\dfrac{-100}{9} = C$

$\qquad C \approx -11.1°$

4. $^{13}\!\not\!65 \times \dfrac{9}{\not\!5} + 32 = F$

$117 + 32 = F$

$\qquad F = 149°$

5. $-8 \times \dfrac{9}{5} + 32 = F$

$\dfrac{-72}{5} + 32 = F$

$-14.4 + 32 = F$

$\qquad F = 17.6°$

6. $212°F$

7.
$$7^2 - 11^2 + 4(Y+3) = -3Y + 2(1.5)$$
$$7^2 - 11^2 + 4Y + (4)(3) = -3Y + 2(1.5)$$
$$49 - 121 + 4Y + (4)(3) = -3Y + 2(1.5)$$
$$49 - 121 + 4Y + 12 = -3Y + 3$$
$$-60 + 4Y = -3Y + 3$$
$$4Y = -3Y + 3 + 60$$
$$4Y = -3Y + 63$$
$$4Y + 3Y = 63$$
$$7Y + 63$$
$$Y = 9$$

8.
$$7^2 - 11^2 + 4(Y+3) = -3Y + 2(1.5) \Rightarrow$$
$$7^2 - 11^2 + 4((9)+3) = -3(9) + 2(1.5)$$
$$7^2 - 11^2 + 4(12) = -3(9) + 2(1.5)$$
$$49 - 121 + 48 = -27 + 3$$
$$-24 = -24$$

9.
$$81B = 9$$
$$\frac{1}{81} \times 81B = \frac{1}{81} \times 9$$
$$B = \frac{9}{81} = \frac{1}{9}$$

10.
$$81B = 9 \Rightarrow \quad 81\left(\frac{1}{9}\right) = 9$$
$$9 = 9$$

11. $4.8 \times 0.3 = 1.44$

12. $0.005 \times 0.45 = 0.00225$

13. $6.7 \times 1.9 = 12.73$

14. $1\frac{5}{8} \times 2\frac{1}{2} \times \frac{2}{5} = \frac{13}{8} \times \frac{5}{2} \times \frac{2}{5} = \frac{13}{8} = 1\frac{5}{8}$

15. $\frac{2}{3} \times \frac{5}{6} \times \frac{3}{5} = \frac{1}{3}$

16. $\frac{3}{4} \times \frac{1}{5} \times \frac{5}{9} = \frac{1}{12}$

17. yes; 96.8°F

18. $25.5 \times 1.6 = 40.8$ mi

19. $\$0.69 \times 3 = \$2.07$
$\$4.00 - \$2.07 = \$1.93$ in change

20.
$$7^2 + 4 \times 7 - 10 =$$
$$49 + 4 \times 7 - 10 =$$
$$49 + 28 - 10 = 67$$

Systematic Review 17F

1. $(194 - 32)\frac{5}{9} = C$ or: $(194 - 32)(0.56) = C$
$$(162)\frac{5}{9} = C \qquad\qquad (162)(0.56) = C$$
$$C = 90° \qquad\qquad\qquad C \approx 90.7°$$

2. 0°C

3. $(100 - 32)\frac{5}{9} = C$ or: $(100 - 32)(0.56) = C$
$$(68)\frac{5}{9} = C \qquad\qquad (68)(0.56) = C$$
$$\frac{340}{9} = C \qquad\qquad\qquad C \approx 38.1°$$
$$C \approx 37.8°$$

4. $60 \times \frac{9}{5} + 32 = F$
$$108 + 32 = F$$
$$F = 140°$$

5. $25 \times \frac{9}{5} + 32 = F$
$$45 + 32 = F$$
$$F = 77°$$

6. $-21 \times \frac{9}{5} + 32 = F$
$$\frac{-189}{5} + 32 = F$$
$$-37.8 + 32 = F$$
$$F = -5.8°$$

7.
$$7 - B + 4 + 2B = 2B + 8 - 7$$
$$11 + B = 2B + 1$$
$$B = 2B + 1 - 11$$
$$B = 2B - 10$$
$$B - 2B = -10$$
$$-B = -10$$
$$B = 10$$

8.
$$7 - B + 4 + 2B = 2B + 8 - 7 \Rightarrow$$
$$7 - (10) + 4 + 2(10) = 2(10) + 8 - 7$$
$$7 - 10 + 4 + 20 = 20 + 8 - 7$$
$$21 = 21$$

9.
$$16X = 5$$
$$\left(\frac{1}{16}\right)(16X) = \left(\frac{1}{16}\right)(5)$$
$$X = \frac{5}{16}$$

10. $16X = 5 \Rightarrow 16\left(\dfrac{5}{16}\right) = 5$

$$\dfrac{80}{16} = 5$$

$$5 = 5$$

11. $1.9 \times 0.24 = 0.456$

12. $0.123 \times 0.6 = 0.0738$

13. $0.21 \times 0.35 = 0.0735$

14. $2\dfrac{2}{5} \times 2\dfrac{3}{4} = \dfrac{\overset{3}{\cancel{12}}}{5} \times \dfrac{11}{\cancel{4}} = \dfrac{33}{5} = 6\dfrac{3}{5}$

15. $1\dfrac{1}{3} \times 1\dfrac{1}{2} = \dfrac{\overset{2}{\cancel{4}}}{\cancel{3}} \times \dfrac{3}{\cancel{2}} = \dfrac{2}{1} = 2$

16. $1\dfrac{2}{5} \times 6\dfrac{3}{7} = \dfrac{\cancel{7}}{\cancel{5}} \times \dfrac{\overset{9}{\cancel{45}}}{\cancel{7}} = \dfrac{9}{1} = 9$

17. Celsius; 3° Fahrenheit is below freezing.

18. $1.5 \times 3.5 = 5.25$ buckets

19. $\$0.99 \times 4 = \$3.96$

$$\$10.00 - \$3.96 = \$6.04$$

20. $4\left(12 \times 14 \times \dfrac{1}{2}\right) + (12 \times 12) =$

$$4(84) + (144) =$$

$$336 + 144 = 480 \text{ in}^2$$

Lesson Practice 18A

1. done

2. done

3. $\left|5 \cdot 6^2\right| =$

$$\left|5 \cdot 36\right| =$$

$$\left|180\right| = 180$$

4. $\left|-10 - 13\right| =$

$$\left|-23\right| = 23$$

5. $\left|-6 + 8\right| =$

$$\left|2\right| = 2$$

6. $\left|18 + 2^3\right| =$

$$\left|18 + 8\right| =$$

$$\left|26\right| = 26$$

7. $\left|3^2 - 5^2\right| - (15 \div 3)^3 + 18 =$

$$\left|9 - 25\right| - (5)^3 + 18 =$$

$$\left|-16\right| - 125 + 18 =$$

$$16 - 125 + 18 = -91$$

8. $\left|10^2 - 5^2\right| + \left|-8^2 + 2^2\right| =$

$$\left|100 - 25\right| + \left|-64 + 4\right| =$$

$$\left|75\right| + \left|-60\right| =$$

$$75 + 60 = 135$$

9. $\left|6 \div (-2)\right| \times 5 + 3^2 =$

$$\left|-3\right| \times 5 + 9 =$$

$$3 \times 5 + 9 =$$

$$15 + 9 = 24$$

10. $-4(P - 6) + 2P = \left|5 - 3 + 6\right|$

$$-4P - (-4)(6) + 2P = \left|8\right|$$

$$-4P - (-24) + 2P = 8$$

$$-2P + 24 = 8$$

$$-2P = 8 - 24$$

$$-2P = -16$$

$$P = 8$$

11. $-5X + \left|9^2 - 3^2\right| + 13 = 12X$

$$-5X + \left|81 - 9\right| + 13 = 12X$$

$$-5X + \left|72\right| + 13 = 12X$$

$$-5X + 85 = 12X$$

$$85 = 12X + 5X$$

$$85 = 17X$$

$$X = 5$$

12. $3(3G + 5G) - \left|3 - 12\right| = 18G + 5(-G - 4)$

$$3(8G) - \left|-9\right| = 18G - 5G - 20$$

$$24G - 9 = 13G - 20$$

$$24G - 13G = -20 + 9$$

$$11G = -11$$

$$G = -1$$

13. $\left(11^2 - 1\right) \div 12 = 3X + \left|-2\right|X$

$$(121 - 1) \div 12 = 3X + 2X$$

$$120 \div 12 = 5X$$

$$10 = 5X$$

$$X = 2$$

14. $(3 - 5)^2 + \left|6 - 4\right| + X = 3X$

$$(-2)^2 + \left|2\right| + X = 3X$$

$$4 + 2 + X = 3X$$

$$6 = 3X - X$$

$$6 = 2X$$

$$X = 3$$

15. $|-4|^2 = (-1)^2 + B(-1) \div 1$

$\qquad 4^2 = 1 - B$

$\qquad 16 = 1 - B$

$\qquad 16 - 1 = -B$

$\qquad 15 = -B$

$\qquad B = -15$

16. $163 - 75 = 88$ mi

Lesson Practice 18B

1. $|25 - 16| =$

$\qquad |9| = 9$

2. $|8 - 3| =$

$\qquad |5| = 5$

3. $|4^2 - 2^2| =$

$\qquad |16 - 4| =$

$\qquad |12| = 12$

4. $|-45 + 11| =$

$\qquad |-34| = 34$

5. $|-1 \times -1 \times -1| =$

$\qquad |-1| = 1$

6. $|3^2 - 8^2| =$

$\qquad |9 - 64| =$

$\qquad |-55| = 55$

7. $|18 - 36| + \left(\left|3 - 5^2\right| - 15\right)^2 =$

$\qquad |-18| + \left(|3 - 25| - 15\right)^2 =$

$\qquad 18 + \left(|-22| - 15\right)^2 =$

$\qquad 18 + (22 - 15)^2 =$

$\qquad 18 + (7)^2 =$

$\qquad 18 + 49 = 67$

8. $|4^2 - 9| + (8 \div 4)^2 =$

$\qquad |16 - 9| + (2)^2 =$

$\qquad |7| + 4 =$

$\qquad 7 + 4 = 11$

9. $|8 - 12| + |6^2 \cdot 5^2| =$

$\qquad |-4| + |36 \cdot 25| =$

$\qquad 4 + |900| =$

$\qquad 4 + 900 = 904$

10. $(3 + 5)^2 + |8 - 11| + Z = 4(Z - 2)$

$\qquad 8^2 + |-3| + Z = 4Z - 8$

$\qquad 64 + 3 + Z = 4Z - 8$

$\qquad 64 + 3 + 8 = 4Z - Z$

$\qquad 75 = 3Z$

$\qquad Z = 25$

11. $(6 + 6)^2 + |100 - 1| - 14^2 = 5 \cdot 9 + B$

$\qquad 12^2 + |99| - 196 = 45 + B$

$\qquad 144 + 99 - 196 - 45 = B$

$\qquad B = 2$

12. $|-3 + 7| - 4^2 + (-4)^2 = 2R$

$\qquad |4| - 16 + 16 = 2R$

$\qquad 4 = 2R$

$\qquad R = 2$

13. $|-3 - 4 - 5 + 2| + W = 3W$

$\qquad |-10| + W = 3W$

$\qquad 10 = 3W - W$

$\qquad 10 = 2W$

$\qquad W = 5$

14. $|-8 - 4| - 8Y = 32 \div |-8|$

$\qquad |-12| - 8Y = 32 \div 8$

$\qquad 12 - 8Y = 4$

$\qquad 12 - 4 = 8Y$

$\qquad 8 = 8Y$

$\qquad Y = 1$

15. $|-9 - 3| = -5C + 3 + 8C$

$\qquad |-12| = 3C + 3$

$\qquad 12 = 3C + 3$

$\qquad 12 - 3 = 3C$

$\qquad 9 = 3C$

$\qquad C = 3$

16. $|40 - 50| =$

$\qquad |-10| = 10$ mi

Lesson Practice 18C

1. $|(-1)(-4)| =$

$\qquad |4| = 4$

2. $|10 \div 2 - 8| =$

$\qquad |5 - 8| =$

$\qquad |-3| = 3$

3. $\left|5^2 - 6^2\right| =$
$\left|25 - 36\right| =$
$\left|-11\right| = 11$

4. $\left|4 - 8 + 1\right| =$
$\left|-3\right| = 3$

5. $\left|-3 \cdot 4\right| =$
$\left|-12\right| = 12$

6. $\left|6^2 - 5^2\right| =$
$\left|36 - 25\right| =$
$\left|11\right| = 11$

7. $\left|(-10)^2 - 9\right| - \left|2^2 - 5^2\right| =$
$\left|100 - 9\right| - \left|4 - 25\right| =$
$\left|91\right| - \left|-21\right| =$
$91 - 21 = 70$

8. $\left|(9^2 \div 9) \div 3\right| =$
$\left|(81 \div 9) \div 3\right| =$
$\left|(9) \div 3\right| =$
$\left|3\right| = 3$

9. $-4^2 + (7 - 3)^2 - \left|-2\right| =$
$-16 + (4)^2 - 2 =$
$-16 + 16 - 2 = -2$

10. $\left|15 + 2\right| = 3A - 6 + 3 + A$
$\left|17\right| = 4A - 3$
$17 = 4A - 3$
$20 = 4A$
$A = 5$

11. $2 \cdot \left|-6 - 6\right| = -5 + 10K - K + 2$
$2 \cdot \left|-12\right| = -3 + 9K$
$2 \cdot 12 = -3 + 9K$
$24 + 3 = 9K$
$27 = 9K$
$K = 3$

12. $\left|-1 - 3 \cdot 4\right| = 3D - 8 + 7 - D$
$\left|-1 - 12\right| = 2D - 1$
$\left|-13\right| = 2D - 1$
$13 = 2D - 1$
$13 + 1 = 2D$
$14 = 2D$
$D = 7$

13. $\left|7 \cdot 11\right| = 10F - 5F + 1 - F$
$77 = 4F + 1$
$77 - 1 = 4F$
$76 = 4F$
$F = 19$

14. $\left|-(2)^2 - 4\right| = -7R + (-2)^2 - 2 + 5R$
$\left|-4 - 4\right| = -7R + 4 - 2 + 5R$
$\left|-8\right| = -2R + 2$
$8 = -2R + 2$
$8 - 2 = -2R$
$6 = -2R$
$R = -3$

15. $3Q + 5Q - 2 - 1 = \left|10^2 - 7\right|$
$8Q - 3 = \left|100 - 7\right|$
$8Q - 3 = \left|93\right|$
$8Q - 3 = 93$
$8Q = 96$
$Q = 12$

16. $\left|50 - 80\right| =$
$\left|-30\right| = 30$ ft

Systematic Review 18D

1. $\left|42 \div 6 - 2\right| \times 11 =$
$\left|7 - 2\right| \times 11 =$
$\left|5\right| \times 11 =$
$5 \times 11 = 55$

2. $(192 \div 8) \times 4 - \left|67 - 200\right| =$
$24 \times 4 - \left|-133\right| =$
$96 - 133 = -37$

3. $\left|(10 + 3)^2 - 9\right| \div 20 =$
$\left|13^2 - 9\right| \div 20 =$
$\left|169 - 9\right| \div 20 =$
$\left|160\right| \div 20 =$
$160 \div 20 = 8$

4. $-(3^2)K - 6 + 3 = |-2^2 \cdot 3| + 3$

$-9K - 3 = |-4 \cdot 3| + 3$

$-9K - 3 = |-12| + 3$

$-9K - 3 = 12 + 3$

$-9K = 12 + 3 + 3$

$-9K = 18$

$K = -2$

5. $-(3^2)K - 6 + 3 = |-2^2 \cdot 3| + 3 \Rightarrow$

$-(3^2)(-2) - 6 + 3 = |-2^2 \cdot 3| + 3$

$(-9)(-2) - 6 + 3 = |-4 \cdot 3| + 3$

$18 - 6 + 3 = |-12| + 3$

$15 = 12 + 3$

$15 = 15$

6. $3B + 6 = 6B + 5 - 2B + |3^2|$

$3B + 6 = 4B + 5 + 9$

$3B + 6 = 4B + 14$

$6 - 14 = 4B - 3B$

$B = -8$

7. $3B + 6 = 6B + 5 - 2B + |3^2| \Rightarrow$

$3(-8) + 6 = 6(-8) + 5 - 2(-8) + |3^2|$

$-24 + 6 = -48 + 5 + 16 + |9|$

$-18 = -18$

8. $^3\!15 \times \dfrac{9}{5} + 32 = F$

$27 + 32 = F$

$F = 59^\circ$

9. $28 \times \dfrac{9}{5} + 32 = F$

$\dfrac{252}{5} + 32 = F$

$50.4 + 32 = F$

$F = 82.4^\circ$

10. $(65 - 32)\dfrac{5}{9} = C$ or: $(65 - 32)(0.56) = C$

$\qquad (33)\dfrac{5}{9} = C$ $(33)(0.56) = C$

$\qquad \dfrac{165}{9} = C$ $C \approx 18.5^\circ$

$\qquad C \approx 18.3^\circ$

11. $(212 - 32)\dfrac{5}{9} = C$ or: $(212 - 32)(0.56) = C$

$\left(^{20}\!180\right)\dfrac{5}{9} = C$ $(180)(0.56) = C$

$\qquad\qquad\qquad C = 100.8^\circ$

12.
$$
\begin{array}{r}
0.057 \\
5\overline{)0.285} \\
\underline{25} \\
35 \\
\underline{35} \\
0
\end{array}
$$

13.
$$
\begin{array}{r}
1.2 \\
9\overline{)10.8} \\
\underline{9} \\
18 \\
\underline{18} \\
0
\end{array}
$$

14.
$$
\begin{array}{r}
2.75 \\
4\overline{)11.00} \\
\underline{8} \\
30 \\
\underline{28} \\
20 \\
\underline{20} \\
0
\end{array}
$$

15.
$$
\begin{array}{r}
7.85 \\
4\overline{)31.40} \\
\underline{28} \\
3\,4 \\
\underline{3\,2} \\
20 \\
\underline{20} \\
0
\end{array}
$$

16. 37°C; 98.6°F

17. $16^2 + 30^2 = 34^2$

$256 + 900 = 1{,}156$

$1{,}156 = 1{,}156$; yes

18. $C + 2 = 2C + 1$

$C = 2C + 1 - 2$

$C = 2C - 1$

$C - 2C = -1$

$-C = -1$

$C = 1$ car

Systematic Review 18E

1. $4^2 - 2(4) + |3 - 4| =$

$16 - 8 + |-1| =$

$16 - 8 + 1 = 9$

2. $3(2)^2 - 2 \div |4 - 3| =$

$3(4) - 2 \div |1| =$

$12 - 2 = 10$

3. $|30 - 46| + |-5| - 9 - 2^2 =$

$|-16| + 5 - 9 - 4 =$

$16 + 5 - 9 - 4 = 8$

4.
$$3Q - 2 + Q = |-3(2+2)+2|$$
$$4Q - 2 = |-3(4)+2|$$
$$4Q - 2 = |-12+2|$$
$$4Q - 2 = |-10|$$
$$4Q - 2 = 10$$
$$4Q = 10 + 2$$
$$4Q = 12$$
$$Q = 3$$

5.
$$3Q - 2 + Q = |-3(2+2)+2| \Rightarrow$$
$$3(3) - 2 + (3) = |-3(2+2)+2|$$
$$9 - 2 + 3 = |-3(4)+2|$$
$$10 = |-12+2|$$
$$10 = |-10|$$
$$10 = 10$$

6.
$$3Y + 8 - 2 - 2Y = |9 - 2^2 + 5|$$
$$Y + 6 = |9 - 4 + 5|$$
$$Y + 6 = |10|$$
$$Y + 6 = 10$$
$$Y = 4$$

7.
$$3Y + 8 - 2 - 2Y = |9 - 2^2 + 5| \Rightarrow$$
$$3(4) + 8 - 2 - 2(4) = |9 - 2^2 + 5|$$
$$12 + 8 - 2 - 8 = |9 - 4 + 5|$$
$$10 = |10|$$
$$10 = 10$$

8.
$$^5\!25 \times \frac{9}{5} + 32 = F$$
$$45 + 32 = F$$
$$F = 77°$$

9.
$$108 \times \frac{9}{5} + 32 = F$$
$$\frac{972}{5} + 32 = F$$
$$194.4 + 32 = F$$
$$F = 226.4°$$

10.
$$(41 - 32)\frac{5}{9} = C \quad \text{or:} \quad (41 - 32)(0.56) = C$$
$$(^1\!9)\frac{5}{9} = C \qquad\qquad (9)(0.56) = C$$
$$C = 5° \qquad\qquad\qquad C \approx 5°$$

11.
$$(86 - 32)\frac{5}{9} = C \quad \text{or:} \quad (86 - 32)(0.56) = C$$
$$(^6\!54)\frac{5}{9} = C \qquad\qquad (54)(0.56) = C$$
$$C = 30° \qquad\qquad\qquad C \approx 30.2°$$

12.
```
     2.44
  3)7.32
     6
     13
     12
      12
      12
       0
```

13.
```
      9.6
  7)67.2
    63
     4 2
     4 2
       0
```

14.
```
     2.4
  5)12.0
    10
     2 0
     2 0
       0
```

15.
```
     1.25   or    1.25
  6)7.50        6)7.50
    7 2           6
     30           15
     30           12
      0           30
                  30
                   0
```

16. Convert 1 ft to 12 in first.
$$2(5 \times 12) + 2(12 \times 4) + 2(4 \times 5) =$$
$$2(60) + 2(48) + 2(20) =$$
$$120 + 96 + 40 = 256 \text{ in}^2$$

17.
$$4\left(12 \times 15 \times \frac{1}{2}\right) + (15 \times 15) =$$
$$4(90) + (225) =$$
$$360 + 225 = 585 \text{ in}^2$$

18. $15.4 \div 2 = 7.7$ days

19. $1.5 \times 1.5 = 2.25 \text{ mi}^2$

20.
$$2N - 2 - N = 16$$
$$N - 2 = 16$$
$$N = 16 + 2$$
$$N = 18 \text{ cards}$$

Systematic Review 18F

1. $(7-3)^2 \times |3-7| =$

$\quad\quad 4^2 \times |-4| =$

$\quad\quad 16 \times 4 = 64$

2. $8 + |4 + 1^2| =$

$\quad\quad 8 + |5| =$

$\quad\quad 8 + 5 = 13$

3. $|8 - 3^2| + |-8| =$

$\quad\quad |8 - 9| + 8 =$

$\quad\quad |-1| + 8 =$

$\quad\quad 1 + 8 = 9$

4. $(8 + 2)^2 = 10X$

$\quad\quad (10)^2 = 10X$

$\quad\quad 100 = 10X$

$\quad\quad X = 10$

5. $(8 + 2)^2 = 10X \Rightarrow (8+2)^2 = 10(10)$

$\quad\quad\quad\quad\quad\quad 10^2 = 100$

$\quad\quad\quad\quad\quad\quad 100 = 100$

6. $-6(Y - 5 + 9) + 7(2Y + 9) = -1$

$\quad\quad -6(Y + 4) + 7(2Y + 9) = -1$

$\quad -6Y + (-6)(4) + 7(2Y) + 7(9) = -1$

$\quad\quad -6Y - 24 + 14Y + 63 = -1$

$\quad\quad\quad\quad\quad\quad 8Y + 39 = -1$

$\quad\quad\quad\quad\quad\quad 8Y = -1 - 39$

$\quad\quad\quad\quad\quad\quad 8Y = -40$

$\quad\quad\quad\quad\quad\quad Y = -5$

7. $-6(Y - 5 + 9) + 7(2Y + 9) = -1 \Rightarrow$

$\quad -6((-5) - 5 + 9) + 7(2(-5) + 9) = -1$

$\quad\quad -6(-1) + 7(-10 + 9) = -1$

$\quad\quad\quad\quad 6 + 7(-1) = -1$

$\quad\quad\quad\quad\quad 6 - 7 = -1$

$\quad\quad\quad\quad\quad\quad -1 = -1$

8. $0°C = 32°F$

9. $48 \times \dfrac{9}{5} + 32 = F$

$\quad\quad \dfrac{432}{5} + 32 = F$

$\quad\quad 86.4 + 32 = F$

$\quad\quad\quad F = 118.4°$

10. $(104 - 32)\dfrac{5}{9} = C$ or: $(104 - 32)(0.56) = C$

$\quad\quad (^8\cancel{72})\dfrac{5}{\cancel{9}} = C \quad\quad\quad\quad (72)(0.56) = C$

$\quad\quad\quad\quad\quad\quad\quad\quad\quad\quad C \approx 40.3°$

$\quad\quad\quad C = 40°$

11. $(58 - 32)\dfrac{5}{9} = C$ or: $(58 - 32)(0.56) = C$

$\quad\quad (26)\dfrac{5}{9} = C \quad\quad\quad\quad\quad (26)(0.56) = C$

$\quad\quad\quad\quad\quad\quad\quad\quad\quad\quad C \approx 14.6°$

$\quad\quad \dfrac{130}{9} = C$

$\quad\quad\quad C \approx 14.4°$

12.
```
    0.84
 4)3.36
   32
   ‾‾
   16
   16
   ‾‾
    0
```

13.
```
    3.8
 9)34.2
   27
   ‾‾
    7 2
    7 2
    ‾‾‾
      0
```

14.
```
    2.5
 6)15.0
   12
   ‾‾
   3 0
   3 0
   ‾‾‾
     0
```

15.
```
      2.94
 5)14.70
   10
   ‾‾
    4 7
    4 5
    ‾‾‾
     20
     20
     ‾‾
      0
```

16. $2(8 \times 20) + 2(20 \times 9) + 2(9 \times 8) =$

$\quad\quad 2(160) + 2(180) + 2(72) =$

$\quad\quad\quad 320 + 360 + 144 = 824 \text{ in}^2$

17. $4\left(9 \times 14 \times \dfrac{1}{2}\right) + (9 \times 9) =$

$\quad\quad\quad 4(63) + (81) =$

$\quad\quad\quad 252 + 81 = 333 \text{ in}^2$

18. $6.3 \div 3 = 2.1$ hr

$\quad 2.1 \times 60 = 126$ min

19. $37°C = 98.6°F =$ normal body temperature

20. $100°C = 212°F =$ boiling point of water; too warm!

Lesson Practice 19A

1. done

2. done

3. done

4. $$\frac{F}{9} = \frac{12}{27}$$
$$27F = 9(12)$$
$$27F = 108$$
$$\frac{1}{27} \times 27F = \frac{1}{27} \times 108$$
$$F = 4$$

5. $$\frac{8}{X} = \frac{72}{81}$$
$$72X = 8(81)$$
$$72X = 648$$
$$X = 9$$

6. $$\frac{10}{12} = \frac{34}{R}$$
$$10R = 12(34)$$
$$10R = 408$$
$$R = 40.8$$

7. $$\frac{P}{9} = \frac{40}{45}$$
$$45P = 9(40)$$
$$45P = 360$$
$$P = 8$$

8. $$\frac{15}{17} = \frac{A}{51}$$
$$17A = 15(51)$$
$$17A = 765$$
$$A = 45$$

9. $$\frac{8}{13} = \frac{D}{26}$$
$$13D = 8(26)$$
$$13D = 208$$
$$D = 16$$

10. done

11. $$\frac{3}{8} = \frac{9}{C}$$
$$3C = 9(8)$$
$$3C = 72$$
$$C = 24 \text{ camels}$$

12. $$\frac{2}{7} = \frac{M}{63}$$
$$7M = 2(63)$$
$$7M = 126$$
$$M = 18 \text{ with mustaches}$$

Lesson Practice 19B

1. $$\frac{D}{6} = \frac{35}{42}$$
$$42D = 6(35)$$
$$42D = 210$$
$$D = 5$$

2. $$\frac{39}{45} = \frac{13}{X}$$
$$39X = 45(13)$$
$$39X = 585$$
$$X = 15$$

3. $$\frac{10}{12} = \frac{G}{6}$$
$$12G = 10(6)$$
$$12G = 60$$
$$G = 5$$

4. $$\frac{36}{E} = \frac{4}{5}$$
$$4E = 36(5)$$
$$4E = 180$$
$$E = 45$$

5. $$\frac{Y}{11} = \frac{24}{33}$$
$$33Y = 11(24)$$
$$33Y = 264$$
$$Y = 8$$

6. $$\frac{14}{38} = \frac{R}{57}$$
$$38R = 14(57)$$
$$38R = 798$$
$$R = 21$$

7. $$\frac{16}{24} = \frac{3}{H}$$
$$16H = 24(3)$$
$$16H = 72$$
$$H = 4.5$$

8. $\dfrac{K}{75} = \dfrac{2}{3}$

$3K = 75(2)$

$3K = 150$

$K = 50$

9. $\dfrac{7}{9} = \dfrac{D}{81}$

$9D = 7(81)$

$9D = 567$

$D = 63$

10. $\dfrac{11}{Z} = \dfrac{55}{75}$

$55Z = 11(75)$

$55Z = 825$

$Z = 15$

11. $\dfrac{7}{8} = \dfrac{R}{56}$

$8R = 7(56)$

$8R = 392$

$R = 49$

12. $\dfrac{2}{T} = \dfrac{8}{12}$

$8T = 2(12)$

$8T = 24$

$T = 3$

13. $\dfrac{25}{100}$

14. $100 - 25 = 75$ are not red

$\dfrac{25}{75}$

15. $\dfrac{1}{10} = \dfrac{80}{E}$

$1E = 10(80)$

$E = \$800$ earned

Lesson Practice 19C

1. $\dfrac{1}{2} = \dfrac{A}{10}$

$2A = 1(10)$

$2A = 10$

$A = 5$

2. $\dfrac{2}{3} = \dfrac{6}{X}$

$2X = 3(6)$

$2X = 18$

$X = 9$

3. $\dfrac{4}{5} = \dfrac{16}{Q}$

$4Q = 5(16)$

$4Q = 80$

$Q = 20$

4. $\dfrac{1}{3} = \dfrac{7}{D}$

$1D = 3(7)$

$D = 21$

5. $\dfrac{3}{5} = \dfrac{B}{45}$

$5B = 3(45)$

$5B = 135$

$B = 27$

6. $\dfrac{W}{6} = \dfrac{45}{54}$

$54W = 6(45)$

$54W = 270$

$W = 5$

7. $\dfrac{5}{8} = \dfrac{35}{J}$

$5J = 8(35)$

$5J = 280$

$J = 56$

8. $\dfrac{10}{25} = \dfrac{R}{4}$

$25R = 10(4)$

$25R = 40$

$R = 1.6$

9. $\dfrac{9}{Y} = \dfrac{3}{5}$

$3Y = 9(5)$

$3Y = 45$

$Y = 15$

10. $\dfrac{T}{4} = \dfrac{33}{44}$

$44T = 4(33)$

$44T = 132$

$T = 3$

11. $\dfrac{120}{180} = \dfrac{2}{D}$

$120D = 180(2)$

$120D = 360$

$D = 3$

12. $\dfrac{2}{4} = \dfrac{15}{Z}$

$2Z = 4(15)$

$2Z = 60$

$Z = 30$

13. $\dfrac{1}{8} = \dfrac{4}{X}$

$1X = 8(4)$

$X = 32 \text{ mi}$

14. $\dfrac{2}{3} = \dfrac{C}{24}$

$3C = 2(24)$

$3C = 48$

$C = 16 \text{ chickadees}$

15. $\dfrac{100}{75} = \dfrac{20}{M}$

$100M = 75(20)$

$100M = 1,500$

$M = 15 \text{ customers}$

Systematic Review 19D

1. $\dfrac{3}{4} = \dfrac{15}{X}$

$3X = 4(15)$

$3X = 60$

$X = 20$

2. $\dfrac{7}{10} = \dfrac{14}{D}$

$7D = 10(14)$

$7D = 140$

$D = 20$

3. $\dfrac{E}{7} = \dfrac{18}{42}$

$42E = 7(18)$

$42E = 126$

$E = 3$

4. $-10G + 30 - 21 + 4G = 7G - 17$

$-6G + 9 = 7G - 17$

$9 + 17 = 7G + 6G$

$26 = 13G$

$G = 2$

5. $-10G + 30 - 21 + 4G = 7G - 17 \Rightarrow$

$-10(2) + 30 - 21 + 4(2) = 7(2) - 17$

$-20 + 30 - 21 + 8 = 14 - 17$

$-3 = -3$

6. $6(B - 4) = 3(5 - 4) + |-3|$

$6B - 24 = 3(1) + 3$

$6B = 3 + 3 + 24$

$6B = 30$

$B = 5$

7. $6(B - 4) = 3(5 - 4) + |-3| \Rightarrow$

$6((5) - 4) = 3(5 - 4) + |-3|$

$6(1) = 3(1) + 3$

$6 = 3 + 3$

$6 = 6$

8. $(100 - 32)\dfrac{5}{9} = C$ or: $(100 - 32)(0.56) = C$

$(68)\dfrac{5}{9} = C$ $\qquad (68)(0.56) = C$

$\qquad\qquad\qquad\qquad C \approx 38.1°$

$\dfrac{340}{9} = C$

$C \approx 37.8°$

9. $53 \times \dfrac{9}{5} + 32 = F$

$\dfrac{477}{5} + 32 = F$

$95.4 + 32 = F$

$F = 127.4°$

10.
```
    2.38
  2)4.76
    4
    0 7
      6
      16
      16
       0
```

11.
```
     50
  3)150
    15
    00
```

12.

$$\begin{array}{r} 400 \\ 7\overline{)2800} \\ \underline{28} \\ 000 \end{array}$$

13.

$$\begin{array}{r} 24 \\ 4\overline{)96} \\ \underline{8} \\ 16 \\ \underline{16} \\ 0 \end{array}$$

14. $5.2 \div 1.3 = 4$ days

15. $\dfrac{5}{8} = \dfrac{B}{40}$

$8B = 5(40)$

$8B = 200$

$B = 25$ blue flowers

16. $212°F$

17. $\sqrt{144} = 12$

18. $\dfrac{1}{3} \times \dfrac{1}{2} \times \dfrac{1}{4} = \dfrac{1}{24}$ of the original price

Systematic Review 19E

1. $\dfrac{2}{5} = \dfrac{Y}{40}$

$5Y = 2(40)$

$5Y = 80$

$Y = 16$

2. $\dfrac{R}{8} = \dfrac{9}{24}$

$24R = 8(9)$

$24R = 72$

$R = 3$

3. $\dfrac{F}{11} = \dfrac{15}{55}$

$55F = 11(15)$

$55F = 165$

$F = 3$

4. $2 - 9X + 8X + 8 = 7X + 49 - 7X$

$10 - X = 49$

$10 - 49 = X$

$X = -39$

5. $2 - 9X + 8X + 8 = 7X + 49 - 7X \Rightarrow$

$2 - 9(-39) + 8(-39) + 8 = 7(-39) + 49 - 7(-39)$

$2 + 351 - 312 + 8 = -273 + 49 + 273$

$49 = 49$

6. $|6 - 2| \cdot 5^2 - Z = 10^2 - 10$

$|6 - 2| \cdot 5^2 - Z = 10^2 - 10$

$|4| \cdot 25 - Z = 100 - 10$

$100 - Z = 100 - 10$

$100 - 100 + 10 = Z$

$Z = 10$

7. $|6 - 2| \cdot 5^2 - Z = 10^2 - 10 \Rightarrow$

$|6 - 2| \cdot 5^2 - (10) = 10^2 - 10$

$|4| \cdot 5^2 - 10 = 100 - 10$

$4 \cdot 25 - 10 = 100 - 10$

$100 - 10 = 100 - 10$

$90 = 90$

8. $(41 - 32)\dfrac{5}{9} = C$ or: $(41 - 32)(0.56) = C$

$(9)\dfrac{5}{9} = C$ $\qquad (9)(0.56) = C$

$C = 5°$ $\qquad C \approx 5.0°$

9. $10 \times \dfrac{9}{5} + 32 = F$

$18 + 32 = F$

$F = 50°$

10. $3.4 \times 0.04 = 0.136$

11. $1.09 + 0.3 = 1.39$

12. $6.9 - 0.13 = 6.77$

13.

$$\begin{array}{r} 0.80 \\ 4\overline{)3.20} \\ \underline{3.20} \\ 0 \end{array}$$

14.

$$\begin{array}{r} 700 \\ 6\overline{)4200} \\ \underline{4200} \\ 0 \end{array}$$

15.

$$\begin{array}{r} 6.4 \\ 2\overline{)12.8} \\ \underline{12} \\ 8 \\ \underline{8} \\ 0 \end{array}$$

16. $4.2 \div 6 = 0.7$ of a pizza

17. $6.9 \div 2.3 = 3$ people

18. $\dfrac{15}{18}$

19. $\dfrac{1}{5} = \dfrac{25}{F}$

$1F = 5(25)$

$F = 125$ flies

20. $7E - 7 = 8 \cdot 7$

$7E - 7 = 56$

$7E = 56 + 7$

$7E = 63$

$E = 9$ eggs

Systematic Review 19F

1. $\dfrac{8}{11} = \dfrac{A}{66}$

$11A = 8(66)$

$11A = 528$

$A = 48$

2. $\dfrac{7}{12} = \dfrac{56}{T}$

$7T = 12(56)$

$7T = 672$

$T = 96$

3. $\dfrac{6}{G} = \dfrac{48}{80}$

$48G = 6(80)$

$48G = 480$

$G = 10$

4. $3A - 10 + 8 + A = 2A + 5A - 11$

$4A - 2 = 7A - 11$

$-2 + 11 = 7A - 4A$

$9 = 3A$

$A = 3$

5. $3A - 10 + 8 + A = 2A + 5A - 11 \Rightarrow$

$3(3) - 10 + 8 + (3) = 2(3) + 5(3) - 11$

$9 - 10 + 8 + 3 = 6 + 15 - 11$

$10 = 10$

6. $\left| -6^2 \right| = 6X$

$\left| -36 \right| = 6X$

$36 = 6X$

$X = 6$

7. $\left| -6^2 \right| = 6X \Rightarrow \left| -6^2 \right| = 6(6)$

$\left| -36 \right| = 36$

$36 = 36$

8. $(77 - 32)\dfrac{5}{9} = C$ or: $(77 - 32)(0.56) = C$

$\left({}^5\!45 \right)\dfrac{5}{9} = C$ $(45)(0.56) = C$

$C = 25°$ $C \approx 25.2°$

9. $13 \times \dfrac{9}{5} + 32 = F$

$\dfrac{117}{5} + 32 = F$

$23.4 + 32 = F$

$F = 55.4°$

10. $1.9 \times 0.12 = 0.228$

11. $0.345 + 1.6 = 1.945$

12. $10.1 - 5.4 = 4.7$

13.
$$\begin{array}{r} 3 \\ 7\overline{)21} \\ 2\underline{]1} \\ 0 \end{array}$$

14.
$$\begin{array}{r} 1{,}600 \\ 4\overline{)6{,}400} \\ \underline{6{,}400} \\ 0 \end{array}$$

15.
$$\begin{array}{r} 2 \\ 2\overline{)4} \\ \underline{4} \\ 0 \end{array}$$

16. $6 \div 0.15 = 40$ times

17. $\dfrac{1}{8} = \dfrac{R}{24}$

$8R = 1(24)$

$R = 3$ redheads

18. $6\dfrac{1}{5} \div 3\dfrac{1}{10} = \dfrac{31}{5} \div \dfrac{31}{10} =$

$\dfrac{\cancel{31}}{\cancel{5}} \times \dfrac{\cancel{10}^2}{\cancel{31}} = \dfrac{2}{1} = 2$ pieces

19. $\dfrac{1}{\cancel{36}} \times \dfrac{\cancel{2}}{3} = \dfrac{1}{9}$ of the work

20. $3(2N - 1) = 15$

$3(2N) - 3(1) = 15$

$6N - 3 = 15$

$6N = 15 + 3$

$6N = 18$

$N = 3$

Lesson Practice 20A

1. done

2. $\dfrac{X}{24} = \dfrac{18}{36}$

$36X = 24(18)$

$36X = 432$

$X = 12$ units

3. $\dfrac{B}{9} = \dfrac{18}{6}$

$6B = 9(18)$

$6B = 162$

$B = 27$ units

4. $\dfrac{D}{48} = \dfrac{6}{24}$

$24D = 48(6)$

$24D = 288$

$D = 12$ units

5. $\dfrac{Y}{10} = \dfrac{32}{8}$

$8Y = 10(32)$

$8Y = 320$

$Y = 40$ units

6. $\dfrac{G}{48} = \dfrac{12}{72}$

$72G = 48(12)$

$72G = 576$

$G = 8$ units

7. $\dfrac{Q}{45} = \dfrac{40}{60}$

$60Q = 45(40)$

$60Q = 1,800$

$Q = 30$ units

8. $\dfrac{F}{120} = \dfrac{5}{100}$

$100F = 5(120)$

$100F = 600$

$F = 6$ units

9. $\dfrac{V}{10} = \dfrac{81}{9}$

$9V = 10(81)$

$9V = 810$

$V = 90$ units

10. $\dfrac{R}{2} = \dfrac{54}{6}$

$6R = 2(54)$

$6R = 108$

$R = 18$ units

Lesson Practice 20B

1. $\dfrac{A}{16} = \dfrac{49}{14}$

$14A = 16(49)$

$14A = 784$

$A = 56$ units

2. $\dfrac{X}{36} = \dfrac{15}{54}$

$54X = 36(15)$

$54X = 540$

$X = 10$ units

3. $\dfrac{B}{20} = \dfrac{115}{23}$

$23B = 20(115)$

$23B = 2,300$

$B = 100$ units

4. $\dfrac{D}{5} = \dfrac{64}{8}$

$8D = 5(64)$

$8D = 320$

$D = 40$ units

5. $\dfrac{Y}{9} = \dfrac{75}{15}$

$15Y = 9(75)$

$15Y = 675$

$Y = 45$ units

6. $\dfrac{G}{10} = \dfrac{56}{16}$

$16G = 10(56)$

$16G = 560$

$G = 35$ units

7. $\dfrac{Q}{80} = \dfrac{21}{48}$

$48Q = 80(21)$

$48Q = 1,680$

$Q = 35$ units

8. $\dfrac{F}{8} = \dfrac{17}{4}$

$4F = 8(17)$

$4F = 136$

$F = 34$ units

9. $\dfrac{V}{20} = \dfrac{21}{15}$

$15V = 20(21)$

$15V = 420$

$V = 28$ units

10. $\dfrac{A}{54} = \dfrac{4}{8}$

$8A = 54(4)$

$8A = 216$

$A = 27$ units

Lesson Practice 20C

1. $\dfrac{A}{54} = \dfrac{22}{44}$

$44A = 54(22)$

$44A = 1,188$

$A = 27$ units

2. $\dfrac{X}{40} = \dfrac{12}{96}$

$96X = 40(12)$

$96X = 480$

$X = 5$ units

3. $\dfrac{B}{68} = \dfrac{13}{52}$

$52B = 68(13)$

$52B = 884$

$B = 17$ units

4. $\dfrac{D}{8} = \dfrac{36}{12}$

$12D = 8(36)$

$12D = 288$

$D = 24$ units

5. $\dfrac{Y}{15} = \dfrac{22}{10}$

$10Y = 15(22)$

$10Y = 330$

$Y = 33$ units

6. $\dfrac{G}{72} = \dfrac{23}{92}$

$92G = 72(23)$

$92G = 1,656$

$G = 18$ units

7. $\dfrac{Q}{22} = \dfrac{60}{15}$

$15Q = 22(60)$

$15Q = 1,320$

$Q = 88$ units

8. $\dfrac{F}{9} = \dfrac{55}{11}$

$11F = 9(55)$

$11F = 495$

$F = 45$ units

9. $\dfrac{V}{38} = \dfrac{11}{19}$

$19V = 38(11)$

$19V = 418$

$V = 22$ units

10. $\dfrac{A}{20} = \dfrac{15}{75}$

$75A = 20(15)$

$75A = 300$

$A = 4$ units

Systematic Review 20D

1. $\dfrac{X}{1} = \dfrac{6}{3}$

$X = \dfrac{6}{3}$

$X = 2$ units

2. $\dfrac{E}{24} = \dfrac{10}{20}$

$20E = 24(10)$

$20E = 240$

$E = 12$ units

3. $\dfrac{8}{Y} = \dfrac{72}{81}$

$72Y = 8(81)$

$72Y = 648$

$Y = 9$

4. $\dfrac{7}{B} = \dfrac{42}{48}$

$42B = 7(48)$

$42B = 336$

$B = 8$

5. $\dfrac{5}{G} = \dfrac{30}{36}$

$30G = 5(36)$

$30G = 180$

$G = 6$

6.
$$25\overline{)7.00} = 0.28$$

$\underline{5\,0}$

200

$\underline{200}$

0

7. $\dfrac{49}{100} = 0.49$

8.
$$40\overline{)9.000} = 0.225$$

$\underline{8\,0}$

$1\,00$

$\underline{80}$

200

$\underline{200}$

0

9.
$$8\overline{)1.000} = 0.125$$

$\underline{8}$

20

$\underline{16}$

40

$\underline{40}$

0

10. $0.82 = \dfrac{82}{100} = \dfrac{41}{50}$

11. $0.444 = \dfrac{444}{1,000} = \dfrac{111}{250}$

12. $0.16 = \dfrac{16}{100} = \dfrac{4}{25}$

13. $0.105 = \dfrac{105}{1,000} = \dfrac{21}{200}$

14. $(98-32)\dfrac{5}{9} = C$ or: $(98-32)(0.56) = C$

$(66)\dfrac{5}{9} = C$ $\qquad (66)(0.56) = C$

$\dfrac{330}{9} = C$ $\qquad\qquad C \approx 37°$

$C \approx 36.7°$

15. $2(10 \times 12) + 2(12 \times 8) + 2(8 \times 10) =$

$2(120) + 2(96) + 2(80) =$

$240 + 192 + 160 = 592 \text{ in}^2$

16. $1.00 - 0.63 = 0.37$ of a pie

17. $1 \times 10^3 + 6 \times 10^2 + 7 \times 10^1 +$

$8 \times 10^0 + 3 \times \dfrac{1}{10^1} + 9 \times \dfrac{1}{10^2}$

18. $2(B+2) = 42$

$2B + 2(2) = 42$

$2B + 4 = 42$

$2B = 42 - 4$

$2B = 38$

$B = 19 \text{ in}$

Systematic Review 20E

1. $\dfrac{X}{4} = \dfrac{6}{3}$

$3X = 4(6)$

$3X = 24$

$X = 8 \text{ units}$

2. $\dfrac{R}{3} = \dfrac{6}{2}$

$2R = 3(6)$

$2R = 18$

$R = 9 \text{ units}$

3. $\dfrac{4}{9} = \dfrac{24}{Y}$

$4Y = 9(24)$

$4Y = 216$

$Y = 54$

4. $\dfrac{B}{10} = \dfrac{70}{100}$

$100B = 10(70)$

$100B = 700$

$B = 7$

5. $\dfrac{15}{G} = \dfrac{3}{7}$

$3G = 15(7)$

$3G = 105$

$G = 35$

6.
$$8\overline{)7.000} = 0.875$$

$\underline{6\,4}$

60

$\underline{56}$

40

$\underline{40}$

0

7. $\dfrac{743}{1,000} = 0.743$

8. $0.115 = \dfrac{115}{1,000} = \dfrac{23}{200}$

9. $0.08 = \dfrac{8}{100} = \dfrac{2}{25}$

10. $2W - 6 = |-66|$

   $2W - 6 = 66$

   $2W = 66 + 6$

   $2W = 72$

   $W = 36$

11. $2W - 6 = |-66| \quad 2(36) - 6 = |-66|$

   $72 - 6 = 66$

   $66 = 66$

12. $(7+1)^2 \div 4 + 7 \cdot 3 - X = 25 + X$

   $8^2 \div 4 + 21 - X = 25 + X$

   $64 \div 4 + 21 - X = 25 + X$

   $16 + 21 - X = 25 + X$

   $16 + 21 - 25 = 2X$

   $12 = 2X$

   $X = 6$

13. $(7+1)^2 \div 4 + 7 \cdot 3 - X = 25 + X$

   $(7+1)^2 \div 4 + 7 \cdot 3 - (6) = 25 + (6)$

   $(8)^2 \div 4 + 7 \cdot 3 - 6 = 25 + 6$

   $64 \div 4 + 7 \cdot 3 - 6 = 25 + 6$

   $16 + 21 - 6 = 25 + 6$

   $31 = 31$

14. $10^2 + 15^2 = 20^2$

   $100 + 225 = 400$

   $325 \neq 400; \text{ no}$

15. $24 \times \dfrac{9}{5} + 32 = F$

   $\dfrac{216}{5} + 32 = F$

   $43.2 + 32 = F$

   $75.2° = F$

   no

16. $\dfrac{8}{16}$ in $= \dfrac{1}{2}$ in

17. $2(16 \times 8) + 2(22 \times 8) =$

   $2(128) + 2(176) =$

   $256 + 352 = 608 \text{ ft}^2$

   $608 \div 400 = 1.52$

   rounded up: 2 gallons

18. $(16 \times 22) = 352 \text{ ft}^2$

   $352 \div 400 = 0.88$ gal per coat

   $2(0.88) = 1.76$ gal

   rounded up: 2 gallons needed

19. $60 \div 3 = 20$

   $20 \times 2 = 40$ hours worked

   $60 - 40 = 20$ hours left

20. $95 \div 0.05 = 1,900$ pieces

Systematic Review 20F

1. $\dfrac{X}{24} = \dfrac{10}{30}$

   $30X = 24(10)$

   $30X = 240$

   $X = 8$ units

2. $\dfrac{R}{5} = \dfrac{7}{5}$

   $5R = 5(7)$

   $R = 7$ units

3. $\dfrac{5}{8} = \dfrac{35}{Y}$

   $5Y = 8(35)$

   $5Y = 280$

   $Y = 56$

4. $\dfrac{63}{72} = \dfrac{B}{8}$

   $72B = 63(8)$

   $72B = 504$

   $B = 7$

5. $\dfrac{7}{G} = \dfrac{35}{50}$

   $35G = 7(50)$

   $35G = 350$

   $G = 10$

6.

$$\begin{array}{r} 0.625 \\ 8\overline{)5.000} \\ \underline{4\,8} \\ 20 \\ \underline{16} \\ 40 \\ \underline{40} \\ 0 \end{array}$$

7. $\dfrac{8}{100} = 0.08$

8. $0.28 = \dfrac{28}{100} = \dfrac{7}{25}$

9. $0.96 = \dfrac{96}{100} = \dfrac{24}{25}$

10. $-2X - X + 12 = X - |-3 \cdot 4|$
$-3X + 12 = X - 12$
$12 + 12 = X + 3X$
$24 = 4X$
$X = 6$

11. $-2X - X + 12 = X - |-3 \cdot 4| \Rightarrow$
$-2(6) - (6) + 12 = (6) - |-3 \cdot 4|$
$-12 - 6 + 12 = 6 - |-12|$
$-12 - 6 + 12 = 6 - 12$
$-6 = -6$

12. $-(6+7)^2 + (10+5)^2 = 5M + 1$
$-(13)^2 + 15^2 = 5M + 1$
$-169 + 225 = 5M + 1$
$56 = 5M + 1$
$55 = 5M$
$M = 11$

13. $-(6+7)^2 + (10+5)^2 = 5M + 1 \Rightarrow$
$-(6+7)^2 + (10+5)^2 = 5(11) + 1$
$-(13)^2 + (15)^2 = 55 + 1$
$-169 + 225 = 56$
$56 = 56$

14. $3.5 + 6.12 + 2.1 = 11.72$ in

15. $10\dfrac{3}{6} - 5\dfrac{1}{6} = 5\dfrac{2}{6} = 5\dfrac{1}{3}$ years

16. $\dfrac{3}{8} \times \dfrac{1}{2} = \dfrac{3}{16}$ of a pizza

17. 32°F

18. $\dfrac{600}{1,000} = \dfrac{C}{2,000}$
$1,000C = 600(2,000)$
$C = 600(2)$
$C = \$1,200$

19. $\$1,200 \div 5 = \$240$
$\$240 \times 3 = \$720$

20. $2A + 3 - 15 = A$
$2A - 12 = A$
$2A = A + 12$
$2A - A = 12$
$A = 12$ apples

Lesson Practice 21A

1. done

2. 3, 6, 9, 12, <u>15</u>, 18, 21, 24, 27, <u>30</u>
5, 10, <u>15</u>, 20, 25, <u>30</u>, 35, 40, <u>45</u>, 50
LCM = 15

3. 4, 8, 12, 16, <u>20</u>, 24, 28, 32, 36, <u>40</u>
5, 10, 15, <u>20</u>, 25, 30, 35, <u>40</u>, 45, 50
LCM = 20

4. done

5. $12 = 2 \times 2 \times 3$

6. $15 = 3 \times 5$

7. $8 = 2 \times 2 \times 2$

8. done

9. $2 \times 2 \times 3 = 12;\ 3 \times 5 = 15$
LCM $= 2 \times 2 \times 3 \times 5 = 60$

10. $2 \times 2 \times 2 = 8;\ 2 \times 3 \times 3 = 18$
LCM $= 2 \times 2 \times 2 \times 3 \times 3 = 72$

11. $2 \times 2 \times 2 = 8;\ 3 \times 5 = 15$
LCM $= 2 \times 2 \times 2 \times 3 \times 5 = 120$

12. $1 \times 5 = 5;\ 2 \times 2 \times 2 = 8$
LCM $= 2 \times 2 \times 2 \times 5 = 40$

13. $2 \times 2 = 4$; $3 \times 3 = 9$
LCM $= 2 \times 2 \times 3 \times 3 = 36$

14. $1 \times 3 = 3$; $2 \times 2 = 4$
LCM $= 2 \times 2 \times 3 = 12$

15. $2 \times 2 \times 2 = 8$; $2 \times 2 \times 3 = 12$
LCM $= 2 \times 2 \times 2 \times 3 = 24$

Lesson Practice 21B

1. 4, <u>8</u>, 12, <u>16</u>, 20, <u>24</u>, 28, <u>32</u>, 36, <u>40</u>
<u>8</u>, <u>16</u>, <u>24</u>, <u>32</u>, <u>40</u>, <u>48</u>, <u>56</u>, <u>64</u>, <u>72</u>, <u>80</u>
LCM $= 8$

2. 6, 12, 18, 24, <u>30</u>, 36, 42, 48, 54, <u>60</u>
10, 20, <u>30</u>, 40, 50, <u>60</u>, 70, 80, <u>90</u>, 100
LCM $= 30$

3. 4, 8, 12, 16, <u>20</u>, 24, 28, 32, 36, <u>40</u>
5, 10, 15, <u>20</u>, 25, 30, 35, <u>40</u>, 45, 50
LCM $= 20$

4. $24 = 2 \times 2 \times 2 \times 3$

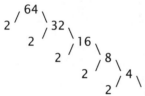

5. $64 = 2 \times 2 \times 2 \times 2 \times 2 \times 2$

6. $32 = 2 \times 2 \times 2 \times 2 \times 2$

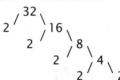

7. $20 = 2 \times 2 \times 5$

8. $2 \times 2 \times 5 = 20$; $2 \times 2 \times 2 \times 3 = 24$
LCM $= 2 \times 2 \times 2 \times 3 \times 5 = 120$

9. $2 \times 2 \times 5 = 20$; $2 \times 2 \times 2 \times 2 \times 2 = 32$
LCM $= 2 \times 2 \times 2 \times 2 \times 2 \times 5 = 160$

10. $2 \times 2 \times 2 \times 3 = 24$; $2 \times 2 \times 2 \times 2 \times 2 \times 2 = 64$
LCM $= 2 \times 2 \times 2 \times 2 \times 2 \times 2 \times 3 = 192$

11. $2 \times 2 \times 2 \times 2 \times 2 = 32$; $2 \times 2 \times 2 \times 2 \times 2 \times 2 = 64$
LCM $= 2 \times 2 \times 2 \times 2 \times 2 \times 2 = 64$

12. $2 \times 3 = 6$; $2 \times 2 \times 3 = 12$
LCM $= 2 \times 2 \times 3 = 12$

13. $1 \times 5 = 5$; $1 \times 7 = 7$
LCM $= 5 \times 7 = 35$

14. $3 \times 3 = 9$; $2 \times 2 \times 3 = 12$
LCM $= 2 \times 2 \times 3 \times 3 = 36$

15. $1 \times 3 = 3$; $3 \times 5 = 15$
LCM $= 3 \times 5 = 15$

Lesson Practice 21C

1. 7, 14, 21, 28, 35, 42, 49, <u>56</u>, 63, 70
8, 16, 24, 32, 40, 48, <u>56</u>, 64, 72, 80
LCM $= 56$

2. 3, 6, 9, 12, 15, 18, <u>21</u> 24, 27, 30
7, 14, <u>21</u> 28, 35, <u>42</u> 49, 56, <u>63</u> 70
LCM $= 21$

3. 2, 4, 6, 8, <u>10</u> 12, 14, 16, 18, <u>20</u>
5, <u>10</u>, 15, <u>20</u>, 25, <u>30</u>, 35, <u>40</u>, 45, <u>50</u>
LCM $= 10$

4. $2 \times 2 \times 2 \times 2 = 16$

5. $5 \times 5 = 25$

6. $2 \times 5 = 10$

7. $2 \times 2 \times 7 = 28$

8. $2 \times 2 \times 2 \times 2 = 16$; $5 \times 5 = 25$
LCM $= 2 \times 2 \times 2 \times 2 \times 5 \times 5 = 400$

9. $5 \times 5 = 25$; $2 \times 5 = 10$
LCM $= 2 \times 5 \times 5 = 50$

10. $2 \times 5 = 10$; $2 \times 2 \times 7 = 28$
LCM $= 2 \times 2 \times 5 \times 7 = 140$

11. $2 \times 2 \times 2 \times 2 = 16$; $2 \times 2 \times 7 = 28$
    $LCM = 2 \times 2 \times 2 \times 2 \times 7 = 112$

12. $2 \times 5 = 10$; $3 \times 5 = 15$
    $LCM = 2 \times 3 \times 5 = 30$

13. $2 \times 2 \times 2 = 8$; $3 \times 3 = 9$
    $LCM = 2 \times 2 \times 2 \times 3 \times 3 = 72$

14. $1 \times 11 = 11$; $2 \times 2 \times 3 = 12$
    $LCM = 2 \times 2 \times 3 \times 11 = 132$

15. $1 \times 5 = 5$; $2 \times 3 = 6$
    $LCM = 2 \times 3 \times 5 = 30$

16. $\dfrac{25}{100} = 25 \div 100 = 0.25 = 25\%$

17. $\dfrac{2}{8} = \dfrac{X}{32}$
    $8X = 2(32)$
    $8X = 64$
    $X = \$8$

18. $8^2 + 10^2 = 15^2$
    $64 + 100 = 225$
    $164 \neq 225$; no

Systematic Review 21D

1. $22 = 2 \times 11$
2. $28 = 2 \times 2 \times 7$
3. $45 = 3 \times 3 \times 5$
4. $36 = 2 \times 2 \times 3 \times 3$
5. $2 \times 3 = 6$; $3 \times 3 = 9$
   $LCM = 2 \times 3 \times 3 = 18$
6. $2 \times 2 = 4$; $3 \times 5 = 15$
   $LCM = 2 \times 2 \times 3 \times 5 = 60$
7. $1 \times 5 = 5$; $3 \times 3 = 9$
   $LCM = 3 \times 3 \times 5 = 45$
8. $2 \times 2 \times 2 = 8$; $2 \times 2 \times 5 = 20$
   $LCM = 2 \times 2 \times 2 \times 5 = 40$
9. $\dfrac{X}{6} = \dfrac{16}{8}$
   $8X = 6(16)$
   $8X = 96$
   $X = 12$ units
10. $\dfrac{Y}{18} = \dfrac{5}{15}$
    $15Y = 18(5)$
    $15Y = 90$
    $Y = 6$ units
11. $|3 - 8| = |-5| = 5$
12. $-|-6 + 9| = -|3| = -3$
13. $\dfrac{1}{5} = 1 \div 5 = 0.20 = 20\%$
14. $\dfrac{6}{100} = 6 \div 100 = 0.06 = 6\%$
15. $\dfrac{5}{8} = 5 \div 8 = 0.62\dfrac{1}{2} = 62\dfrac{1}{2}\%$

Systematic Review 21E

1. $6 = 2 \times 3$
2. $14 = 2 \times 7$
3. $48 = 2 \times 2 \times 2 \times 2 \times 3$
4. $2 \times 5 = 10$; $2 \times 2 \times 3 = 12$
   $LCM = 2 \times 2 \times 3 \times 5 = 60$
5. $3 \times 3 = 9$; $2 \times 5 = 10$
   $LCM = 2 \times 3 \times 3 \times 5 = 90$
6. $2 \times 2 \times 3 = 12$; $2 \times 2 \times 2 \times 3 = 48$
   $LCM = 2 \times 2 \times 2 \times 2 \times 3 = 48$
7. $\dfrac{A}{12} = \dfrac{5}{10}$
   $10A = 12(5)$
   $10A = 60$
   $A = 6$ units
8. $\dfrac{B}{6} = \dfrac{9}{6}$
   $6B = 6(9)$
   $B = 9$ units
9. done

10.
$$\begin{array}{r} 7.12 \\ 5\overline{)35.6} \\ \underline{35} \\ 0\,6 \\ \underline{0\,5} \\ 10 \\ \underline{10} \\ 0 \end{array}$$

11.
$$\begin{array}{r} 0.857 \approx 0.86 \\ 7\overline{)6.00} \\ \underline{5\,6} \\ 40 \\ \underline{35} \\ 50 \\ \underline{49} \\ 1 \end{array}$$

12. $\dfrac{2}{5} = 2 \div 5 = 0.40 = 40\%$

13. $\dfrac{1}{4} = 1 \div 4 = 0.25 = 25\%$

14. $\frac{1}{3} = 1 \div 3 \approx 0.33 \frac{1}{3} = 33\frac{1}{3}\%$

15. $25 \times \frac{9}{5} + 32 = F$

$\quad\quad 45 + 32 = F$

$\quad\quad\quad\quad F = 77°$

16. $(30-32)\frac{5}{9} = C$ or: $(30-32)(0.56) = C$

$\quad\quad (-2)\frac{5}{9} = C$ $\quad\quad\quad (-2)(0.56) = C$

$\quad\quad\quad\quad\quad\quad\quad\quad\quad\quad C \approx -1.1°$

$\quad\quad \frac{-10}{9} = C$

$\quad\quad\quad C \approx -1.1°$

17. $\$65.35 + \$4.50 = \$69.85$

18. $\frac{1}{4} = \frac{X}{2}$

$\quad 4X = 1(2)$

$\quad 4X = 2$

$\quad\quad X = \frac{1}{2}$ gal or 0.5 gal

19. $6^2 + 8^2 = H^2$

$\quad 36 + 64 = H^2$

$\quad\quad 100 = H^2$

$\quad\quad\quad H = 10$ ft

20. $20 \div 5 = 4$

$\quad 4 \times 4 = 16$ brought umbrellas.

$\quad 20 - 16 = 4$ did not bring umbrellas.

Systematic Review 21F

1. $2 \times 17 = 34$

2. $2 \times 2 \times 2 \times 5 = 40$

3. $2 \times 3 \times 3 \times 3 = 54$

4. $2 \times 7 = 14; 2 \times 2 \times 7 = 28$

$\quad$ LCM $= 2 \times 2 \times 7 = 28$

5. $1 \times 3 = 3; 2 \times 2 \times 2 \times 2 = 16$

$\quad$ LCM $= 2 \times 2 \times 2 \times 2 \times 3 = 48$

6. $3 \times 5 = 15; 2 \times 2 \times 2 \times 5 = 40$

$\quad$ LCM $= 2 \times 2 \times 2 \times 3 \times 5 = 120$

7. $\frac{R}{2.3} = \frac{4.5}{1.5}$

$\quad 1.5R = 2.3(4.5)$

$\quad 1.5R = 10.35$

$\quad\quad R = 6.9$ units

8. $\frac{E}{0.5} = \frac{1}{0.2}$

$\quad 0.2E = 0.5(1)$

$\quad 0.2E = 0.5$

$\quad\quad E = 2.5$ units

9.
```
   0.875
 8)7.000
   6 4
    60
    56
     40
     40
```

10.
```
      5.2
 25)130.0
    125
     50
     50
      0
```

11.
```
   33.333 ≈ 33.33
 3)100.000
    9
    10
     9
     10
      9
      10
       9
       10
```

12. $\frac{3}{4} = 3 \div 4 = 0.75 = 75\%$

13. $\frac{5}{8} = 5 \div 8 = 0.62\frac{1}{2} = 62\frac{1}{2}\%$ or 62.5%

14. $\frac{1}{100} = 1 \div 100 = 0.01 = 1\%$

15.
$$\frac{20}{1} = \frac{X}{5}$$
$$1X = 20(5)$$
$$X = 100 \text{ mi}$$

16.
$$12^2 + 9^2 = H^2$$
$$144 + 81 = H^2$$
$$225 = H^2$$
$$H = 15 \text{ mi}$$
$$15 + 12 = 27 \text{ mi total}$$

17. $\frac{1}{4} + \frac{3}{5} + \frac{3}{20} = \frac{5}{20} + \frac{12}{20} + \frac{3}{20} = \frac{20}{20}$; yes

18. $(230 - 32)\dfrac{5}{9} = C$ or: $(230 - 32)(0.56) = C$
     $\left(^{22}\cancel{198}\right)\dfrac{5}{\cancel{9}} = C$ $(198)(0.56) = C$
          $C = 110°$ $C \approx 110.9°$

19.
$$4\left(6 \times 4 \times \frac{1}{2}\right) + (4 \times 4) =$$
$$4(12) + (16) =$$
$$48 + 16 = 64 \text{ ft}^2$$

20.
$$2X + 8 = 16$$
$$2X = 16 - 8$$
$$2X = 8$$
$$X = 4 \text{ chores}$$

Lesson Practice 22A

1. done
2. $12 : \underline{1}, \underline{2}, 3, \underline{4}, 6, 12$
   $16 : \underline{1}, \underline{2}, \underline{4}, 8, 16$
   GCF = 4
3. $6 : \underline{1}, \underline{2}, \underline{3}, \underline{6}$
   $18 : \underline{1}, \underline{2}, \underline{3}, \underline{6}, 9, 18$
   GCF = 6
4. done
5. $36 = 2 \times 2 \times 3 \times 3$
6. $28 = 2 \times 2 \times 7$
7. $42 = 2 \times 3 \times 7$
8. done
9. $28 = \underline{2} \times \underline{2} \times 7; \ 36 = \underline{2} \times \underline{2} \times 3 \times 3$
   GCF = $2 \times 2 = 4$
10. $28 = \underline{2} \times 2 \times \underline{7}; \ 42 = \underline{2} \times 3 \times \underline{7}$
    GCF = $2 \times 7 = 14$

11. $24 = \underline{2} \times 2 \times 2 \times \underline{3}; \ 42 = \underline{2} \times \underline{3} \times 7$
    GCF = $2 \times 3 = 6$
12. $30 = 2 \times \underline{3} \times \underline{5}; \ 75 = \underline{3} \times \underline{5} \times 5$
    GCF = $3 \times 5 = 15$
13. $28 = \underline{2} \times 2 \times \underline{7}; \ 14 = \underline{2} \times \underline{7}$
    GCF = $2 \times 7 = 14$
14. $8 = \underline{2} \times \underline{2} \times \underline{2}; \ 24 = \underline{2} \times \underline{2} \times \underline{2} \times 3$
    GCF = $2 \times 2 \times 2 = 8$
15. $21 = 3 \times \underline{7}; \ 35 = 5 \times \underline{7}$
    GCF = 7

Lesson Practice 22B

1. $4 : \underline{1}, \underline{2}, 4$
   $6 : \underline{1}, \underline{2}, 3, 6$
   GCF = 2
2. $16 : \underline{1}, \underline{2}, \underline{4}, \underline{8}, \underline{16}$
   $32 : \underline{1}, \underline{2}, \underline{4}, \underline{8}, \underline{16}, 32$
   GCF = 16
3. $10 : \underline{1}, \underline{2}, \underline{5}, \underline{10}$
   $20 : \underline{1}, \underline{2}, 4, \underline{5}, \underline{10}, 20$
   GCF = 10
4. $25 = 5 \times 5$
5. $35 = 5 \times 7$
6. $40 = 2 \times 2 \times 2 \times 5$
7. $56 = 2 \times 2 \times 2 \times 7$
8. $25 = \underline{5} \times 5; \ 35 = \underline{5} \times 7$
   GCF = 5
9. $35 = \underline{5} \times 7; \ 40 = 2 \times 2 \times 2 \times \underline{5}$
   GCF = 5
10. $35 = 5 \times \underline{7}; \ 56 = 2 \times 2 \times 2 \times \underline{7}$
    GCF = 7
11. $40 = \underline{2} \times \underline{2} \times \underline{2} \times 5; \ 56 = \underline{2} \times \underline{2} \times \underline{2} \times 7$
    GCF = $2 \times 2 \times 2 = 8$
12. $36 = \underline{2} \times \underline{2} \times \underline{3} \times 3; \ 48 = \underline{2} \times \underline{2} \times 2 \times 2 \times \underline{3}$
    GCF = $2 \times 2 \times 3 = 12$
13. $66 = \underline{2} \times \underline{3} \times 11; \ 90 = \underline{2} \times \underline{3} \times 3 \times 5$
    GCF = $2 \times 3 = 6$
14. $75 = 3 \times \underline{5} \times \underline{5}; \ 100 = 2 \times 2 \times \underline{5} \times \underline{5}$
    GCF = $5 \times 5 = 25$
15. $12 = 2 \times 2 \times \underline{3}; \ 15 = \underline{3} \times 5$
    GCF = 3

Lesson Practice 22C

1. $6:\underline{1},\underline{2},3,6$
   $8:\underline{1},\underline{2},4,8$
   GCF = 2

2. $15:\underline{1},3,\underline{5},15$
   $20:\underline{1},2,4,\underline{5},10,20$
   GCF = 5

3. $7:\underline{1},\underline{7}$
   $14:\underline{1},2,\underline{7},14$
   GCF = 7

4. $8 = 2\times2\times2$

5. $64 = 2\times2\times2\times2\times2\times2$

6. $16 = 2\times2\times2\times2$

7. $20 = 2\times2\times5$

8. $8 = \underline{2}\times\underline{2}\times\underline{2}$; $64 = \underline{2}\times\underline{2}\times\underline{2}\times2\times2\times2$
   GCF = $2\times2\times2 = 8$

9. $8 = \underline{2}\times\underline{2}\times2$; $20 = \underline{2}\times\underline{2}\times5$
   GCF = $2\times2 = 4$

10. $64 = \underline{2}\times\underline{2}\times\underline{2}\times\underline{2}\times2\times2$; $16 = \underline{2}\times\underline{2}\times\underline{2}\times\underline{2}$
    GCF = $2\times2\times2\times2 = 16$

11. $16 = \underline{2}\times\underline{2}\times2\times2$; $20 = \underline{2}\times\underline{2}\times5$
    GCF = $2\times2 = 4$

12. $27 = \underline{3}\times\underline{3}\times3$; $45 = \underline{3}\times\underline{3}\times5$
    GCF = $3\times3 = 9$

13. $99 = 3\times3\times\underline{11}$; $110 = 2\times5\times\underline{11}$
    GCF = 11

14. $15 = 3\times\underline{5}$; $100 = 2\times2\times\underline{5}\times5$
    GCF = 5

15. $32 = \underline{2}\times\underline{2}\times\underline{2}\times\underline{2}\times2$; $48 = \underline{2}\times\underline{2}\times\underline{2}\times\underline{2}\times3$
    GCF = $2\times2\times2\times2 = 16$

Systematic Review 22D

1. $36 = \underline{2}\times\underline{2}\times3\times3$; $40 = \underline{2}\times\underline{2}\times2\times5$
   GCF = $2\times2 = 4$

2. $15 = 3\times\underline{5}$; $20 = 2\times2\times\underline{5}$
   GCF = 5

3. $32 = \underline{2}\times\underline{2}\times\underline{2}\times2\times2$; $56 = \underline{2}\times\underline{2}\times\underline{2}\times7$
   GCF = $2\times2\times2 = 8$

4. $3 = 1\times3$; $4 = 2\times2$
   LCM = $2\times2\times3 = 12$

5. $8 = 2\times2\times2$; $36 = 2\times2\times3\times3$
   LCM = $2\times2\times2\times3\times3 = 72$

6. $10 = 2\times5$; $45 = 3\times3\times5$
   LCM = $2\times3\times3\times5 = 90$

7. $\dfrac{Q}{2.5} = \dfrac{28}{7}$
   $7Q = 2.5(28)$
   $7Q = 70$
   $Q = 10$ units

8. $\dfrac{R}{3} = \dfrac{20}{5}$
   $5R = 3(20)$
   $5R = 60$
   $R = 12$ units

9. $\sqrt{25} = 5$

10. $\sqrt{100} = 10$

11. $\sqrt{X^2} = X$

12. $0.47\times85 = 39.95$

13. $0.05\times100 = 5$

14. $0.69\times12.8 = 8.832$

15. $\$58\times0.10 = \$5.80$

16. $140.1+153.09+106.25 = 399.44$ ft^2

17. $\dfrac{20}{D} = \dfrac{1}{2}$
    $D = 2(20)$
    $D = 40$ dry days

18. $\dfrac{3}{8}\times\dfrac{1}{6} = \dfrac{3}{48} = \dfrac{1}{16}$ of Dad's check
    $\$320\div16 = \$20$

Systematic Review 22E

1. $12 = \underline{2}\times2\times\underline{3}$; $18 = \underline{2}\times\underline{3}\times3$
   GCF = $2\times3 = 6$

2. $30 = 2\times\underline{3}\times\underline{5}$; $45 = 3\times\underline{3}\times\underline{5}$
   GCF = $3\times5 = 15$

3. $10 = \underline{2}\times\underline{5}$; $100 = \underline{2}\times2\times\underline{5}\times5$
   GCF = $2\times5 = 10$

4. $15 = 3\times5$; $18 = 2\times3\times3$
   LCM = $2\times3\times3\times5 = 90$

5. $6 = 2\times3$; $10 = 2\times5$
   LCM = $2\times3\times5 = 30$

6. $3 = 1\times3$; $5 = 1\times5$
   LCM = $3\times5 = 15$

7. $\dfrac{X}{8} = \dfrac{10}{5}$

$5X = 8(10)$

$5X = 80$

$X = 16$ units

8. $\dfrac{Y}{2} = \dfrac{6}{2}$

$2Y = 2(6)$

$Y = 6$ units

9. $2.861 + 4.4 = 7.261$

10. $0.78 - 0.09 = 0.69$

11. $0.75 \div 0.3 = 2.5$

12. $0.13 \times 61 = 7.93$

13. $0.06 \times 2.45 = 0.147$

14. $0.10 \times 950 = 95$

15. $16X + 5X - 8 = |{-97}|$

$21X - 8 = 97$

$21X = 97 + 8$

$21X = 105$

$X = 5$

16. $3(2Y - 11) = 3(7)$

$3(2Y) - 3(11) = 3(7)$

$6Y - 33 = 21$

$6Y = 21 + 33$

$6Y = 54$

$Y = 9$

17. $A - 8 + 6(3) = 13$

$A - 8 + 18 = 13$

$A + 10 = 13$

$A = 13 - 10$

$A = 3$

18. $0.25 \times \$49.00 = \$12.25$ off

$\$49.00 - \$12.25 = \$36.75$

19. $\$36.75 \times 1.06 \approx \$38.96$

20. $\dfrac{4}{5} = \dfrac{8}{B}$

$4B = 5(8)$

$4B = 40$

$B = 10$ biographies

Systematic Review 22F

1. $75 = \underline{3} \times \underline{5} \times 5; \; 45 = \underline{3} \times 3 \times \underline{5}$

$GCF = 3 \times 5 = 15$

2. $7 = 1 \times \underline{7}; \; 21 = 3 \times \underline{7}$

$GCF = 7$

3. $33 = 3 \times \underline{11}; \; 55 = 5 \times \underline{11}$

$GCF = 11$

4. $5 = 1 \times 5; \; 10 = 2 \times 5$

$LCM = 2 \times 5 = 10$

5. $8 = 2 \times 2 \times 2; \; 24 = 2 \times 2 \times 2 \times 3$

$LCM = 2 \times 2 \times 2 \times 3 = 24$

6. $7 = 1 \times 7; \; 5 = 1 \times 5$

$LCM = 7 \times 5 = 35$

7. $\dfrac{F}{28} = \dfrac{92}{46}$

$46F = 28(92)$

$46F = 2{,}576$

$F = 56$ units

8. $\dfrac{D}{36} = \dfrac{10}{15}$

$15D = 36(10)$

$15D = 360$

$D = 24$ units

9. $(-11) + (-23) = -34$

10. $(-45) \times 12 = -540$

11. $17 - 21 = -4$

12. $0.75 \times 100 = 75$

13. $0.03 \times 14.6 = 0.438$

14. $0.11 \times 0.67 = 0.0737$

15. $3 \div 4 = 0.75 = 75\%$

16. $1 \div 2 = 0.50 = 50\%$

17. $7 \div 9 = 0.77\dfrac{7}{9} = 77\dfrac{7}{9}\%$

18. $10 \div 2 = 5$ answers

19. $\$25.56 + \$6.78 = \$32.34$

$\$32.34 - \$16.16 = \$16.18$

20. $\dfrac{S}{28} = \dfrac{3}{7}$

$7S = 28(3)$

$7S = 84$

$S = 12$ snowy days

$28 - 12 = 16$ not snowy

Lesson Practice 23A

1. 3
2. 2
3. C : $X^2 + 2X + 2$
4. D : $X^2 + 3X + 4$
5. B : $3X^2 + 6$
6. A : $X^2 + 4X + 3$
7. done

8.
$$X^2 + 3X + 6$$
$$+ X^2 - 2X + 8$$
$$2X^2 + X + 14$$

9.
$$4X^2 - 6X - 6$$
$$+ 2X^2 + 2X - 3$$
$$6X^2 - 4X - 9$$

10.
$$8X^2 + 2X - 15$$
$$+ X^2 - 7X + 20$$
$$9X^2 - 5X + 5$$

11.
$$6X^2 - 10X + 3$$
$$+ 2X^2 - 5X - 8$$
$$8X^2 - 15X - 5$$

12.
$$X^2 - 3X + 9$$
$$+ 2X^2 - 6X - 11$$
$$3X^2 - 9X - 2$$

Lesson Practice 23B

1. trinomial
2. X
3. C : $X^2 + 9X + 2$
4. D : $3X^2 + 2X + 5$
5. A : $X^2 + 4X + 1$
6. B : $2X^2 + 3X$

7.
$$5X^2 + 3X - 2$$
$$+ 3X^2 + 2X - 5$$
$$8X^2 + 5X - 7$$

8.
$$4X^2 - 6X + 8$$
$$+ X^2 - 2X - 10$$
$$5X^2 - 8X - 2$$

9.
$$5X^2 - 2X + 4$$
$$+ 2X^2 - 3X + 6$$
$$7X^2 - 5X + 10$$

10.
$$7X^2 - 3X + 6$$
$$+ {-3X^2} - 6X + 2$$
$$4X^2 - 9X + 8$$

11.
$$8X^2 - 3X + 2$$
$$+ {-4X^2} - 5X - 8$$
$$4X^2 - 8X - 6$$

12.
$$-3X^2 - 6X + 8$$
$$+ 4X^2 + 5X - 7$$
$$X^2 - X + 1$$

Lesson Practice 23C

1. 2
2. 3
3. C : $X^2 + X$
4. D : $X^2 + 1$
5. A : $2X^2 + 3X + 4$
6. B : $2X^2 + 5X + 2$

7.
$$-4X^2 + 3X - 3$$
$$+ 4X^2 + 2X - 7$$
$$5X - 10$$

8.
$$2X^2 + X - 6$$
$$+ {-2X^2} - 3X + 2$$
$$-2X - 4$$

9.
$$3X^2 - 4X - 7$$
$$+ {-2X^2} + 2X + 10$$
$$X^2 - 2X + 3$$

10.
$$6X^2 + 3X - 3$$
$$+ {-7X^2} + 2X + 5$$
$$-X^2 + 5X + 2$$

11.
$$4X^2 + X - 3$$
$$+ \ \underline{-6X^2 - 3X + 2}$$
$$-2X^2 - 2X - 1$$

12.
$$-2X^2 + 8X + 1$$
$$+ \ \underline{5X^2 - 3X - 9}$$
$$3X^2 + 5X - 8$$

Systematic Review 23D

1.
$$-3X^2 + 6X - 8$$
$$+ \ \underline{2X^2 - 2X + 9}$$
$$-X^2 + 4X + 1$$

2.
$$5X^2 + 3X + 2$$
$$+ \ \underline{-3X^2 - 2X + 5}$$
$$2X^2 + X + 7$$

3.
$$8X^2 - 3X + 9$$
$$+ \ \underline{-3X^2 - 2X - 2}$$
$$5X^2 - 5X + 7$$

4. $3 = 1 \times 3; \ 6 = 2 \times 3$
$\text{LCM} = 2 \times 3 = 6$

5. $11 = 1 \times \underline{11}; \ 22 = 2 \times \underline{11}$
$\text{GCF} = 11$

6. $\dfrac{X}{31} = \dfrac{42}{93}$
$93X = 31(42)$
$93X = 1{,}302$
$X = 14$

7. $\dfrac{16}{19} = \dfrac{64}{A}$
$16A = 19(64)$
$16A = 1{,}216$
$A = 76$

8. $\dfrac{5}{9} = \dfrac{55}{Y}$
$5Y = 9(55)$
$5Y = 495$
$Y = 99$

9. $\dfrac{1}{9} = 1 \div 9 \approx 0.11$

10. $\dfrac{3}{7} = 3 \div 7 \approx 0.43$

11. $\dfrac{25}{100} = 25 \div 100 = 0.25$

12. $2(5 \times 15) + 2(15 \times 5) + 2(5 \times 5) =$
$2(75) + 2(75) + 2(25) =$
$150 + 150 + 50 = 350 \text{ in}^2$

13. $4\left(20 \times 12.8 \times \dfrac{1}{2}\right) + (20 \times 20) =$
$4(128) + (400) =$
$512 + 400 = 912 \text{ in}^2$

14. done

15. $4.00 \times 25 = 100$

16. $2.25 \times 0.5 = 1.125$

17. $\$75.00 \times 1.30 = \$97.50$
$\$97.50 - \$75.00 = \$22.50$

18. $^{20}\cancel{100} \times \dfrac{9}{\cancel{5}} + 32 = F$
$180 + 32 = F$
$F = 212°$

Systematic Review 23E

1.
$$-4X^2 + 7X - 7$$
$$+ \ \underline{3X^2 - 3X + 8}$$
$$-X^2 + 4X + 1$$

2.
$$X^2 + 2X + 3$$
$$+ \ \underline{-4X^2 - 6X + 5}$$
$$-3X^2 - 4X + 8$$

3.
$$5X^2 - 3X + 3$$
$$+ \ \underline{-2X^2 - 2X - 3}$$
$$3X^2 - 5X$$

4. $4 = 2 \times 2; \ 12 = 2 \times 2 \times 3$
$\text{LCM} = 2 \times 2 \times 3 = 12$

5. $6 = \underline{2} \times 3; \ 16 = \underline{2} \times 2 \times 2 \times 2$
$\text{GCF} = 2$

6. $\dfrac{D}{6} = \dfrac{25}{30}$
$30D = 6(25)$
$30D = 150$
$D = 5$

7. $\dfrac{9}{F} = \dfrac{18}{48}$

$18F = 9(48)$

$18F = 432$

$F = 24$

8. $\dfrac{15}{20} = \dfrac{G}{44}$

$20G = 15(44)$

$20G = 660$

$G = 33$

9. $3.25 = 3\dfrac{25}{100} = 3\dfrac{1}{4}$

10. $0.45 = \dfrac{45}{100} = \dfrac{9}{20}$

11. $0.08 = \dfrac{8}{100} = \dfrac{2}{25}$

12. $2(4 \times 8.5) + 2(8.5 \times 6) + 2(6 \times 4) =$

$2(34) + 2(51) + 2(24) =$

$68 + 102 + 48 = 218$ in^2

13. $4\left(3.5 \times 1.2 \times \dfrac{1}{2}\right) + (1.2 \times 1.2) =$

$4(2.1) + (1.44) =$

$8.4 + 1.44 = 9.84$ ft^2

14. $2.10 \times 50 = 105$

15. $1.20 \times 4.5 = 5.4$

16. $5.00 \times 38 = 190$

17. $2\dfrac{1}{3} + 1\dfrac{1}{6} = 2\dfrac{2}{6} + 1\dfrac{1}{6} = 3\dfrac{3}{6} = 3\dfrac{1}{2}$ hours

18. $\$45.99 \times 0.30 \approx \$13.80$

$\$45.99 - \$13.80 = \$32.19$

19. $5X - 5 - 9 = 36$

$5X - 14 = 36$

$5X = 36 + 14$

$5X = 50$

$X = 10$

20. $\$45.98 \times 1.20 \approx \$55.18$

Systematic Review 23F

1. $8X^2 - 5X - 2$

$+ \quad -7X^2 + 4X + 5$

$\overline{\quad X^2 - \ X + 3}$

2. $4X^2 + 2X + 5$

$+ \quad 2X^2 + 7X - 2$

$\overline{\quad 6X^2 + 9X + 3}$

3. $-X^2 - X - 1$

$+ \quad -2X^2 + X + 3$

$\overline{\quad -3X^2 \quad + 2}$

4. $10 = 2 \times 5; \ 100 = 2 \times 2 \times 5 \times 5$

$LCM = 2 \times 2 \times 5 \times 5 = 100$

5. $16 = 2 \times \underline{2} \times \underline{2} \times \underline{2}; \ 64 = \underline{2} \times \underline{2} \times \underline{2} \times \underline{2} \times 2 \times 2$

$GCF = 2 \times 2 \times 2 \times 2 = 16$

6. $\dfrac{3}{4} = \dfrac{R}{100}$

$4R = 3(100)$

$4R = 300$

$R = 75$

7. $\dfrac{36}{X} = \dfrac{2}{3}$

$2X = 36(3)$

$2X = 108$

$X = 54$

8. $\dfrac{98}{100} = \dfrac{49}{Q}$

$98Q = 100(49)$

$98Q = 4,900$

$Q = 50$

9. $15^2 = 225$

10. $12^2 = 144$

11. $9^2 = 81$

12. $2(8 \times 13.4) + 2(13.4 \times 10) + 2(10 \times 8) =$

$2(107.2) + 2(134) + 2(80) =$

$214.4 + 268 + 160 = 642.4$ in^2

13. $4\left(0.4 \times 0.8 \times \dfrac{1}{2}\right) + (0.4 \times 0.4) =$

$4(0.16) + (0.16) =$

$0.64 + 0.16 = 0.8$ ft^2

14. $6.00 \times 1.2 = 7.2$

15. $1.50 \times 22.4 = 33.6$

16. $2.25 \times 80 = 180$

17. $(350 - 32)\dfrac{5}{9} = C$ or: $(350 - 32)(0.56) = C$

$(318)\dfrac{5}{9} = C$ $(318)(0.56) = C$

$C \approx 178.1°$

$\dfrac{1,590}{9} = C$

$C \approx 176.7°$

18. $\dfrac{1}{6} + \dfrac{1}{5} + \dfrac{1}{4} = \dfrac{5}{30} + \dfrac{6}{30} + \dfrac{1}{4} = \dfrac{11}{30} + \dfrac{1}{4} =$

$\dfrac{44}{120} + \dfrac{30}{120} = \dfrac{74}{120} = \dfrac{37}{60}$ of the job

19. $16^2 + L^2 = 20^2$

$256 + L^2 = 400$

$L^2 = 400 - 256$

$L^2 = 144$

$L = 12 \text{ ft}$

20. $2.00 \times 5 = 10 \text{ in}$

Lesson Practice 24A

1. done

2. $V \approx 3.14(9)^2 \times 25$

$V \approx 3.14 \times 81 \times 25 \approx 6{,}358.5 \text{ ft}^3$

3. $V \approx 3.14(14)^2(12)$

$V \approx 3.14(196)(12) \approx 7{,}385.28 \text{ in}^3$

4. $V \approx 3.14(8)^2(16)$

$V \approx 3.14(64)(16) \approx 3{,}215.36 \text{ ft}^3$

5. $V \approx 3.14(5)^2(6)$ because 18 ft = 6 yd

$V \approx 3.14(25)(6) \approx 471 \text{ yd}^3$

6. $V \approx 3.14(4)^2(12)$

$V \approx 3.14(16)(12) \approx 602.88 \text{ ft}^3$

7. $V \approx 3.14(1)^2(3)$

$V \approx 3.14(1)(3) \approx 9.42 \text{ in}^3$

8. $V \approx 3.14(40)^2(2)$

$V \approx 3.14(1{,}600)(2) \approx 10{,}048 \text{ mi}^3$

Lesson Practice 24B

1. $V \approx 3.14(8)^2(10)$

$V \approx 3.14(64)(10) \approx 2{,}009.6 \text{ in}^3$

2. $V \approx 3.14(5)^2(13)$

$V \approx 3.14(25)(13) \approx 1{,}020.5 \text{ ft}^3$

3. $V \approx 3.14(5)^2(6)$

$V \approx 3.14(25)(6) \approx 471 \text{ in}^3$

4. $V \approx 3.14(4)^2(11)$

$V \approx 3.14(16)(11) \approx 552.64 \text{ ft}^3$

5. $V \approx 3.14(7)^2(3)$

$V \approx 3.14(49)(3) \approx 461.58 \text{ yd}^3$

6. $V \approx 3.14(3)^2(20)$

$V \approx 3.14(9)(20) \approx 565.2 \text{ ft}^3$

7. $V \approx 3.14(1.5)^2(6.5)$

$V \approx 3.14(2.25)(6.5) \approx 45.9 \text{ in}^3$

8. $V \approx 3.14(10)^2(40)$

$V \approx 3.14(100)(40) \approx 12{,}560 \text{ ft}^3$

Lesson Practice 24C

1. $V \approx 3.14(3)^2(8)$

$V \approx 3.14(9)(8) \approx 226.08 \text{ in}^3$

2. $V \approx 3.14(10)^2(20)$

$V \approx 3.14(100)(20) \approx 6{,}280 \text{ ft}^3$

3. $V \approx 3.14(15)^2(8)$

$V \approx 3.14(225)(8) \approx 5{,}652 \text{ in}^3$

4. $V \approx 3.14(4)^2(7.5)$

$V \approx 3.14(16)(7.5) \approx 376.8 \text{ ft}^3$

5. $V \approx 3.14(1)^2(2.2)$

$V \approx 3.14(1)(2.2) \approx 6.9 \text{ yd}^3$

6. $V \approx 3.14(2.5)^2(13)$

$V \approx 3.14(6.25)(13) \approx 255.1 \text{ ft}^3$

7. $V \approx 3.14(2)^2(7.5)$

$V \approx 3.14(4)(7.5) \approx 94.2 \text{ in}^3$

8. $V \approx 3.14(3.5)^2(9)$

$V \approx 3.14(12.25)(9) \approx 346.2$ ft^3

Systematic Review 24D

1. $V \approx 3.14(5)^2(9)$

$V \approx 3.14(25)(9) \approx 706.5$ in^3

2. $V \approx 3.14(4)^2(8)$

$V \approx 3.14(16)(8) \approx 401.9$ ft^3

3.
$$\begin{array}{r} X^2 + 5X + 3 \\ + 2X^2 + 7X + 9 \\ \hline 3X^2 + 12X + 12 \end{array}$$

4.
$$\begin{array}{r} 4X^2 + 6X + 2 \\ + 7X^2 + 3X + 8 \\ \hline 11X^2 + 9X + 10 \end{array}$$

5.
$$\begin{array}{r} 5X^2 + 9X + 8 \\ + -4X^2 - 6X - 1 \\ \hline X^2 + 3X + 7 \end{array}$$

6. $2 = 1 \times 2; 7 = 1 \times 7$

$LCM = 2 \times 7 = 14$

7. $15 = 3 \times \underline{5}; 20 = 2 \times 2 \times \underline{5}$

$GCF = 5$

8. $\frac{3}{4} = 3 \div 4 = 0.75 = 75\%$

9. $\frac{1}{2} = 1 \div 2 = 0.50 = 50\%$

10. $\frac{4}{5} = 4 \div 5 = 0.80 = 80\%$

11. plane

12. point

13. line segment

14. line

15. 32°F; 0°C

16. $V \approx 3.14(5)^2(25)$

$V \approx 3.14(25)(25) \approx 1,962.5$ ft^3

17. $200\% = \frac{200}{100} = 2$

$2 \times 45 = 90$ people needed

18. $\frac{25}{G} = \frac{5}{7}$

$5G = 25(7)$

$5G = 175$

$G = 35$ green

Systematic Review 24E

1. $V \approx 3.14(10)^2(14.6)$

$V \approx 3.14(100)(14.6) \approx 4,584.4$ in^3

2. $V \approx 3.14(8)^2(22)$

$V \approx 3.14(64)(22) \approx 4,421.1$ ft^3

3.
$$\begin{array}{r} 4X^2 + X - 3 \\ + -6X^2 - 3X + 2 \\ \hline -2X^2 - 2X - 1 \end{array}$$

4.
$$\begin{array}{r} -2X^2 + 8X + 1 \\ + 5X^2 - 3X - 9 \\ \hline 3X^2 + 5X - 8 \end{array}$$

5.
$$\begin{array}{r} 2X^2 + 4X + 3 \\ + 3X^2 + 6X + 8 \\ \hline 5X^2 + 10X + 11 \end{array}$$

6. $3 = 1 \times 3; 15 = 3 \times 5$

$LCM = 3 \times 5 = 15$

7. $12 = \underline{2} \times \underline{2} \times 3; 16 = \underline{2} \times \underline{2} \times 2 \times 2;$

$GCF = 2 \times 2 = 4$

8. $20\% = 0.20 = \frac{20}{100} = \frac{1}{5}$

9. $10\% = 0.10 = \frac{10}{100} = \frac{1}{10}$

10. $35\% = 0.35 = \frac{35}{100} = \frac{7}{20}$

11. b

12. c

13. d

14. a

15. Units must be the same, so convert 3 yards to 9 feet.

$V \approx 3.14(9)^2(5)$

$V \approx 3.14(81)(5) \approx 1,272$ ft^3

16. $1,272 \times 7 = 8,904$ gal

17. $(82-32)\dfrac{5}{9}=C$ or $(82-32)(0.56)=C$

$(50)\dfrac{5}{9}=C$ $(50)(0.56)=C$

$\dfrac{250}{9}=C$ $C=28^\circ$

$C\approx 27.8^\circ$

18. $\dfrac{1}{2}=50\%$; $60\%>50\%$, so 60% is better.

19. $0.60\times\$25=\$15$;
$\$25-\$15=\$10$

20. $\dfrac{1}{40}=\dfrac{6}{X}$

$1X=40(6)$

$X=240$ mi

11. line

12. line segment

13. point

14. plane

15. $\dfrac{9}{10}=\dfrac{90}{100}=0.90=90\%$

16. $\dfrac{19}{20}=\dfrac{95}{100}=0.95=95\%$

17. no; 37°C = normal body temperature

18. yes; 0°C = freezing point of water

19. $2(8\times16)+2(8\times22)=$
$2(128)+2(176)=$
$256+352=608$ ft^2

20. $600\div400=1.5$
1.5 rounded up = 2 gal

Systematic Review 24F

1. $V\approx 3.14(6)^2(15)$
$V\approx 3.14(36)(15)\approx 1{,}695.6$ in^3

2. $V\approx 3.14(7)^2(25)$
$V\approx 3.14(49)(25)\approx 3{,}846.5$ ft^3

3. $6X^2+\ 5X+10$
$+\ \ \ 5X^2+\ 9X+\ 2$
$\overline{\ \ 11X^2+14X+12}$

4. $7X^2+7X+5$
$+\ \ -5X^2-4X-3$
$\overline{\ \ \ 2X^2+\ 3X+2}$

5. $4X^2+\ 8X+9$
$+\ \ -3X^2+\ 2X-6$
$\overline{\ \ \ \ X^2+10X+3}$

6. $4=2\times2$; $9=3\times3$
LCM $=2\times2\times3\times3=36$

7. $6=2\times\underline{3}$; $21=\underline{3}\times7$
GCF $=3$

8. $300\%=\dfrac{300}{100}=3$

9. $250\%=\dfrac{250}{100}=2\dfrac{1}{2}$

10. $175\%=\dfrac{175}{100}=1\dfrac{3}{4}$

Lesson Practice 25A

1. done

2. done

3. $(2X+1)(X+2)=2X^2+5X+2$

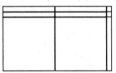

4. $(X+2)(X+4)=X^2+6X+8$

5. $(X+2)(X+6)=X^2+8X+12$

6. $(X)(X+4)=X^2+4X$

7. done

8.
$$X + 7$$
$$\times \quad X + 1$$
$$X + 7$$
$$X^2 + 7X$$
$$X^2 + 8X + 7$$

9.
$$X + \ 6$$
$$\times \quad X + \ 3$$
$$3X + 18$$
$$X^2 + 6X$$
$$X^2 + 9X + 18$$

10. done

11. done

12. $(X + 7)(X + 3) = X^2 + 10X + 21$

$$X^2 + 10X + 21 \Rightarrow (10)^2 + 10(10) + 21 =$$
$$100 + 100 + 21 =$$
$$221 \text{ units}^2$$

Lesson Practice 25B

1. $(X + 6)(X + 1) = X^2 + 7X + 6$

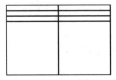

2. $(2X)(X + 3) = 2X^2 + 6X$

3. $(X + 7)(X + 2) = X^2 + 9X + 14$

4. $(2X + 2)(X + 1) = 2X^2 + 4X + 2$

5. $(X + 1)(X + 4) = X^2 + 5X + 4$

6. $(X)(X + 5) = X^2 + 5X$

7.
$$X + \ 8$$
$$\times \quad X + \ 2$$
$$2X + 16$$
$$X^2 + \ 8X$$
$$X^2 + 10X + 16$$

8.
$$X + \ 9$$
$$\times \quad X + \ 6$$
$$6X + 54$$
$$X^2 + 9X$$
$$X^2 + 15X + 54$$

9.
$$2X + \ 9$$
$$\times \quad X + \ 2$$
$$4X + 18$$
$$2X^2 + \ 9X$$
$$2X^2 + 13X + 18$$

10.
$$X + \ 7$$
$$\times \quad X + \ 6$$
$$6X + 42$$
$$X^2 + \ 7X$$
$$X^2 + 13X + 42$$

11.
$$X + \ 5$$
$$\times \quad X + \ 4$$
$$4X + 20$$
$$X^2 + 5X$$
$$X^2 + 9X + 20$$

12.
$$2X + 3$$
$$\times \quad X + 1$$
$$2X + 3$$
$$2X^2 + 3X$$
$$2X^2 + 5X + 3$$

13. $(X+3)(X+4) = X^2 + 7X + 12$

14. $X^2 + 7X + 12 \Rightarrow (4)^2 + 7(4) + 12 =$
$16 + 28 + 12 = 56$

$X^2 + 7X + 12 \Rightarrow (8)^2 + 7(8) + 12 =$
$64 + 56 + 12 = 132$

15. $(R)(R+5) = R^2 + 5R$ rabbits

Lesson Practice 25C

1. $(X)(X+1) = X^2 + X$

2. $(X+1)(X+5) = X^2 + 6X + 5$

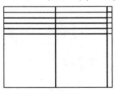

3. $(X+4)(X+5) = X^2 + 9X + 20$

4. $(2X+1)(X+5) = 2X^2 + 11X + 5$

5. $(X+2)(X+8) = X^2 + 10X + 16$

6. $(X+1)(X+1) = X^2 + 2X + 1$

7.
$$
\begin{array}{r}
2X+1 \\
\times \quad X+4 \\
\hline
8X+4 \\
2X^2 + \ X \\
\hline
2X^2 + 9X + 4
\end{array}
$$

8.
$$
\begin{array}{r}
X+ 8 \\
\times \quad X+ 3 \\
\hline
3X+24 \\
X^2 + \ 8X \\
\hline
X^2 + 11X + 24
\end{array}
$$

9.
$$
\begin{array}{r}
X+ 9 \\
\times \quad X+ 7 \\
\hline
7X+63 \\
X^2 + \ 9X \\
\hline
X^2 + 16X + 63
\end{array}
$$

10.
$$
\begin{array}{r}
2X+ 5 \\
\times \quad X+ 3 \\
\hline
6X+15 \\
2X^2 + \ 5X \\
\hline
2X^2 + 11X + 15
\end{array}
$$

11.
$$
\begin{array}{r}
X+ 6 \\
\times \quad X+ 4 \\
\hline
4X+24 \\
X^2 + \ 6X \\
\hline
X^2 + 10X + 24
\end{array}
$$

12.
$$
\begin{array}{r}
X+ 11 \\
\times \quad X+ 3 \\
\hline
3X+33 \\
X^2 + 11X \\
\hline
X^2 + 14X + 33
\end{array}
$$

13. $(X+8)(X+5) = X^2 + 13X + 40$

14. $X^2 + 13X + 40 \Rightarrow (3)^2 + 13(3) + 40 =$
$9 + 39 + 40 = 88$

$X^2 + 13X + 40 \Rightarrow (6)^2 + 13(6) + 40 =$
$36 + 78 + 40 = 154$

15. $A = (F+10)(F+5)$
$A = F^2 + 15F + 50$ ft^2

Systematic Review 25D

1. $(X+7)(X+1) = X^2 + 8X + 7$

2. $(X+2)(X+8) = X^2 + 10X + 16$

3. $(X+3)(X+6) = X^2 + 9X + 18$

4.
$$
\begin{array}{r}
X + 5 \\
\times \quad X + 3 \\
\hline
3X + 15 \\
X^2 + 5X \quad\quad \\
\hline
X^2 + 8X + 15
\end{array}
$$

5.
$$
\begin{array}{r}
2X + 5 \\
\times \quad X + 4 \\
\hline
8X + 20 \\
2X^2 + 5X \quad\quad \\
\hline
2X^2 + 13X + 20
\end{array}
$$

6.
$$
\begin{array}{r}
X + 9 \\
\times \quad X + 2 \\
\hline
2X + 18 \\
X^2 + 9X \quad\quad \\
\hline
X^2 + 11X + 18
\end{array}
$$

7. $V \approx 3.14(4)^2(13.7)$
$V \approx 3.14(16)(13.7) \approx 688.3$ in^3

8. $V \approx 3.14(5.5)^2(14)$
$V \approx 3.14(30.25)(14) \approx 1{,}329.8$ ft^3

9.
$$
\begin{array}{r}
4X^2 + 8X + 9 \\
+ \quad -3X^2 + 2X - 6 \\
\hline
X^2 + 10X + 3
\end{array}
$$

10.
$$
\begin{array}{r}
5X^2 + 5X + 4 \\
+ \quad 7X^2 - 1X - 2 \\
\hline
12X^2 + 4X + 2
\end{array}
$$

11.
$$
\begin{array}{r}
8X^2 + 3X - 2 \\
+ \quad -4X^2 + 7X + 11 \\
\hline
4X^2 + 10X + 9
\end{array}
$$

12. $3.4 \times 6.5 = 22.1$

13. $12.8 \div 3.2 = 4$

14. $10 - 3.9 = 6.1$

15. line segment

16. $3.50 \times 150 = 525$ lb needed in all
$525 - 150 = 375$ lb needed to buy

17. $25 = \underline{5} \times 5$; $45 = 3 \times 3 \times \underline{5}$
GCF = 5

18. $3 - \dfrac{19}{8} = \dfrac{24}{8} - \dfrac{19}{8} = \dfrac{5}{8}$ of a pizza

Systematic Review 25E

1. $(X+1)(X+5) = X^2 + 6X + 5$

2. $(X+2)(X+1) = X^2 + 3X + 2$

3. $(X+3)(X+4) = X^2 + 7X + 12$

4.
$$
\begin{array}{r}
2X + 2 \\
\times \quad X + 5 \\
\hline
10X + 10 \\
2X^2 + 2X \quad\quad \\
\hline
2X^2 + 12X + 10
\end{array}
$$

5.

$$\begin{array}{r} X+7 \\ \times\quad X+1 \\ \hline X+7 \\ X^2+7X \\ \hline X^2+8X+7 \end{array}$$

6.

$$\begin{array}{r} 2X+8 \\ \times\quad X+2 \\ \hline 4X+16 \\ 2X^2+8X \\ \hline 2X^2+12X+16 \end{array}$$

7. $V \approx 3.14(5)^2(10.6)$
$V \approx 3.14(25)(10.6) = 832.1 \text{ in}^3$

8. $V \approx 3.14(6.5)^2(17.3)$
$V \approx 3.14(42.25)(17.3) \approx 2,295.1 \text{ ft}^3$

9.

$$\begin{array}{r} 9X^2+5X+4 \\ +\quad -8X^2+3X-7 \\ \hline X^2+8X-3 \end{array}$$

10.

$$\begin{array}{r} 9X^2-2X+9 \\ +\quad -2X^2-6X-4 \\ \hline 7X^2-8X+5 \end{array}$$

11.

$$\begin{array}{r} X^2-X+6 \\ +\quad 5X^2-8X-3 \\ \hline 6X^2-9X+3 \end{array}$$

12. $7(3+4) =$
$7(3)+7(4) =$
$21+28 = 49$

13. $8(X+5) =$
$8X+8(5) = 8X+40$

14. $Y(Y+9) = Y^2+9Y$

15. $\dfrac{S}{12} = \dfrac{14}{21}$
$21S = 12(14)$
$21S = 168$
$S = 8 \text{ units}$

16. $\dfrac{D}{4} = \dfrac{10}{2}$
$2D = 4(10)$
$2D = 40$
$D = 20 \text{ units}$

17. angle

18. $9 = 3 \times 3; \ 5 = 1 \times 5$
$LCM = 3 \times 3 \times 5 = 45$

19. $\dfrac{1}{4}$ of $32 = $8 off
$32.00 - $8.00 = $24.00 new price
$24.00 \times 0.04 = $0.96 tax
$24.00 + $0.96 = $24.96 total cost

20. $25.00 - $24.96 = $0.04 in change

Systematic Review 25F

1. $(X+4)(X+6) = X^2+10X+24$

2. $(X+9)(X+1) = X^2+10X+9$

3. $(2X+3)(X+1) = 2X^2+5X+3$

4.

$$\begin{array}{r} 3X+4 \\ \times\quad X+3 \\ \hline 9X+12 \\ 3X^2+4X \\ \hline 3X^2+13X+12 \end{array}$$

5.

$$\begin{array}{r} X+3 \\ \times\quad X+2 \\ \hline 2X+6 \\ X^2+3X \\ \hline X^2+5X+6 \end{array}$$

6.

$$\begin{array}{r} 2X+5 \\ \times\quad X+1 \\ \hline 2X+5 \\ 2X^2+5X \\ \hline 2X^2+7X+5 \end{array}$$

7. $V \approx 3.14(3)^2(8)$

$V \approx 3.14(9)(8) \approx 226.1 \text{ in}^3$

8. $V \approx 3.14(5)^2(5.3)$

$V \approx 3.14(25)(5.3) \approx 416.1 \text{ ft}^3$

9.
$$6X^2 + 5X + 2$$
$$+ \quad 7X^2 + 3X - 1$$
$$13X^2 + 8X + 1$$

10.
$$4X^2 - X - 9$$
$$+ \quad 8X^2 - 8X - 2$$
$$12X^2 - 9X - 11$$

11.
$$-3X^2 + 6X - 5$$
$$+ \quad 8X^2 - 4X + 7$$
$$5X^2 + 2X + 2$$

12. $5^2 + 2 \cdot 3 - 8 + |-5| =$
$25 + 2 \cdot 3 - 8 + 5 =$
$25 + 6 - 8 + 5 = 28$

13. $|16 \div 4 + (3)(4) - 19| =$
$|4 + 12 - 19| =$
$|-3| = 3$

14. $(2^2 - 6) + 11 - 9 =$
$(4 - 6) + 11 - 9 =$
$-2 + 11 - 9 = 0$

15. $\frac{W}{21} = \frac{15}{25}$
$25W = 21(15)$
$25W = 315$
$W = 12.6 \text{ units}$

16. $\frac{X}{1} = \frac{4.5}{1.5}$
$X = \frac{4.5}{1.5}$
$X = 3 \text{ units}$

17. $24 = 2 \times 2 \times 2 \times 3; \ 32 = 2 \times 2 \times 2 \times 2 \times 2$
$GCF = 2 \times 2 \times 2 = 8$

18. $0.40 \times \$12.00 = \$4.80 \text{ discount}$
$\$12.00 - \$4.80 = \$7.20$

19. $\$12.00 - \$7.20 = \$4.80 \text{ profit}$

20. $\$2.98 \times 1.77 \approx \$5.27$

Lesson Practice 26A

1. done
2. done
3.
$$5:30$$
$$+ \quad 1:30$$
$$7:00$$
4.
$$1:15$$
$$+ \quad 2:40$$
$$3:55$$
5.
$$0:55$$
$$+ \quad 1:15$$
$$2:10$$
6.
$$4:43$$
$$+ \quad 3:22$$
$$8:05$$
7. done
8.
$$3:75$$
$$- \quad 3:40$$
$$:35$$
9.
$$6:95$$
$$- \quad 2:58$$
$$4:37$$
10. done
11. $11:24 + 12 = 11:36$
$- \quad 5:48 + 12 = 6:00$
$5:36$
12. $4:15 + 14 = 4:29$
$- \quad 2:46 + 14 = 3:00$
$1:29$
13. $8:35 \text{ a.m.} + 1:25 = 10:00 \text{ a.m.}$
14. $11:20 + 1:50 = 13 \text{ hr } 10 \text{ min}$
15. $4:30 \text{ p.m.} - 2:45 = 1:45 \text{ p.m.}$

Lesson Practice 26B

1.
$$5:18$$
$$+ \quad 6:36$$
$$11:54$$
2.
$$7:36$$
$$+ \quad 3:40$$
$$11:16$$

3.
$$
\begin{array}{r}
9:50 \\
+\ \ 6:30 \\
\hline
16:20
\end{array}
$$

4.
$$
\begin{array}{r}
3:38 \\
+\ \ 2:21 \\
\hline
5:59
\end{array}
$$

5.
$$
\begin{array}{r}
2:56 \\
+\ \ 1:18 \\
\hline
4:14
\end{array}
$$

6.
$$
\begin{array}{r}
3:45 \\
+\ \ 1:19 \\
\hline
5:04
\end{array}
$$

7.
$$
\begin{array}{r}
8:75 \\
-\ \ 6:30 \\
\hline
2:45
\end{array}
$$

8.
$$
\begin{array}{r}
7:85 \\
-\ \ 6:45 \\
\hline
1:40
\end{array}
$$

9.
$$
\begin{array}{r}
2:81 \\
-\ \ 1:50 \\
\hline
1:31
\end{array}
$$

10.
$$
\begin{array}{r}
8:19+37=8:56 \\
-\ \ 4:23+37=5:00 \\
\hline
3:56
\end{array}
$$

11.
$$
\begin{array}{r}
5:27+31=5:58 \\
-\ \ 2:29+31=3:00 \\
\hline
2:58
\end{array}
$$

12.
$$
\begin{array}{r}
5:37+5=5:42 \\
-\ \ 1:55+5=2:00 \\
\hline
3:42
\end{array}
$$

13. $6:20\ \text{a.m.}+5:15=11:35\ \text{a}$
14. $2:20+1:40=4\ \text{hr}\ 00\ \text{min}$
15. $6:15\ \text{p.m.}-4:13=2:02\ \text{p}$

3.
$$
\begin{array}{r}
3:14 \\
+\ \ 5:17 \\
\hline
8:31
\end{array}
$$

4.
$$
\begin{array}{r}
8:30 \\
+\ \ 6:40 \\
\hline
15:10
\end{array}
$$

5.
$$
\begin{array}{r}
2:43 \\
+\ \ 3:16 \\
\hline
5:59
\end{array}
$$

6.
$$
\begin{array}{r}
8:25 \\
+\ \ 1:46 \\
\hline
10:11
\end{array}
$$

7.
$$
\begin{array}{r}
8:90 \\
-\ \ 2:40 \\
\hline
6:50
\end{array}
$$

8.
$$
\begin{array}{r}
3:\overset{6\ \ 1}{75} \\
-\ \ 2:46 \\
\hline
1:29
\end{array}
$$

9.
$$
\begin{array}{r}
11:78 \\
-\ \ 3:55 \\
\hline
8:23
\end{array}
$$

10.
$$
\begin{array}{r}
8:36+13=8:49 \\
-\ \ 2:47+13=3:00 \\
\hline
5:49
\end{array}
$$

11.
$$
\begin{array}{r}
5:30+5=5:35 \\
-\ \ 1:55+5=2:00 \\
\hline
3:35
\end{array}
$$

12.
$$
\begin{array}{r}
4:25+17=4:42 \\
-\ \ 1:43+17=2:00 \\
\hline
2:42
\end{array}
$$

13. $7:45\ \text{a.m.}+2:55=10:40\ \text{a.}$
14. $5:10+6:42=11\ \text{hr}\ 52\ \text{min}$
15. $5:25\ \text{p.m.}-4:30\ \text{p.m.}=55$

Lesson Practice 26C

1.
$$
\begin{array}{r}
6:17 \\
+\ \ 5:55 \\
\hline
12:12
\end{array}
$$

2.
$$
\begin{array}{r}
3:18 \\
+\ \ 1:53 \\
\hline
5:11
\end{array}
$$

Systematic Review 26D

1.
$$
\begin{array}{r}
7:\overset{8\ \ 1}{90} \\
-\ \ 2:55 \\
\hline
5:35
\end{array}
$$

2.
$$
\begin{array}{r}
2:58 \\
+\ \ 5:12 \\
\hline
8:10
\end{array}
$$

3.

$$\begin{array}{r} 5:38 \\ +\ 2:25 \\ \hline 8:03 \end{array}$$

4.

$$\begin{array}{r} X+\ 3 \\ \times\quad X+\ 9 \\ \hline 9X+27 \\ X^2+\ 3X \\ \hline X^2+12X+27 \end{array}$$

5.

$$\begin{array}{r} 3X+\ 2 \\ \times\quad X+\ 5 \\ \hline 15X+10 \\ 3X^2+\ 2X \\ \hline 3X^2+17X+10 \end{array}$$

6.

$$\begin{array}{r} 3X+\ 6 \\ \times\quad X+\ 7 \\ \hline 21X+42 \\ 3X^2+\ 6X \\ \hline 3X^2+27X+42 \end{array}$$

7. $2(1.2\times3)+2(3\times0.5)+2(0.5\times1.2)=$
$\qquad 2(3.6)+2(1.5)+2(0.6)=$
$\qquad\qquad 7.2+3+1.2=11.4 \text{ ft}^2$

8. $4\left(4\times6\times\dfrac{1}{2}\right)+(4\times4)=$
$\qquad\qquad 4(12)+(16)=$
$\qquad\qquad\quad 48+16=64 \text{ ft}^2$

9. $2(4\times6.4)+2(6.4\times2)+2(2\times4)=$
$\qquad 2(25.6)+2(12.8)+2(8)=$
$\qquad\qquad 51.2+25.6+16=92.8 \text{ in}^2$

10. $2(3+X)-16=|2\cdot3|$
$\quad 2(3)+2X-16=|6|$
$\quad\ \ 6+2X-16=6$
$\qquad\ \ 2X-10=6$
$\qquad\qquad 2X=6+10$
$\qquad\qquad 2X=16$
$\qquad\qquad\ \ X=8$

11. $A(4+8)-5=2(3+A)-1^2$
$\quad A(12)-5=2(3)+2A-1$
$\quad\ \ 12A-5=6+2A-1$
$\quad\ \ 12A-5=2A+5$
$\qquad 12A=2A+5+5$
$\qquad 12A=2A+10$
$\quad 12A-2A=10$
$\qquad 10A=10$
$\qquad\ \ A=1$

12. $(-3)^2-10-X=-|6-12|$
$\quad 9-10-X=-|-6|$
$\qquad -1-X=-6$
$\qquad\quad -X=-6+1$
$\qquad\quad -X=-5$
$\qquad\quad\ \ X=5$

13. acute

14. obtuse

15. $8:30 \text{ a.m.}-5:20=3:10 \text{ a.m.}$
3 hours and 10 minutes

16. $V\approx3.14(12)^2(50)$
$V\approx3.14(144)(50)\approx22{,}608 \text{ ft}^3$

17. $V\approx3.14(6)^2(25)$
$V\approx3.14(36)(25)\approx2{,}826 \text{ ft}^3$, no

18. $V\approx3.14(12)^2(25)$
$V\approx3.14(144)(25)\approx11{,}304 \text{ ft}^3$
$22{,}608 \text{ ft}^3\div2=11{,}304 \text{ ft}^3$, yes
or $11{,}304\times2=22{,}608$; yes

Systematic Review 26E

1.

$$\begin{array}{r} 2:25 \\ +\ 4:45 \\ \hline 7:10 \end{array}$$

2.

$$\begin{array}{r} 3:79 \\ -\ 1:36 \\ \hline 2:43 \end{array}$$

3.

$$\begin{array}{r} 8:23 \\ +\ 2:46 \\ \hline 11:09 \end{array}$$

4.
$$
\begin{array}{r}
2X+1 \\
\times \quad X+6 \\
\hline
12X+6 \\
2X^2+\ \ X \\
\hline
2X^2+13X+6
\end{array}
$$

5.
$$
\begin{array}{r}
X+\ 12 \\
\times \quad X+\ \ 9 \\
\hline
9X+108 \\
X^2+12X \\
\hline
X^2+21X+108
\end{array}
$$

6.
$$
\begin{array}{r}
X+10 \\
\times \quad X+\ 8 \\
\hline
8X+80 \\
X^2+10X \\
\hline
X^2+18X+80
\end{array}
$$

7. $2(0.4\times1.3)+2(1.3\times0.2)+2(0.2\times0.4)=$
$2(0.52)+2(0.26)+2(0.08)=$
$1.04+0.52+0.16\approx1.7\,\text{yd}^2$

8. $4\left(3.4\times10\times\dfrac{1}{2}\right)+(3.4\times3.4)=$
$4(17)+(11.56)=$
$68+11.56\approx79.6\ \text{in}^2$

9. $2(10\times20)+2(20\times5)+2(5\times10)=$
$2(200)+2(100)+2(50)=$
$400+200+100=700\ \text{ft}^2$

10. $75\div100=0.75$

11. $5\div6\approx0.83$

12. $1\div4=0.25$

13. $4\div7\approx0.57$

14. $180°$

15. $90°$

16. $7:05\ \text{p.m.}-5:45=1:20\ \text{p.m.}$

17. $(425-32)\dfrac{5}{9}=C$ \quad or: \quad $(425-32)(0.56)=C$
$(393)\dfrac{5}{9}=C$ \qquad\qquad $(393)(0.56)=C$
$\dfrac{1{,}965}{9}=C$ \qquad\qquad\quad $C\approx220.1°$
$C\approx218.3°$

18. $325\%=3\dfrac{25}{100}=3.25$

19. $10\times2=20$
$2.00=200\%$

20. $\dfrac{3}{8}\times\dfrac{1}{2}=\dfrac{3}{16}$ of a pizza

Systematic Review 26F

1.
$$
\begin{array}{r}
5:27 \\
-\ \ 1:19 \\
\hline
4:08
\end{array}
$$

2.
$$
\begin{array}{r}
3:86 \\
-\ \ 3:50 \\
\hline
:36
\end{array}
$$

3.
$$
\begin{array}{r}
3:19 \\
+\ \ 1:55 \\
\hline
5:14
\end{array}
$$

4.
$$
\begin{array}{r}
X+\ 6 \\
\times \quad X+\ 5 \\
\hline
5X+30 \\
X^2+\ \ 6X \\
\hline
X^2+11X+\ 30
\end{array}
$$

5.
$$
\begin{array}{r}
X+\ 9 \\
\times \quad X+\ 2 \\
\hline
2X+18 \\
X^2+\ 9X \\
\hline
X^2+11X+18
\end{array}
$$

6.
$$
\begin{array}{r}
3X+1 \\
\times \quad X+5 \\
\hline
15X+5 \\
3X^2+\ \ X \\
\hline
3X^2+16X+5
\end{array}
$$

7. $2(6\times12)+2(12\times3)+2(3\times6)=$
$2(72)+2(36)+2(18)=$
$144+72+36=252\ \text{yd}^2$

8. $4\left(4.8\times6.5\times\dfrac{1}{2}\right)+(4.8\times4.8)=$
$4(15.6)+(23.04)=$
$62.4+23.04\approx85.4\ \text{in}^2$

9. $2(5 \times 8) + 2(8 \times 2.1) + 2(2.1 \times 5) =$
$\qquad 2(40) + 2(16.8) + 2(10.5) =$
$\qquad\qquad 80 + 33.6 + 21 = 134.6 \text{ ft}^2$

10. $0.5 = \dfrac{5}{10} = \dfrac{1}{2}$

11. $0.35 = \dfrac{35}{100} = \dfrac{7}{20}$

12. $0.875 = \dfrac{875}{1,000} = \dfrac{175}{200} = \dfrac{35}{40} = \dfrac{7}{8}$

13. $0.4 = \dfrac{4}{10} = \dfrac{2}{5}$

14. obtuse

15. acute

16. $3:30 \text{ p.m.} + 1:45 = 5:15 \text{ p.m.}$

17. $(230 - 32)\dfrac{5}{9} = C \quad \text{or:} \quad (230 - 32)(0.56) = C$
$\qquad (^{22}\cancel{198})\dfrac{5}{9} = C \qquad\qquad (198)(0.56) = C$
$\qquad\qquad\qquad\qquad\qquad\qquad\qquad C \approx 110.9^\circ$
$\qquad\qquad C = 110^\circ$

18. $25 - 15 = 10$ away games
$\qquad 0.40 \times 10 = 4$ away games lost

19. $V \approx 3.14(28)^2(48)$
$\qquad V \approx 3.14(784)(48) \approx 118,164.5 \text{ ft}^3$

20. $9\dfrac{5}{8} + 6\dfrac{4}{8} = 15\dfrac{9}{8} = 16\dfrac{1}{8}$ pies

Lesson Practice 27A

1. done

2. done

3. $V = \dfrac{1}{3}Bh$
$\qquad V = \dfrac{1}{3}(10 \times 10)(40)$
$\qquad V \approx 1,333.33 \text{ in}^3$

4. $V = \dfrac{1}{3}Bh$
$\qquad V \approx \dfrac{1}{3}(3.14)(2)^2(6)$
$\qquad V \approx 25.12 \text{ in}^3$

5. $V = \dfrac{1}{3}Bh$
$\qquad V = \dfrac{1}{3}(3.6 \times 3.6)(4)$
$\qquad V = 17.28 \text{ ft}^3$

6. $V = \dfrac{1}{3}Bh$
$\qquad V \approx \dfrac{1}{3}(3.14)(4.2)^2(9)$
$\qquad V \approx 166.17 \text{ ft}^3$

7. $V = \dfrac{1}{3}Bh$
$\qquad V = \dfrac{1}{3}(2.5 \times 2.5)(4.5)$
$\qquad V \approx 9.38 \text{ yd}^3$
or :
$\qquad \dfrac{1}{3} \times \dfrac{5}{2} \times \dfrac{5}{2} \times \dfrac{9}{2} = \dfrac{225}{24} = 9\dfrac{3}{8} \text{ yd}^3$

8. $V \approx \dfrac{1}{3}(3.14)(4.5)^2(8)$
$\qquad V \approx 169.56 \text{ in}^3$

Lesson Practice 27B

1. $V = \dfrac{1}{3}Bh$
$\qquad V = \dfrac{1}{3}(3.5 \times 3.5)(3.5)$
$\qquad V \approx 14.29 \text{ yd}^3$

2. $V = \dfrac{1}{3}Bh$
$\qquad V \approx \dfrac{1}{3}(3.14)(3.4)^2(7)$
$\qquad V \approx 84.70 \text{ in}^3$

3. $V = \dfrac{1}{3}Bh$
$\qquad V = \dfrac{1}{3}(5 \times 5)(6)$
$\qquad V = 50 \text{ in}^3$

4. $V = \dfrac{1}{3}Bh$
$\qquad V \approx \dfrac{1}{3}(3.14)(5)^2(8)$
$\qquad V \approx 209.33 \text{ in}^3$

5. $V = \dfrac{1}{3}Bh$
$\qquad V = \dfrac{1}{3}(2.4 \times 2.4)(5)$
$\qquad V = 9.6 \text{ ft}^3$

6. $V = \frac{1}{3}Bh$

$V \approx \frac{1}{3}(3.14)(11)^2(14.3)$

$V \approx 1{,}811.05 \text{ ft}^3$

7. $V = \frac{1}{3}(3 \times 3)(4)$

$V = 12 \text{ yd}^3$

8. $V \approx \frac{1}{3}(3.14)(2)^2(5)$

$V \approx 20.93 \text{ in}^3$

Lesson Practice 27C

1. $V = \frac{1}{3}Bh$

$V = \frac{1}{3}(60 \times 60)(50)$

$V = 60{,}000 \text{ ft}^3$

2. $V = \frac{1}{3}Bh$

$V \approx \frac{1}{3}(3.14)(62)^2(52)$

$V \approx 209{,}216.11 \text{ ft}^3$

3. $V = \frac{1}{3}Bh$

$V = \frac{1}{3}(1 \times 1)(9)$

$V = 3 \text{ in}^3$

4. $V = \frac{1}{3}Bh$

$V \approx \frac{1}{3}(3.14)(8)^2(12.4)$

$V \approx 830.63 \text{ in}^3$

5. $V = \frac{1}{3}Bh$

$V = \frac{1}{3}(4.8 \times 4.8)(10)$

$V = 76.8 \text{ ft}^3$

6. $V = \frac{1}{3}Bh$

$V \approx \frac{1}{3}(3.14)(14)^2(17)$

$V \approx 3{,}487.49 \text{ ft}^3$

7. $(6)(6)(6) = 216 \text{ in}^3$

$\frac{1}{3}(6 \times 6)(6) = 72 \text{ in}^3$

8. $\frac{1}{3}(3.14)(3)^2(6) \approx 56.52 \text{ in}^3$

Systematic Review 27D

1. $V \approx \frac{1}{3}(3 \times 3)(12) \approx 36 \text{ in}^3$

2. $V \approx \frac{1}{3}(3.14)(3)^2(12) \approx 113.04 \text{ in}^3$

3.
$$2:52 + 0:05 = 2:57$$
$$-\ \ 1:55 + 0:05 = 2:00$$
$$:57$$

4.
$$\begin{array}{r} 4\!:\!56 \\ +\ 2:49 \\ \hline 4:45 \end{array}$$

5.
$$\begin{array}{r} 3:52 \\ +\ 1:40 \\ \hline 5:32 \end{array}$$

6. $(X + 3)(X + 7) = X^2 + 10X + 21$

7. $(2X + 1)(X + 4) = 2X^2 + 9X + 4$

8. $(X + 6)(X + 1) = X^2 + 7X + 6$

9. $\frac{3.2}{8} = \frac{12.8}{P}$

$3.2P = 8(12.8)$

$3.2P = 102.4$

$P = 32$

10. $\frac{15}{75} = \frac{X}{12}$

$75X = 15(12)$

$75X = 180$

$X = 2.4$

11. $\frac{5}{8} = \frac{R}{32}$

$8R = 5(32)$

$8R = 160$

$R = 20$

12. $\text{mean} = \frac{1 + 2 + 3 + 6 + 7 + 8 + 8}{7} = \frac{35}{7} = 5$

$\text{median} = 6$

$\text{mode} = 8$

13. $\text{mean} = \dfrac{2+4+4}{3} = \dfrac{10}{3} = 3.33$

median = 4

mode = 4

14. $\text{mean} = \dfrac{9+9+11+12+13}{5} = \dfrac{54}{5} = 10.8$

median = 11

mode = 9

15. $\text{mean} = \dfrac{15+25+25}{3} = \dfrac{65}{3} = 21.67$

median = 25

mode = 25

16. $9^2 + 12^2 = H^2$

$81 + 144 = H^2$

$H = 15$ ft

17. 90°

18. infinity

Systematic Review 27E

1. $V \approx \dfrac{1}{3}(8 \times 8)(5.3) \approx 113.07 \text{ ft}^3$

2. $V \approx \dfrac{1}{3}(3.14)(4.5)^2(15) \approx 317.93 \text{ in}^3$

3. $\begin{aligned} 6:15 + 0:20 &= 6:35 \\ - \quad 2:40 + 0:20 &= 3:00 \\ &\quad \; 3:35 \end{aligned}$

4. $\begin{aligned} 5:10 + 0:27 &= 5:37 \\ - \quad 3:33 + 0:27 &= 4:00 \\ &\quad \; 1:37 \end{aligned}$

5. $\begin{aligned} 6&:^118 \\ + \quad 2&:29 \\ \hline 8&:47 \end{aligned}$

6. $(X+3)(X+4) = X^2 + 7X + 12$

7. $(X+4)(X+5) = X^2 + 9X + 20$

8. $(X+1)(X+8) = X^2 + 9X + 8$

9. $\dfrac{P}{7} = \dfrac{72}{84}$

$84P = 7(72)$

$84P = 504$

$P = 6$

10. $\dfrac{3}{10} = \dfrac{18}{A}$

$3A = 10(18)$

$3A = 180$

$A = 60$

11. $\dfrac{B}{4} = \dfrac{24}{32}$

$32B = 4(24)$

$32B = 96$

$B = 3$

12. $\dfrac{4}{5} \div \dfrac{1}{7} = \dfrac{4}{5} \times \dfrac{7}{1} = \dfrac{28}{5} = 5\dfrac{3}{5}$

13. $\dfrac{\cancel{4}}{\cancel{9}} \times \dfrac{\overset{2}{\cancel{18}}}{\underset{5}{\cancel{20}}} = \dfrac{2}{5}$

14. $\dfrac{2}{3} \div \dfrac{1}{9} = \dfrac{2}{\cancel{3}} \times \dfrac{\overset{3}{\cancel{9}}}{1} = \dfrac{6}{1} = 6$

15. $\text{mean} = \dfrac{5+5+6+6+6+7+7+10+11}{9} = 7$

median = 6

mode = 6

16. $\sqrt{144} = 12$

17. acute

18. $\dfrac{2}{3} + \dfrac{3}{4} + \dfrac{1}{2} = \dfrac{8}{12} + \dfrac{9}{12} + \dfrac{6}{12} = \dfrac{23}{12} = 1\dfrac{11}{12}$ cups

19. 1:16 p.m. + 7:51 = 9:07 p.m.

20. $\begin{aligned} X^2 + 3X + \;\; &9 \\ + \quad 5X^2 + 2X + \;\; &4 \\ \hline 6X^2 + 5X + 13 \; &\text{units}^2 \end{aligned}$

Systematic Review 27F

1. $V = \dfrac{1}{3}(5 \times 5)(12) = 100 \text{ ft}^3$

2. $V \approx \dfrac{1}{3}(3.14)(10)^2(30) \approx 3{,}140 \text{ in}^3$

3. $\begin{aligned} ^13&:17 \\ + \quad 5&:50 \\ \hline 9&:07 \end{aligned}$

4. $\begin{aligned} ^15&:32 \\ + \quad 1&:40 \\ \hline 7&:12 \end{aligned}$

5. $\begin{aligned} 8&:16 \\ + \quad 3&:30 \\ \hline 11&:46 \end{aligned}$

6. $(X+2)(X+2) = X^2+4X+4$

7. $(2X+6)(X+2) = 2X^2+10X+12$

8. $(X)(X+5) = X^2+5X$

9. $\dfrac{2}{9} = \dfrac{6}{X}$

$2X = 9(6)$

$2X = 54$

$X = 27$

10. $\dfrac{T}{8} = \dfrac{5}{40}$

$40T = 8(5)$

$40T = 40$

$T = 1$

11. $\dfrac{4}{7} = \dfrac{12}{Q}$

$4Q = 7(12)$

$4Q = 84$

$Q = 21$

12. b

13. c

14. a

15. $\text{mean} = \dfrac{15+20+25+30+35+35+40}{7} =$

$\dfrac{200}{7} \approx 28.57$

median $= 30$

mode $= 35$

16. 11:13 p.m. $-8:25 = 2:48$ p.m.

17. straight

18. no; he was saying that 51° was the average high temperature for the month.

19. a point

20. $\$3.60 \times 6 = \$21.60$

$\$21.60 \times 1.04 \approx \$22.46$

Lesson Practice 28A

1. done

2. done

3. $9:25 + 12:00 = 2125$

4. 0810

5. done

6. done

7. $1921 - 12:00 = 7:21$ p.m.

8. 9:48 a.m.

9. done

10.
```
  ¹1215
+ 0842
  2057
```

11.
```
  ⁰³28
+ 1950
  2318
```

12. done

13.
```
  16⁸90
- 1345
  0345
```

14.
```
  2102        ¹ ¹2062
- 0320      - 0320
              1742
```

15. done

16.
```
  2314+35 = 2349
- 2125+35 = 2200
            0149
```

17.
```
  1630+15 = 1645
- 1245+15 = 1300
            0345
```

18. $1100 + 4:00 = 1500$

Lesson Practice 28B

1. $10:10 + 12:00 = 2210$

2. $7:15 + 12:00 = 1915$

3. 0100

4. 0620

5. $2310 - 12:00 = 11:10$ p.m

6. $1314 - 12:00 = 1:14$ p.m.

7. 6:50 a.m.

8. 1:05 a.m.

9.
```
  0⁵54
+ 0325
  0919
```

10.
```
  1315
+ 1040
  2355
```

11.
$$\begin{array}{r} 1\overset{1}{1}48 \\ + \ 0152 \\ \hline 1340 \end{array}$$

12.
$$\begin{array}{r} 0935 \\ - \ 0251 \\ \hline \end{array} \quad \begin{array}{r} 0895 \\ - \ 0251 \\ \hline 0644 \end{array}$$

13.
$$\begin{array}{r} 1220 \\ - \ 1130 \\ \hline \end{array} \quad \begin{array}{r} 1180 \\ - \ 1130 \\ \hline 0050 \end{array}$$

14.
$$\begin{array}{r} 2315 \\ - \ 2045 \\ \hline \end{array} \quad \begin{array}{r} 2275 \\ - \ 2045 \\ \hline 0230 \end{array}$$

15.
$$\begin{array}{r} 1821 + 25 = 1846 \\ - \ 1535 + 25 = 1600 \\ \hline 0246 \end{array}$$

16.
$$\begin{array}{r} 2206 + 40 = 2246 \\ - \ 0420 + 40 = 0500 \\ \hline 1746 \end{array}$$

17.
$$\begin{array}{r} 0938 + 10 = 0948 \\ - \ 0250 + 10 = 0300 \\ \hline 0648 \end{array}$$

18. $3:14 + 12:00 = 1514$

Lesson Practice 28C

1. $8:20 + 12:00 = 2020$
2. 0540
3. 0000
4. 1200
5. $1730 - 12:00 = 5:30$ p.m.
6. $2150 - 12:00 = 9:50$ p.m.
7. $5:35$ a.m.
8. $12:15$ a.m.

9.
$$\begin{array}{r} 10\overset{1}{2}8 \\ + \ 0035 \\ \hline 1103 \end{array}$$

10.
$$\begin{array}{r} 20\overset{1}{4}5 \\ + \ 0245 \\ \hline 2330 \end{array}$$

11.
$$\begin{array}{r} \overset{1}{1}6\overset{1}{1}8 \\ + \ 0718 \\ \hline 2336 \end{array}$$

12.
$$\begin{array}{r} 2400 \\ - \ 1128 \\ \hline \end{array} \quad \begin{array}{r} 23\overset{5}{\cancel{6}}\overset{1}{0} \\ - \ 1128 \\ \hline 1232 \end{array}$$

13.
$$\begin{array}{r} 1355 \\ - \ 0655 \\ \hline 0700 \end{array}$$

14.
$$\begin{array}{r} 0714 \\ - \ 0045 \\ \hline \end{array} \quad \begin{array}{r} 06\overset{6}{\cancel{7}}\overset{1}{4} \\ - \ 0045 \\ \hline 0629 \end{array}$$

15.
$$\begin{array}{r} 1700 + 18 = 1718 \\ - \ 0342 + 18 = 0400 \\ \hline 1318 \end{array}$$

16.
$$\begin{array}{r} 2319 + 01 = 2320 \\ - \ 1259 + 01 = 1300 \\ \hline 1020 \end{array}$$

17.
$$\begin{array}{r} 1012 + 22 = 1034 \\ - \ 0938 + 22 = 1000 \\ \hline 0034 \end{array}$$

18. $1015 + 1300 = 2315$

Systematic Review 28D

1. $2:15 + 12:00 = 1415$
2. 0820
3. $5:45$ a.m.
4. $1902 - 12:00 = 7:02$ p.m.

5.
$$\begin{array}{r} 0\overset{1}{1}53 \\ + \ 0050 \\ \hline 0243 \end{array}$$

6.
$$\begin{array}{r} 0642 \\ + \ 0310 \\ \hline 0952 \end{array}$$

7.
$$\begin{array}{r} 1836 + 20 = 1856 \\ - \ 1140 + 20 = 1200 \\ \hline 0656 \end{array}$$

8.
$$\begin{array}{r} 2215 + 22 = 2237 \\ - \ 1038 + 22 = 1100 \\ \hline 1137 \end{array}$$

9. $V \approx \dfrac{1}{3}(6.5)(7 \times 7) \approx 106.17 \text{ in}^3$

10. $V \approx \dfrac{1}{3}(3.14)(4)^2(5.4) \approx 90.43 \text{ in}^3$

11. $V \approx (3.14)(3)^2(10) \approx 282.6 \text{ ft}^3$

12. $(2X+3)(X+3) = 2X^2 + 7X + 6$

13. $(X+2)(X+4) = X^2 + 6X + 8$

14. $(2X)(X+3) = 2X^2 + 6X$

15. $\dfrac{5}{100} = \dfrac{1}{20}$

16. $\dfrac{50+45}{100} = \dfrac{95}{100} = \dfrac{19}{20}$

17. $8 = 2 \times 2 \times 2; \ 3 = 1 \times 3$
    LCM $= 2 \times 2 \times 2 \times 3 = 24$

18. $1.50 \times 6 = 9$ ft tall now
    $9 - 6 = 3$ ft of growth

17. mean $= \dfrac{5+6+7+7+10}{5} = \dfrac{35}{5} = 7$
    median $= 7$
    mode $= 7$

18. $\dfrac{2}{350} = \dfrac{1}{175}$

19. $\dfrac{5}{350} = \dfrac{1}{70}$

20. $150 \times \dfrac{9}{5} + 32 = F$
    $270 + 32 = F$
    $F = 302°$

Systematic Review 28E

1. 1205

2. 0135

3. $2305 - 12 = 11:05$ p.m.

4. 3:16 a.m.

5. $\begin{array}{r} 0\overset{1}{1}53 \\ +\ 0050 \\ \hline 0243 \end{array}$

6. $\begin{array}{r} 0642 \\ +\ 0310 \\ \hline 0952 \end{array}$

7. $\begin{array}{r} 2419+15 = 2434 \\ -\ 0645+15 = 0700 \\ \hline 1734 \end{array}$

8. $\begin{array}{r} 1052+05 = 1057 \\ -\ 0855+05 = 0900 \\ \hline 0157 \end{array}$

9. $V = \dfrac{1}{3}(1.5)(1.5)(3) = 2.25 \text{ in}^3$

10. $V \approx \dfrac{1}{3}(3.14)(10)^2(15) \approx 1{,}570 \text{ in}^3$

11. $V \approx 3.14(4)^2(6.5) \approx 326.56 \text{ ft}^3$

12. $(2X+4)(X+4) = 2X^2 + 12X + 16$

13. $(X+6)(X+1) = X^2 + 7X + 6$

14. $(2X+1)(2X+5) = 4X^2 + 12X + 5$

15. $1345 - 12:00 = 1:45$ p.m.

16. $1345 + 0115 = 1500$
    $1500 - 12:00 = 3:00$ p.m.

Systematic Review 28F

1. $6:18 + 12:00 = 1818$

2. 1020

3. 12:16 a.m.

4. $1530 - 12:00 = 3:30$ p.m.

5. $\begin{array}{r} \overset{1}{1}018 \\ +\ 0350 \\ \hline 1408 \end{array}$

6. $\begin{array}{r} 1\overset{1}{0}45 \\ +\ 1045 \\ \hline 2130 \end{array}$

7. $\begin{array}{r} 1825+22 = 1847 \\ -\ 0238+22 = 0300 \\ \hline 1547 \end{array}$

8. $\begin{array}{r} 0921+23 = 0944 \\ -\ 0637+23 = 0700 \\ \hline 0244 \end{array}$

9. $V = \dfrac{1}{3}(8 \times 8)(9.3) = 198.4 \text{ in}^3$

10. $V \approx \dfrac{1}{3}(3.14)(0.1)^2(3.6) \approx 0.04 \text{ yd}^3$

11. $V \approx 3.14(2)^2(8.3) \approx 104.25 \text{ ft}^3$

12. $(X+2)(X+5) = X^2 + 7X + 10$

13. $(3X+1)(X+3) = 3X^2 + 10X + 3$

14. $(X+4)(X+3) = X^2 + 7X + 12$

15. $\dfrac{10}{100} = \dfrac{1}{10}$

16. $24 = \underline{2} \times \underline{2} \times 2 \times \underline{3}; \ 36 = \underline{2} \times \underline{2} \times 3 \times \underline{3}$
    GCF $= 2 \times 2 \times 3 = 12$

17. 80

18. 0000 is midnight, so it was probably dark.

19. obtuse

20. $\dfrac{165}{515} = \dfrac{33}{103} = 33 \div 103 \approx 0.320$

Lesson Practice 29A

1. done

2.
$$\begin{array}{r} 5 \text{ yd } 2 \text{ ft} \\ + \ 2 \text{ yd } 2 \text{ ft} \\ \hline 7 \text{ yd } 4 \text{ ft} \end{array} = 8 \text{ yd } 1 \text{ ft}$$

3.
$$\begin{array}{r} 4 \text{ lb } 10 \text{ oz} \\ + \ 1 \text{ lb } \ 8 \text{ oz} \\ \hline 5 \text{ lb } 18 \text{ oz} \end{array} = 6 \text{ lb } 2 \text{ oz}$$

4. done

5.
$$\begin{array}{rcl} 8 \text{ yd } 1 \text{ ft} & \Rightarrow & 7 \text{ yd } 4 \text{ ft} \\ - \ 3 \text{ yd } 2 \text{ ft} & & - \ 3 \text{ yd } 2 \text{ ft} \\ \hline & & 4 \text{ yd } 2 \text{ ft} \end{array}$$

6.
$$\begin{array}{rcl} 9 \text{ lb } 5 \text{ oz} & \Rightarrow & 8 \text{ lb } 21 \text{ oz} \\ - \ 4 \text{ lb } 7 \text{ oz} & & - \ 4 \text{ lb } \ 7 \text{ oz} \\ \hline & & 4 \text{ lb } 14 \text{ oz} \end{array}$$

7. done

8.
$$\begin{array}{r} 9 \text{ yd } 1 \text{ ft} + 1 \text{ ft} = 9 \text{ yd } 2 \text{ ft} \\ - \ 4 \text{ yd } 2 \text{ ft} + 1 \text{ ft} = 5 \text{ yd } 0 \text{ ft} \\ \hline 4 \text{ yd } 2 \text{ ft} \end{array}$$

9.
$$\begin{array}{r} 14 \text{ lb } 9 \text{ oz} + 4 \text{ oz} = 14 \text{ lb } 13 \text{ oz} \\ - \ 10 \text{ lb } 12 \text{ oz} + 4 \text{ oz} = 11 \text{ lb } \ 0 \text{ oz} \\ \hline 3 \text{ lb } 13 \text{ oz} \end{array}$$

10.
$$\begin{array}{r} 7'5" + 6" = 7'11" \\ - \ 2'6" + 6" = 3' \ 0" \\ \hline 4'11" \end{array}$$

11.
$$\begin{array}{r} 10 \text{ yd } 2 \text{ ft} \\ - \ 8 \text{ yd } 2 \text{ ft} \\ \hline 2 \text{ yd } 0 \text{ ft} \end{array}$$

12.
$$\begin{array}{r} 6 \text{ lb } 1 \text{ oz} + 14 \text{ oz} = 6 \text{ lb } 15 \text{ oz} \\ - \ 2 \text{ lb } 2 \text{ oz} + 14 \text{ oz} = 3 \text{ lb } \ 0 \text{ oz} \\ \hline 3 \text{ lb } 15 \text{ oz} \end{array}$$

13.
$$\begin{array}{r} 7' 6" + 1" = 7'7" \\ - \ 5'11" + 1" = 6'0" \\ \hline 1'7" \end{array}$$

14.
$$\begin{array}{r} 3' \ 8" \\ + \ 1' \ 4" \\ \hline 4'12" \end{array} = 5'0"$$

15.
$$\begin{array}{r} 7 \text{ lb } \ 5 \text{ oz} \\ + \ 9 \text{ lb } 12 \text{ oz} \\ \hline 16 \text{ lb } 17 \text{ oz} \end{array} = 17 \text{ lb } 1 \text{ oz}$$

Lesson Practice 29B

1.
$$\begin{array}{r} 5'11" \\ + \ 1' \ 4" \\ \hline 6'15" \end{array} = 7'3"$$

2.
$$\begin{array}{r} 10 \text{ yd } 1 \text{ ft} \\ + \ 6 \text{ yd } 1 \text{ ft} \\ \hline 16 \text{ yd } 2 \text{ ft} \end{array}$$

3.
$$\begin{array}{r} 7 \text{ lb } 12 \text{ oz} \\ + \ 8 \text{ lb } 5 \text{ oz} \\ \hline 15 \text{ lb } 17 \text{ oz} \end{array} = 16 \text{ lb } 1 \text{ oz}$$

4.
$$\begin{array}{rcl} 13'2" & \Rightarrow & 12'14" \\ - \ 8'5" & & - \ 8' \ 5" \\ \hline & & 4' \ 9" \end{array}$$

5.
$$\begin{array}{rcl} 7 \text{ yd } 0 \text{ ft} & \Rightarrow & 6 \text{ yd } 3 \text{ ft} \\ - \ 2 \text{ yd } 1 \text{ ft} & & - \ 2 \text{ yd } 1 \text{ ft} \\ \hline & & 4 \text{ yd } 2 \text{ ft} \end{array}$$

6.
$$\begin{array}{rcl} 18 \text{ lb } 3 \text{ oz} & \Rightarrow & 17 \text{ lb } 19 \text{ oz} \\ - \ 16 \text{ lb } 4 \text{ oz} & & - \ 16 \text{ lb } \ 4 \text{ oz} \\ \hline & & 1 \text{ lb } 15 \text{ oz} \end{array}$$

7.
$$\begin{array}{r} 20'3" + 4" = 20'7" \\ - \ 17'8" + 4" = 18'0" \\ \hline 2'7" \end{array}$$

8.
$$\begin{array}{r} 4 \text{ yd } 2 \text{ ft} \\ - \ 1 \text{ yd } 2 \text{ ft} \\ \hline 3 \text{ yd } 0 \text{ ft} \end{array}$$

9.
$$\begin{array}{r} 9 \text{ lb } 11 \text{ oz} + 1 \text{ oz} = 9 \text{ lb } 12 \text{ oz} \\ - \ 6 \text{ lb } 15 \text{ oz} + 1 \text{ oz} = 7 \text{ lb } \ 0 \text{ oz} \\ \hline 2 \text{ lb } 12 \text{ oz} \end{array}$$

10.
$$\begin{array}{r} 14'2" + 7" = 14'9" \\ - \ 3'5" + 7" = \ 4'0" \\ \hline 10'9" \end{array}$$

11.
$$
\begin{array}{r}
7 \text{ yd } 2 \text{ ft} \\
- \ 4 \text{ yd } 1 \text{ ft} \\
\hline
3 \text{ yd } 1 \text{ ft}
\end{array}
$$

12.
$$
\begin{array}{r}
33 \text{ lb } 8 \text{ oz} + 7 \text{ oz} = 33 \text{ lb } 15 \text{ oz} \\
- \ 21 \text{ lb } 9 \text{ oz} + 7 \text{ oz} = 22 \text{ lb } \ 0 \text{ oz} \\
\hline
11 \text{ lb } 15 \text{ oz}
\end{array}
$$

13.
$$
\begin{array}{r}
5 \text{ lb } \ 8 \text{ oz} \\
+ \ 9 \text{ lb } \ 8 \text{ oz} \\
\hline
14 \text{ lb } 16 \text{ oz}
\end{array}
= 15 \text{ lb}
$$

14.
$$
\begin{array}{r}
14 \text{ lb } 16 \text{ oz} \\
- \ 2 \text{ lb } \ 4 \text{ oz} \\
\hline
12 \text{ lb } 12 \text{ oz}
\end{array}
$$

15.
$$
\begin{array}{r}
15 \text{ yd } 0 \text{ ft} + 1 \text{ ft} = 15 \text{ yd } 1 \text{ ft} \\
- \ \ 0 \text{ yd } 2 \text{ ft} + 1 \text{ ft} = \ 1 \text{ yd } 0 \text{ ft} \\
\hline
14 \text{ yd } 1 \text{ ft}
\end{array}
$$

Lesson Practice 29C

1.
$$
\begin{array}{r}
8'6" \\
+ \ 6'6" \\
\hline
14'12"
\end{array}
= 15'
$$

2.
$$
\begin{array}{r}
14 \text{ yd } 3 \text{ ft} \\
+ \ 9 \text{ yd } 3 \text{ ft} \\
\hline
23 \text{ yd } 6 \text{ ft}
\end{array}
= 25 \text{ yd}
$$

3.
$$
\begin{array}{r}
5 \text{ lb } 2 \text{ oz} \\
+ \ 5 \text{ lb } 8 \text{ oz} \\
\hline
10 \text{ lb } 10 \text{ oz}
\end{array}
$$

4.
$$
\begin{array}{rr}
10' 8" \Rightarrow & 9'20" \\
- \ 4'11" & - \ 4'11" \\
\hline
& 5' \ 9"
\end{array}
$$

5.
$$
\begin{array}{rr}
23 \text{ yd } 0 \text{ ft} \Rightarrow & 22 \text{ yd } 3 \text{ ft} \\
- \ 16 \text{ yd } 2 \text{ ft} & - \ 16 \text{ yd } 2 \text{ ft} \\
\hline
& 6 \text{ yd } 1 \text{ ft}
\end{array}
$$

6.
$$
\begin{array}{rr}
7 \text{ lb } \ 3 \text{ oz} \Rightarrow & 6 \text{ lb } 19 \text{ oz} \\
- \ 5 \text{ lb } 15 \text{ oz} & - \ 5 \text{ lb } 15 \text{ oz} \\
\hline
& 1 \text{ lb } \ 4 \text{ oz}
\end{array}
$$

7.
$$
\begin{array}{r}
16'7" \\
- \ 8'3" \\
\hline
8' 4"
\end{array}
$$

8.
$$
\begin{array}{r}
7 \text{ yd } 1 \text{ ft} + 1 \text{ ft} = 7 \text{ yd } 2 \text{ ft} \\
- \ 2 \text{ yd } 2 \text{ ft} + 1 \text{ ft} = 3 \text{ yd } 0 \text{ ft} \\
\hline
4 \text{ yd } 2 \text{ ft}
\end{array}
$$

9.
$$
\begin{array}{r}
15 \text{ lb } 2 \text{ oz} + 13 \text{ oz} = 15 \text{ lb } 15 \text{ oz} \\
- \ 11 \text{ lb } 3 \text{ oz} + 13 \text{ oz} = 12 \text{ lb } \ 0 \text{ oz} \\
\hline
3 \text{ lb } 15 \text{ oz}
\end{array}
$$

10.
$$
\begin{array}{r}
31'0" + 9" = 31'9" \\
- \ 20'3" + 9" = 21'0" \\
\hline
10'9"
\end{array}
$$

11.
$$
\begin{array}{r}
42 \text{ yd } 1 \text{ ft} + 1 \text{ ft} = 42 \text{ yd } 2 \text{ ft} \\
- \ 17 \text{ yd } 2 \text{ ft} + 1 \text{ ft} = 18 \text{ yd } 0 \text{ ft} \\
\hline
24 \text{ yd } 2 \text{ ft}
\end{array}
$$

12.
$$
\begin{array}{r}
12 \text{ lb } 10 \text{ oz} \\
- \ 7 \text{ lb } \ 8 \text{ oz} \\
\hline
5 \text{ lb } \ 2 \text{ oz}
\end{array}
$$

13.
$$
\begin{array}{r}
5' \ 4" + 1" = 5'5" \\
- \ 4'11" + 1" = 5'0" \\
\hline
5"
\end{array}
$$

14.
$$
\begin{array}{rr}
70 \text{ lb } 0 \text{ oz} \Rightarrow & 69 \text{ lb } 16 \text{ oz} \\
- \ 36 \text{ lb } 5 \text{ oz} & - \ 36 \text{ lb } \ 5 \text{ oz} \\
\hline
& 33 \text{ lb } 11 \text{ oz}
\end{array}
$$

15.
$$
\begin{array}{rr}
7 \text{ yd } 0 \text{ in} \Rightarrow & 6 \text{ yd } 36 \text{ in} \\
- \ 5 \text{ yd } 6 \text{ in} & - \ 5 \text{ yd } \ 6 \text{ in} \\
\hline
& 1 \text{ yd } 30 \text{ in}
\end{array}
$$
$$
\text{or } 1 \text{ yd } 2 \text{ ft } 6 \text{ in}
$$

Systematic Review 29D

1.
$$
\begin{array}{r}
6'4" = 5'16" \\
- \ 2'8" = 2' \ 8" \\
\hline
3' \ 8"
\end{array}
$$

2.
$$
\begin{array}{r}
10 \text{ yd } 2 \text{ ft} \\
+ \ 5 \text{ yd } 1 \text{ ft} \\
\hline
15 \text{ yd } 3 \text{ ft}
\end{array}
= 16 \text{ yd}
$$

3.
$$
\begin{array}{r}
9 \text{ lb } 11 \text{ oz} \\
- \ 7 \text{ lb } \ 8 \text{ oz} \\
\hline
2 \text{ lb } \ 3 \text{ oz}
\end{array}
$$

4. 0745

5. $3:30 + 12:00 = 1530$

6. $2319 - 12:00 = 11:19 \text{ p.m.}$

7. 8:21 a.m.

8.
$$2X^2 + 5X - 1$$
$$+ \;\; -3X^2 + 6X + 10$$
$$-X^2 + 11X + 9$$

9.
$$-6X^2 + 4X - 5$$
$$+ \;\; 2X^2 - 4X + 4$$
$$-4X^2 \qquad -1$$

10.
$$-4X^2 + 5X - 1$$
$$+ \;\; 8X^2 - 2X + 6$$
$$4X^2 + 3X + 5$$

11. $0.34 \times 0.018 = 0.00612$

12. $0.75 \div 0.03 = 25$

13. $0.5 + 1.09 = 1.59$

14. dekagram

15. $\dfrac{1}{100}$

16. $A: \dfrac{50}{200} = \dfrac{1}{4}$

    $B: \dfrac{90}{120} = \dfrac{3}{4}$

    $\dfrac{3}{4} > \dfrac{1}{4}$; B is more dangerous.

17. 6 years

18.
$$1615 + 30 = 1645$$
$$- \;\; 1130 + 30 = 1200$$
$$4{:}45 \quad \text{or 4 hr 45 min}$$

Systematic Review 29E

1.
$$7'10"$$
$$+ \;\; 1'\;\; 7"$$
$$8'17" \;\; = 9'5"$$

2.
$$25 \text{ yd } 1 \text{ ft} + 1 \text{ ft} = 25 \text{ yd } 2 \text{ ft}$$
$$- \;\; 18 \text{ yd } 2 \text{ ft} + 1 \text{ ft} = 19 \text{ yd } 0 \text{ ft}$$
$$6 \text{ yd } 2 \text{ ft}$$

3.
$$2 \text{ lb } \;\; 8 \text{ oz}$$
$$+ \;\; 5 \text{ lb } \;\; 8 \text{ oz}$$
$$7 \text{ lb } 16 \text{ oz} \;\; = 8 \text{ lb}$$

4. 0112

5. $9{:}55 + 12{:}00 = 2155$

6. 10:30 a.m.

7. $1942 - 12{:}00 = 7{:}42$ p.m.

8.
$$-3X^2 + 7X - 2$$
$$+ \;\; 6X^2 - 6X + 9$$
$$3X^2 + \;\; X + 7$$

9.
$$-5X^2 + 9X - 3$$
$$+ \;\; 6X^2 + 4X - 3$$
$$X^2 + 13X - 6$$

10.
$$6X^2 + 8X - 20$$
$$+ \;\; 7X^2 - 5X + 13$$
$$13X^2 + 3X - \;\; 7$$

11. $\dfrac{3}{4} \div \dfrac{3}{16} = \dfrac{\cancel{3}}{4} \times \dfrac{\overset{4}{\cancel{16}}}{\cancel{3}} = \dfrac{4}{1} = 4$

12. $\dfrac{1}{2} \times \dfrac{5}{6} = \dfrac{5}{12}$

13. $\dfrac{7}{8} + \dfrac{1}{5} = \dfrac{35}{40} + \dfrac{8}{40} = \dfrac{43}{40} = 1\dfrac{3}{40}$

14. hectoliter

15. $\dfrac{1}{1,000}$

16. $V \approx 3.14(5)^2(6) = 471 \text{ in}^3$

    $V \approx 3.14(6)^2(5) = 565.2 \text{ in}^3$

    $565.2 > 471$; the second one holds more.

17.
$$11{:}04 + {:}54 = 11{:}58$$
$$- \;\; 5{:}06 + {:}54 = \;\; 6{:}00$$
$$5{:}58 \quad \text{AM}$$

18. a plane

19. $V \approx \dfrac{1}{3}(3.14)(10)^2(20) \approx 2,093.33 \text{ ft}^3$

20. $\dfrac{2}{36} = \dfrac{1}{18}$

Systematic Review 29F

1.
$$38'3" + 8" = 38'11"$$
$$- \;\; 21'4" + 8" = 22'\; 0"$$
$$16'11"$$

2.
$$5 \text{ yd } 3 \text{ ft}$$
$$- \;\; 1 \text{ yd } 2 \text{ ft}$$
$$4 \text{ yd } 1 \text{ ft}$$

3.

$$8 \text{ lb } 7 \text{ oz}$$
$$+ \quad 3 \text{ lb } 12 \text{ oz}$$
$$\overline{11 \text{ lb } 19 \text{ oz}} = 12 \text{ lb } 3 \text{ oz}$$

4. 0527

5. 0000

6. 12:30 a.m.

7. 1550 − 12:00 = 3:50 p.m.

8.

$$-9X^2 + 4X - 8$$
$$+ \quad 3X^2 - 11X + 2$$
$$\overline{-6X^2 - 7X - 6}$$

9.

$$12X^2 - X - 8$$
$$+ \quad 11X^2 + 7X - 9$$
$$\overline{23X^2 + 6X - 17}$$

10.

$$6X^2 + 5X - 18$$
$$+ \quad 9X^2 - 7X + 12$$
$$\overline{15X^2 - 2X - 6}$$

11. $2\frac{1}{2} \div 1\frac{3}{4} = \frac{5}{2} \div \frac{7}{4} = \frac{5}{\cancel{2}} \times \frac{\cancel{4}^2}{7} = \frac{10}{7} = 1\frac{3}{7}$

12. $4\frac{5}{8} \times 3\frac{1}{4} = \frac{37}{8} \times \frac{13}{4} = \frac{481}{32} = 15\frac{1}{32}$

13. $7\frac{1}{3} + 9\frac{4}{5} = 7\frac{5}{15} + 9\frac{12}{15} = 16\frac{17}{15} = 17\frac{2}{15}$

14. 1,000 m

15. 100 cl

16. $\frac{5}{265} = \frac{1}{53}$

17. mean $= \frac{3+4+8+8+12}{5} = \frac{35}{5} = 7$

median = 8

mode = 8

18. 1430 − 1015 = 4 hr 15 min or 4.25 hr

4.25 × $6.50 ≈ $27.63

19. $45.00 × 1.08 = $48.60

$100.00 − $48.60 = $51.40

20. $\frac{3}{4} = \frac{9}{G}$

3G = 4(9)

3G = 36

G = 12 girls

Lesson Practice 30A

1. rational: $5 = \frac{5}{1}$

2. irrational: $\sqrt{5}$ is equal to a non-repeating, non-terminating decimal.

3. rational: $\sqrt{4} = 2$

4. rational

5. rational: $95 = \frac{95}{1}$

6. rational: $0.16 = \frac{16}{100}$

7. rational

8. rational: $\sqrt{25} = 5$

9. irrational: π is equal to a non-repeating, non-terminating decimal.

10. rational: $0.7 = \frac{7}{10}$

11. rational: $15 = \frac{15}{1}$

12. irrational: $\sqrt{2}$ is equal to a non-repeating, non-terminating decimal.

13. false

14. true

15. false

16. true

17. false

18. done

19. see number line

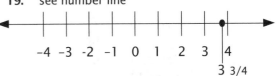

20.

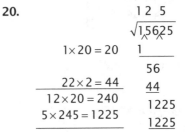

check: $125^2 = 15,625$

Lesson Practice 30B

1. rational
2. rational
3. irrational
4. rational
5. irrational
6. rational
7. rational
8. irrational
9. rational
10. rational
11. rational
12. irrational
13. true
14. false
15. true
16. false
17. false
18. see number line
19. see number line

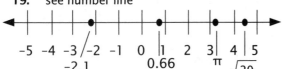

-5 -4 -3 /-2 -1 0 |1 2 3| 4 |5
    -2.1   0.66   π $\sqrt{20}$

20.

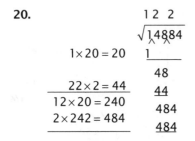

$$\begin{array}{r} 1\ 2\ \ 2 \\ \sqrt{1\,48\,84} \end{array}$$

$1 \times 20 = 20$ 1

               48

$22 \times 2 = 44$ 44

$12 \times 20 = 240$ 484

$2 \times 242 = 484$ 484

check: $122^2 = 14{,}884$

Lesson Practice 30C

1. irrational
2. rational
3. rational
4. rational
5. rational

6. rational
7. irrational
8. rational
9. rational
10. rational
11. irrational
12. rational
13. true
14. false
15. false
16. true
17. false
18. see previous column for number line
19. see previous column for number line

20.

$$\begin{array}{r} 2\ 5\ \ 8 \\ \sqrt{6\,65\,64} \end{array}$$

$2 \times 20 = 40$ 4

               265

$45 \times 5 = 225$ 225

$25 \times 20 = 500$ 4064

$8 \times 508 = 4064$ 4064

check: $258^2 = 66{,}564$

Systematic Review 30D

1. rational
2. irrational
3. rational
4.   6'17"
   – 3' 6"
     3' 11"

5.   9 yd 2 ft
 + 6 yd 2 ft
  15 yd 4 ft $= 16$ yd 1 ft

6.   15 lb 7 oz + 6 oz = 15 lb 13 oz
 – 8 lb 10 oz + 6 oz = 9 lb 0 oz
                       6 lb 13 oz

7. 0815
8. $4:25 + 12:00 = 1625$
9. $1921 - 12:00 = 7:21$ p.m.

10. 6:38 a.m.

11. $(X+1)(X+4) = X^2 + 5X + 4$

12. $(X+7)(X+2) = X^2 + 9X + 14$

13. $(X+5)(X+3) = X^2 + 8X + 15$

14. done

15. done

16. 30

17. 0.35

18. $V \approx \frac{1}{3}(3.14)(10)^2(20) \approx 2{,}093.33 \text{ in}^3$

Systematic Review 30E

1. rational

2. rational

3. irrational

4. $\begin{array}{r} 6:21 \\ -\ 2:19 \\ \hline 4:02 \end{array}$

5. $\begin{array}{r} 3:24+0:30 = 3:54 \\ -\ 1:30+0:30 = 2:00 \\ \hline 1:54 \end{array}$

6. $\begin{array}{r} 4:20 \\ +\ 3:45 \\ \hline 7:65 \ = 8:05 \end{array}$

7. $15 = 3 \times 5; \ 20 = 2 \times 2 \times 5$
   GCF = 5

8. $12 = 2 \times 2 \times 3; \ 18 = 2 \times 3 \times 3$
   GCF $= 2 \times 3 = 6$

9. $75 = 3 \times 5 \times 5; \ 45 = 3 \times 3 \times 5$
   GCF $= 3 \times 5 = 15$

10. $13X = 65$
    $X = 5$

11. $7W + 3 = 4W + 2^2 + 8$
    $7W + 3 = 4W + 4 + 8$
    $7W + 3 = 4W + 12$
    $7W = 4W + 12 - 3$
    $7W = 4W + 9$
    $7W - 4W = 9$
    $3W = 9$
    $W = 3$

12. $\begin{array}{r} -4X + 9 = -51 \\ -4X = -51 - 9 \\ -4X = -60 \\ X = 15 \end{array}$

13. 4,000

14. 5,000

15. 0.21

16. 16

17. $V \approx 3.14(5)^2(4) \approx 314 \text{ in}^3$

18. $\dfrac{20}{500} = \dfrac{1}{25}$

Systematic Review 30F

1. rational

2. irrational

3. irrational

4. $2Q = 50$
   $Q = 25$

5. $4X = 160$
   $X = 40$

6. $7Q = 168$
   $Q = 24$

7. $12 = 2 \times 2 \times 3; \ 48 = 2 \times 2 \times 2 \times 2 \times 3$
   LCM $= 2 \times 2 \times 2 \times 2 \times 3 = 48$

8. $3 = 1 \times 3; \ 16 = 2 \times 2 \times 2 \times 2$
   LCM $= 2 \times 2 \times 2 \times 2 \times 3 = 48$

9. $9 = 3 \times 3; \ 12 = 2 \times 2 \times 3$
   LCM $= 2 \times 2 \times 3 \times 3 = 36$

10. 9,000

11. 0.075

12. $V = \frac{1}{3}(9)^2(14) = 378 \text{ in}^3$

13. mean $= \dfrac{85 + 87 + 90 + 90 + 98}{5} = \dfrac{450}{5} = 90$

    median = 90

    mode = 90

14. obtuse

15. $|-4| = 4$

16. $2(2 \times 6) + 2(6 \times 4) + 2(4 \times 2) =$
    $2(12) + 2(24) + 2(8) =$
    $24 + 48 + 16 = 88 \text{ ft}^2$

17. $\dfrac{358}{1,000} = \dfrac{179}{500}$

18. $\$25.00 \times 2.50 = \$62.50$

Application and Enrichment Solutions

Application and Enrichment Lesson 1

1. $135 \div 9 = 15$;
   $15 \times 2 = 30$ people approved.
   $135 \div 5 = 27$ people disapproved.
   $30 + 27 = 57$ people answered.
   $135 - 57 = 78$ people didn't answer.
   More people didn't answer.

2. $49,170 \div 1,250 = 39$ r.420
   39 times with 420 mi^2 left over

3. $2 \times \$35.99 = \$71.98$
   $\$71.98 + \$15.95 = \$87.93$
   $\$87.93 - \$5.00 = \$82.93$
   $\$100.00 - \$82.93 = \$17.07$ change

4. $\$17.07 - \$10.00 = \$7.07$ change
   $\$7.07 - \$5.00 = \$2.07$;
   $\$2.07 - \$2.00 = \$0.07$;
   $\$0.07 - \$0.05 = \$0.02$;
   a ten, a five, two ones, a nickel,
   and two pennies

5. $24 \times 12 = 288$ per case;
   $900 \div 288 = 3.125$
   rounded to next whole number is 4.

6. $1,260 \div 60 = 21$ hours

7. $15 + (-33) = -18$;
   $-18 + 5 = -13°$

Application and Enrichment Lesson 2

1. Beginning price was $60, and he
   purchased 30 shares, so he spent
   $30 \times \$60$, or about $1,800. Ending
   price was $45, and he sold 30
   shares, so he received $30 \times \$45$, or
   $1,350. $\$1,800 - \$1,350 = \$450$ lost

2. $\frac{3}{8} + \frac{1}{8} + \frac{3}{8} = \frac{7}{8}$ of a mile traveled
   $\frac{8}{8} - \frac{7}{8} = \frac{1}{8}$ of a mile left

3. ran $\frac{3}{8} + \frac{3}{8} = \frac{6}{8} = \frac{3}{4}$ mile
   $5,280 \div 4 = 1,320$; $1,320 \times 3 =$
   3,960 ft running
   jogged $\frac{1}{8}$ mile
   $5,280 \div 8 = 660$ ft jogging
   Distance walking is the same
   as distance jogging, so that is
   also 660 ft.

4. $21 \times 60 = 1,260$ per hour
   $1,260 \times 24 = 30,240$ per day
   $30,240 \times 365 = 11,037,600$ per year

5. $-5 + 4 - 8 + 10 + 5 - 4 - 6 = -4$ gallons

6. $-4 \times 4 = -16$ qt

Application and Enrichment Lesson 3

1. $68 \div 4 = 17$ units on a side
   $17 \times 17 = 289$ $units^2$

2. $8 \times 6 = 48$ $units^2$
   $16 \times 12 = 192$ $units^2$
   $192 \div 48 = 4$ times the original

3. $4 \times 3 = 12$ $units^2$
   $12 \div 48 = \frac{1}{4}$ the original

4.

   area of rectangle A = XY $units^2$
   area of rectangle B = 9XY $units^2$
   $9XY \div XY = 9$
   The area of B is 9 times that of A.

5. 39

6. 13   This can easily be solved
   by drawing a diagram or a
   number line.

7. rectangle : $14 \times 16 = 224$ in^2

   triangle: $\frac{1}{2} \times 14 \times 15 = 105$ in^2

   total: $224 + 105 = 329$ in^2

8. $3.14(15^2) = 706.5$ in^2

   $3.14(12^2) = 452.16$ in^2

   $706.5 - 452.16 = 254.34$ in^2

Application and Enrichment Lesson 4

1. $\frac{1}{2} + \frac{1}{3} = \frac{3}{6} + \frac{2}{6} = \frac{5}{6}$

   $\frac{6}{6} - \frac{5}{6} = \frac{1}{6}$

2. $12:00 - 7:30 = 4:30$

   $4:30 + 3:00 = 7:30$ hours worked

   $7.5 \times 4.65 = \$34.875$, or $34.88 earned

3. $\frac{3}{4} = \frac{15}{20}$ or $\frac{30}{40}$; $\frac{4}{5} = \frac{16}{20}$ or $\frac{32}{40}$

   E. $\frac{31}{40}$

4. $\frac{3}{4} = \frac{18}{24}$; $\frac{5}{6} = \frac{20}{24}$

   A or $\frac{19}{24}$ is an answer.

   Check other fractions by using
   the Rule of Four to compare each
   with the two given fractions.
   E also falls between the
   given fractions.

   $\frac{3}{4} \Leftrightarrow \frac{11}{14}, \frac{42}{56} \Leftrightarrow \frac{44}{56}$

   $\frac{5}{6} \Leftrightarrow \frac{11}{14}, \frac{70}{84} \Leftrightarrow \frac{66}{84}$

   You can also change each fraction
   to a decimal for easy comparison.

5. It will be quadrupled:

   $3.14(5^2) \approx 78.5$ ft^2

   $3.14(10^2) \approx 314$ ft^2

   $314 \div 78.5 = 4$

6. $12 \times 22 = 264$ in^2

7. rectangle :

   $18 \times 30 = 540$ in^2

   paralellogram:

   $8 \times 15 = 120$ in^2

   $540 - 120 = 420$ in^2

8. area of square:

   $36 \times 36 = 1,296$ cm^2

   semicircles:

   $\frac{1}{2}(3.14)(5^2) \approx 39.25$ cm^2

   $39.25 \times 4 \approx 157$ cm^2

   $1,296 - 157 \approx 1,139$ cm^2

Application and Enrichment Lesson 5

1. $1.00

   $5 \times \$1.00 = \$5.00$

2. $2.00 the first day

   $4.00 the second day

   $16.00 the third day

   $256.00 the fourth day

   $65,536.00 the fifth day

   $65,814.00 total

3. $3 \times 2 = 6$ units^2

   $9 \times 4 = 36$ units^2

4. Sketches and dimensions will vary.
   The student should notice that
   when the dimensions are squared,
   the area will be squared.

5. Sketches and dimensions will vary.
   The student should notice that when
   the dimensions are cubed, the area
   will be cubed.

6. Area = base × height, so the area of this rectangle will be ab units^2. If the length and the width of the rectangle are both cubed, the new area will be a^3b^3 units^2, which can also be expressed as $(ab)^3$ units^2.

7. If the radius is doubled, the area will be four times greater.

    Ex : $r = 2$, $A \approx 3.14(4) \approx 12.56$ units^2

    $r^2 = 4$, $A \approx 3.14(16) \approx 50.24$ units^2, new area is 4 times original area

8. If you start with a radius of 3 and square it, the new area will be 9 times the original area. Squaring the radius of a circle causes the area to increase by a factor of r^2.

9. $A = 10\left(\dfrac{20+15}{2}\right)$

    $= 10\left(\dfrac{35}{2}\right)$

    $= \dfrac{350}{2} = 175$ in^2

10. trapezoid:

    $12\left(\dfrac{21+26}{2}\right) = 12\left(\dfrac{47}{2}\right)$

    $= 6(47) = 282$ cm^2

    large semicircle:

    $\dfrac{3.14(6)^2}{2} \approx \dfrac{3.14(36)}{2}$

    $\approx 3.14(18) \approx 56.52$ cm^2

    small semicircle:

    $\dfrac{3.14(2)^2}{2} \approx \dfrac{3.14(4)}{2}$

    $\approx \dfrac{12.56}{2} \approx 6.28$ cm^2

    $282 - 56.52 - 6.28 \approx 219.2$ cm^2

Application and Enrichment Lesson 6

1.

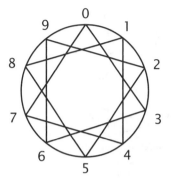

2.

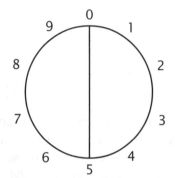

3.

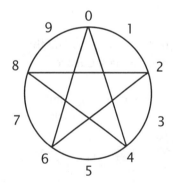

4.

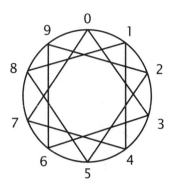

5. 3 facts and 7 facts (or 4 and 6 facts)
6. 12
7. 28
8. 88; multiply one less than the number of weeks by four to get the number of rooms.

Application and Enrichment Lesson 7

There may be alternate ways to describe some patterns. If you get the next number in the sequence correct, your description is valid.

1. Add 7 to the last number to find the next number in the sequence.
2. 35
3. Square the last number to find the next number in the sequence.
4. 65,536
5. Add one more to the last number each time: $1+4=5$; $5+5=10$; $10+6=16$
6. $16+7=23$
   $23+8=31$
7. Add twice as many to the last number each time: $1+2=3$; $3+4=7$; $7+8=15$
8. $15+16=31$
   $31+32=63$
9. Add the last two numbers in the sequence to find the next: $1+2=3$; $2+3=5$; $3+5=8$

10. $5+8=13$
    $8+13=21$

11.

step	1	2	3	4	5	6	7
blocks	1	3	6	10	15	21	28

12. Add one more than the number that was added in the previous step.
    $28+8=36$
13. Take the number of steps times one more than the number of steps and divide by 2.
    $9(9+1)\div 2 =$
    $9(10)\div 2 =$
    $90 \div 2 = 45$

Application and Enrichment Lesson 8

1. sum, how many, total in all
2. difference, how many more, have left
3. times, product, fraction of
4. how many for each, how many parts
5. $2+5=7$
   $7-1=6$
   $(2+5)-1=6$ pencils
6. $K+M+Q$
7. $X+Y$ total treats
   $(X+Y)\div Z$ treats per person
8. $\frac{1}{3}$ completed, $\frac{2}{3}$ to go
   $\frac{2}{3}\times(A+B)$ or $\frac{2(A+B)}{3}$
   $\frac{2A+2B}{3}$ is also correct.

Application and Enrichment Lesson 9

1. $\dfrac{A}{A} + \dfrac{B}{A} = \dfrac{A+B}{A}$

2. $\dfrac{Y}{Z} - \dfrac{X}{Z} = \dfrac{Y-X}{Z}$

3. $\dfrac{A+B}{E} + \dfrac{C}{E} = \dfrac{A+B+C}{E}$

4. $\dfrac{A}{X} + \dfrac{B}{Y} = \dfrac{AY}{XY} + \dfrac{BX}{XY} = \dfrac{AY+BX}{XY}$

5. $\dfrac{EF}{T} - \dfrac{G}{S} = \dfrac{EFS}{TS} - \dfrac{GT}{TS} = \dfrac{EFS-GT}{TS}$

6. $\dfrac{X}{RS} + \dfrac{X}{QS} = \dfrac{XQ}{QRS} + \dfrac{XR}{QRS} = \dfrac{XQ+XR}{QRS}$

7. $\dfrac{A}{B} \times \dfrac{C}{D} = \dfrac{AC}{BD}$

8. $\dfrac{X}{R} \times \dfrac{X}{S} = \dfrac{X^2}{RS}$

9. $\dfrac{DF}{YZ} \times \dfrac{Y}{D} = \dfrac{DFY}{YZD} = \dfrac{F}{Z}$

10. $\dfrac{A}{B} \div \dfrac{A}{B} = \dfrac{A \div B}{A \div B} = \dfrac{1}{1} = 1$

    $\dfrac{A}{B} \div \dfrac{A}{B} = \dfrac{A \div A}{B \div B} = \dfrac{1}{1} = 1$

    $\dfrac{A}{B} \div \dfrac{A}{B} = \dfrac{AB}{AB} = \dfrac{1}{1} = 1$

11. $\dfrac{Q}{Z} \div \dfrac{YZ}{T} = \dfrac{QT}{ZT} \div \dfrac{YZ^2}{ZT} =$

    $\dfrac{QT \div YZ^2}{ZT \div ZT} = \dfrac{QT}{YZ^2}$

12. $\dfrac{X}{R} \div \dfrac{R}{X} = \dfrac{X^2}{RX} \div \dfrac{R^2}{RX} = \dfrac{X^2 \div R^2}{RX \div RX} = \dfrac{X^2}{R^2}$

13. $\dfrac{Q}{X} + \dfrac{R}{P} = \dfrac{QP}{XP} + \dfrac{XR}{XP} = \dfrac{QP+XR}{XP}$

14. $\dfrac{DT}{S} \times \dfrac{C}{D} = \dfrac{DTC}{SD} = \dfrac{TC}{S}$

15. $\dfrac{L}{B} \div \dfrac{U}{B} = \dfrac{L \div U}{B \div B} = \dfrac{L \div U}{1} = \dfrac{L}{U}$

16. $X = A - Y$

17. $Y = A - X$

18. $5X - 4X = B + B$

    $X = B + B$

    $X = 2B$

Application and Enrichment Lesson 10

1. direct route

   $18^2 + 24^2 = H^2$

   $324 + 576 = 900$

   $30 \text{ miles} = H$

   same way he came

   $18 + 24 = 42$ miles

   $42 - 30 = 12$ miles shorter

   by direct route

2. $15^2 + 36^2 = H^2$

   $225 + 1296 = 1521$

   $39 \text{ ft} = H$

   $39 + 3 = 42$ ft

3. $3^2 + 4^2 = H^2$

   $9 + 16 = 25$

   $5 \text{ miles} = H$

   $5 + 5 = 10$ miles

4. $20^2 + 48^2 = H^2$

   $400 + 2304 = 2704$

   $52 \text{ mi} = H$

   $P = 20 + 48 + 20 + 48 = 136$ mi

   $136 + 52 = 188$ miles of fence

5. $\dfrac{A}{B} \div \dfrac{C}{D} = \dfrac{AD}{BD} \div \dfrac{BC}{BD} =$

   $\dfrac{AD \div BC}{BD \div BD} = \dfrac{AD \div BC}{1} = \dfrac{AD}{BC}$

6. $\dfrac{A}{B} \times \dfrac{D}{C} = \dfrac{AD}{BC}$

7. $\dfrac{AD}{BC} = \dfrac{AD}{BC}$

8. $\dfrac{XY}{Z} \div \dfrac{B}{CD} = \dfrac{XYCD}{ZCD} \div \dfrac{ZB}{ZCD} =$

   $\dfrac{XYCD \div ZB}{ZCD \div ZCD} = \dfrac{XYCD \div ZB}{1} = \dfrac{XYCD}{ZB}$

   $\dfrac{XY}{Z} \times \dfrac{CD}{B} = \dfrac{XYCD}{ZB}$

   The answers are equal.

Application and Enrichment Lesson 11

1. Multiply by 3 and add 1.
2. 202
3. Divide by 2.
4. $\frac{5}{8}$
5. Take the square root.
6. 2
7. Subtract half of what was subtracted the previous time.
8. $2\frac{1}{2}$; $2\frac{1}{4}$
9.

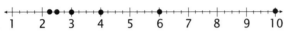

10. 2; no
11.

step	1	2	3	4	5
circles	1	4	9	16	25
squares	4	8	12	16	20

12. The number of circles equals the step number squared.
13. The number of squares equals the step number times 4.
14. $8^2 = 64$ circles
    $8 \times 4 = 32$ squares

Application and Enrichment Lesson 12

1. 1×36
   2×18
   3×12
   4×9
   6×6
2. $P = 2(1) + 2(36) = 2 + 72 = 74$ units
   $P = 2(2) + 2(18) = 4 + 36 = 40$ units
   $P = 2(3) + 2(12) = 6 + 24 = 30$ units
   $P = 2(4) + 2(9) = 8 + 18 = 26$ units
   $P = 2(6) + 2(6) = 12 + 12 = 24$ units
3. 6×6

4. 6×10
5. 1×15
6. 1×5
   2×4
   3×3
7. $1 \times 5 = 5$ units^2
   $2 \times 4 = 8$ units^2
   $3 \times 3 = 9$ units^2
8. $3 \times 3 = 9$ ft^2
9. The shape she chooses would depend on what she intended it to be used for. Some possibilities:
   $5 \times 5 = 25$ ft^2
   $4 \times 6 = 24$ ft^2
   $3 \times 7 = 21$ ft^2
10. They enclose the most space with least exposure.

Application and Enrichment Lesson 13

1. $AX = ABC$
   $X = BC$
2. $XY - B = Q$
   $XY = B + Q$
   $X = \dfrac{B + Q}{Y}$
3. $CDX + E = RD$
   $CDX = RD - E$
   $X = \dfrac{RD - E}{CD}$
4. $YX - YT = YZ$
   $X - T = Z$
   $X = Z + T$
5. $Q(X + B) = R(X + C)$
   $QX + QB = RX + RC$
   $QX - RX = RC - QB$
   $X(Q - R) = RC - QB$
   $X = \dfrac{RC - QB}{Q - R}$

6. $$AX - BX - C = CX + X + E$$
$$AX - BX - CX - X = E + C$$
$$X(A - B - C - 1) = E + C$$
$$X = \frac{E + C}{A - B - C - 1}$$

7. $\sqrt{32} = 5.7$
8. $\sqrt{150} = 12.2$
9. $\sqrt{75} = 8.7$
10. $\sqrt{481} = 21.9$
11. $$L^2 + L^2 = H^2$$
$$9^2 + 7^2 = H^2$$
$$81 + 49 = H^2$$
$$130 = H^2$$
$$H = \sqrt{130}$$

between 11 and 12:
$$11^2 = 121$$
$$12^2 = 144$$

12. 11.4 ft
$0.4 \times 12 = 4.8"$
to the nearest inch
11'5"

Application and Enrichment Lesson 14

1. $$X + X + 20 = 144$$
$$2X + 20 = 144$$
$$2X = 124$$
$$X = 62 \text{ on one shelf}$$
$62 + 20 = 82$ books on the other shelf
$62 + 82 = 144$ books total

2. X boys went out for swimming.
X + 18 boys went out for baseball.
$$X + X + 18 = 48$$
$$2X + 18 = 48$$
$$2X = 30$$
$$X = 15 \text{ boys for swimming}$$
$X + 18 = 33$ boys for baseball
$15 + 33 = 48$ boys total

3. Lisa made X cards.
$$X + 3X = 32$$
$$4X = 32$$
$$X = 8 \text{ cards for Lisa}$$
$8 \cdot 3 = 24$ cards for June
$24 + 8 = 32$ cards total

4. $$P = 2L + 2W$$
$$(40) = 2(16) + 2W$$
$$40 = 32 + 2W$$
$$8 = 2W$$
$$4 = W$$
$$2(4) + 2(16) = 40$$
$$8 + 32 = 40$$
$$40 = 40$$

5. J = number of dollars Jill earned
$$J + 2J + 3J = \$150$$
$$6J = \$150$$
$$J = \$25 \text{ for Jill}$$
$2 \cdot \$25 = \$50$ for Joan
$3 \cdot \$25 = \$75$ for Deb
$\$25 + \$50 + \$75 = \$150$ total

6. $$P = 2L + 2W$$
$$22 = 2(X) + 2(X + 1)$$
$$22 = 2X + 2X + 2$$
$$20 = 4X$$
$$X = 5 \text{ in for the short side}$$
$X + 1 = 6$ in for the long side
$2(5) + 2(6) = 10 + 12 = 22$ in

Application and Enrichment Lesson 15

1. rectangular walls:
$$2(25 \times 12) + 2(18 \times 12) =$$
$$2(300) + 2(216) =$$
$$600 + 432 = 1,032 \text{ ft}^2$$
triangular sections:
$$2\left(\frac{1}{2}\right)(12 \times 18) = 216 \text{ ft}^2$$
total:
$$1,032 + 216 = 1,248 \text{ ft}^2$$

2. $1{,}248 \div 425 = 2.94$ gal (rounded)
2.94×2 coats $= 5.88$ gal, so
6 gal will need to be purchased.
$6 \times 28 = \$168$

3. If 2 5-gallon buckets were purchased:
$2 \times 120 = \$240.00$
In a real-life situation you probably would have purchased one 5-gallon bucket and a 1-gallon bucket.
$120 + 28 = \$148$
$168 - 148 = \$20$ savings

4. $1{,}248 \div 250 = 5$ gal (rounded)
5×2 coats $= 10$ gal
$10 \times 20 = \$200$
The more expensive paint is a better buy because you don't have to buy as much of it.

5. $4(18) + 4(25) =$
$72 + 100 =$
172 ft^2
$1248 - 172 = 1076$ ft^2

6. $1076 \times 1.12 = 1205.12$
13 squares

7. whole rectangle:
$12 \times 18 = 216$ ft^2
closet:
$6 \times 3 = 18$ ft^2
cutout:
$4 \times 8 = 32$ ft^2
$216 - 18 - 32 = 166$ ft^2

8. 9 sq ft in a yd^2
$166 \div 9 = 18.44$ yd^2 (rounded)

9. $18.44 \times 1.10 = 20.28$ (rounded)
21 yd^2 needed
$12 \times 21 = \$252$

10. $166 + 18 = 184$ ft^2
$184 \div 9 = 20.45$ yd^2
$20.45 \times 1.10 = 22.495$ yd^2
23 yd^2 will be needed.
$23 \times 12 = \$276$

Application and Enrichment Lesson 16

1. Each face is a triangle.
$A = \frac{1}{2}(bh)$
$A = \frac{1}{2}(4)(3.5)$
$A = 7$ in^2 per face
$7 \cdot 8 = 56$ in^2

2. Each face is a square.
$5 \times 5 = 25$ in^2 per face
$25 \times 6 = 150$ in^2

3. Each face is a triangle.
$A = \frac{1}{2}(bh)$
$A = \frac{1}{2}(10)(8.7)$
$A = 43.5$ cm^2 per face
$43.5 \times 20 = 870$ cm^2

4. $4 + 4 = 6 + 2$
$8 = 8$

5. $8 + 6 = 12 + 2$
$14 = 14$

6. $12 + 20 = 30 + 2$
$32 = 32$

7. $20 + 12 = 30 + 2$
$32 = 32$

Application and Enrichment Lesson 17

1-2.

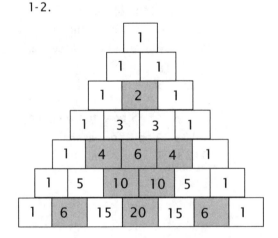

3-4.

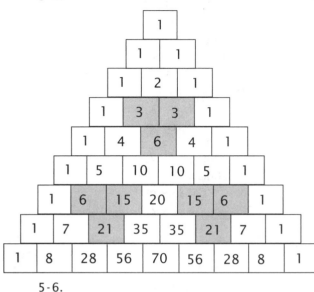

5-6.

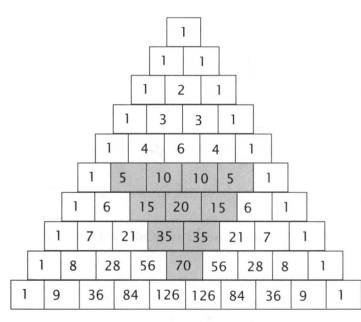

7. $1 + 4 + 10 = 15$

8. Answers will vary, but the sum of the numbers in the "handle" of the "hockey stick" will always equal the number in the smaller rectangle.

Application and Enrichment Lesson 18

1. $M + (M - 11) = 21$
    $$2M - 11 = 21$$
    $$2M = 32$$
    $$M = \$16 \text{ for the meal}$$
    $16 - 11 = \$5.00$ for dessert

2. $6 - 7 + 3 - 4 = -2$ mi east, or 2 miles west. The answer should not be written as a negative number because it is a distance, and distance is always positive.

3. $X + (X - 200) = 300$
    $$2X - 200 = 300$$
    $$2X = 500$$
    $$X = 250$$
    Isaac has $250.

4. Let J = the number of dollars John earned
    $$J + (J - 18) = 60 - 3.50$$
    $$2J - 18 = 56.50$$
    $$2J = 74.50$$
    $$J = \$37.25$$

5. In a square, the perimeter is 4 times the length of one side, so:
    $$S = (S + 57) \div 4$$
    $$4S = S + 57$$
    $$3S = 57$$
    $$S = 19$$

6. Distance is always positive, so he should have reported the distance as 20 ft.

7. $P = W + W + L + L$
    $$52 = W + W + 20 + 20$$
    $$52 = 2W + 40$$
    $$12 = 2W$$
    $$W = 6 \text{ ft}$$

8. using fractions:

$$\left(N \times \frac{9}{5}\right) + 32 = (N-32) \times \frac{5}{9}$$

$$45\left(N \times \frac{9}{5}\right) + 45(32) = (N-32) \times \frac{5}{9}(45)$$

Multiplying each term by 45 cancels the denominator, as 9 and 5 both go into 45 evenly.

$$81N + 1{,}440 = (N-32) \times 25$$
$$81N + 1{,}440 = 25N - 800$$
$$56N = -800 - 1{,}440$$
$$56N = -2{,}240$$
$$N = -40°$$

using decimals:

$$1.8N + 32 = (N-32) \times 0.56 \text{(rounded)}$$
$$1.8N + 32 = 0.56N - 17.92$$
$$1.8N - 0.56N = -17.92 - 32$$
$$1.24N = -49.92$$
$$124N = -4992$$
$$N = -40.26°$$

(In this case, the fractions give the exact value and the decimals give an approximate value because of the rounding.)

Application and Enrichment Lesson 19

1.
$$\frac{8+6}{6} = \frac{180}{F}$$
$$\frac{14}{6} = \frac{180}{F}$$
$$14F = 6(180)$$
$$7F = 3(180)$$
$$7F = 540$$
$$F = 77\frac{1}{7}\text{ gal}$$

2.
$$\frac{40+20}{20} = \frac{135}{S}$$
$$\frac{60}{20} = \frac{135}{S}$$
$$60S = 20(135)$$
$$3S = 135$$
$$S = \$45 \text{ for the son}$$
$$\$135 - \$45 = \$90 \text{ for the father}$$

3.
$$\frac{4}{200} = \frac{T}{575-200}$$
$$\frac{4}{200} = \frac{T}{375}$$
$$200T = 4(375)$$
$$50T = 375$$
$$T = 7\frac{1}{2} = 7 \text{ hrs } 30 \text{ min}$$

4.
$$\frac{8.5}{200} = \frac{G}{575}$$
$$200G = 8.5(575)$$
$$200G = 4887.5$$
$$G = 24.4 \text{ gal}$$
$$\text{(rounded)}$$

5.
$$\frac{3}{2} = \frac{7}{L}$$
$$3L = 14$$
$$L = 4\frac{2}{3} \text{ loaves}$$

She can make 4 whole loaves.

6.
$$\frac{4}{3} = \frac{T}{81}$$
$$3T = 324$$
$$T = 108 \text{ ft}$$

7.
$$6 \times 5 = M \times 3$$
$$30 = 3M$$
$$M = 10 \text{ machines}$$

8.
$$15 \times 36 = (15+9) \times D$$
$$540 = 24D$$
$$D = 22.5 \text{ days}$$

Application and Enrichment Lesson 20

1.
$$\frac{4}{5.2} = \frac{25}{D}$$
$$4D = 5.2(25)$$
$$4D = 130$$
$$D = 32.5$$
$$33 \text{ miles rounded}$$

2.
$$8.2 + 4.5 = 12.7 \text{ cm}$$
$$\frac{4}{12.7} = \frac{25}{D}$$
$$4D = 12.7(25)$$
$$4D = 317.5$$
$$D = 79.375$$
$$79 \text{ miles rounded}$$

3. $\dfrac{5}{7} = \dfrac{14}{D}$

$5D = 7(14)$

$5D = 98$

$D = 19\dfrac{3}{5}$ or 19.6 miles

4. $\dfrac{5}{3} = \dfrac{D}{6}$

$3D = 5(6)$

$3D = 30$

$D = 10$ cm

5. $\dfrac{10}{15} = \dfrac{3,000}{D}$

$10D = 15(3,000)$

$10D = 45,000$

$D = 4,500$ miles

2. yes

3. no

4. 11; it holds true (see diagram)

5. The next prime is 13; see diagram for shading of multiples of ten

Application and Enrichment Lesson 22

1. 20; 35,690
2. 20; 35; 35,690
3. 20; 35,690
4. 0; 105; 75,084
5. 6055; 45,759
6. 792; 1,639; 90,959

Application and Enrichment Lesson 21

1.

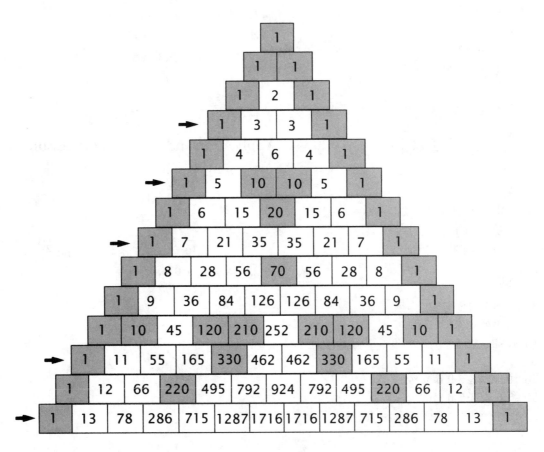

7. Digits add to 33, so it is a multiple of 3. $692,835 \div 3 = 230,945$. It ends in 5, so it is a multiple of 5. $230,945 \div 5 = 46,189$. $4,618 - 18 = 4,600$: not a multiple of 7. $4 + 1 + 9 = 14$; $6 + 8 = 14$: $14 - 14 = 0$, so it is a multiple of 11. $46,189 \div 11 = 4,199$ $4 + 9 = 13$; $1 + 9 = 10$; $13 - 10 = 3$ not a multiple of 11.
Try 13: $4\,199 \div 13 = 323$.
Try 17: $323 \div 17 = 19$.
Prime factors of 692,835 are:
$3 \times 5 \times 11 \times 13 \times 17 \times 19$

Application and Enrichment Lesson 23

1. $P = 2W + 2L$
$2(X - 5) + 2(2X + 9) =$
$2X - 10 + 4X + 18 =$
$6X + 8$

2. $6(8) + 8 = 48 + 8 = 56$
$W = (8) - 5 = 3$
$L = 2(8) + 9 = 25$
$3 + 3 + 25 + 25 = 56$
yes

3. $(X - 3) + (X + 18) + (X^2 - 2) =$
$X^2 + 2X + 13$

4. $(5)^2 - 2 = 25 - 2 = 23$
$(5) - 3 = 2$
$(5) + 18 = 23$

5. $(4X + 3) + (3X + 1) + (X) + (2X) + (X) +$
$((4X + 3) - (2X)) + (3X + 1) =$
$14X + 5 + (2X + 3) = 16X + 8$

6. room:
$4(3) + 3 = 15$ ft
$3(3) + 1 = 10$ ft
The closet is 3 ft x 6 ft.

7. $16(3) + 8 = 56$ ft

8. 6

9. $56 \times 0.10 = 5.6$ ft of waste
$56 + 5.6 = 61.6$ ft total
7 lengths should be purchased.

Application and Enrichment Lesson 24

1. $4 \times 6 \times 0.5 = 12$ ft^3

2. $3 \times 3 \times 3 = 27$ ft^3

3. $27 - 12 = 15$ ft^3

4. $5 \times 6 \times 0.5 = 15$ ft^3

5. $15 + 12 = 27$ ft^3
27 ft^3 = 1 yd^3
No sand will be left over.

6. Mr. Brown:
$\frac{12}{27}$ of $40 = \$17.78$
Mr. White:
$\frac{15}{27}$ of $40 = \$22.22$

7. $12 \times 18 = 216$ ft^2

8. $216 \times 0.5 = 108$ ft^3

9. $108 \div 27 = 4$ yd^3

10. $4 \times 80 = \$320$

11. $\$500 - \$320 = \$180$

12. $12 \times 24 \times 0.5 = 144$ ft^3
$144 \div 27 = 5.33$ yd^3
$10 - 4 = 6$ yd^3
yes

Application and Enrichment Lesson 25

1. $A = \frac{1}{2}(X + 1)(2X - 6) =$
$\frac{1}{2}(2X^2 - 4X - 6) =$
$X^2 - 2X - 3$
$(4)^2 - 2(4) - 3 = 16 - 8 - 3 = 5$ units^2

2. $A = (2X + 1)(X + 7) =$
$2X^2 + 15X + 7$

3. $A = (X)(2X) = 2X^2$

4. closet :

$2(5)^2 = 2(25) = 50 \text{ ft}^2$

bedroom:

$2(5)^2 + 15(5) + 7 = 50 + 75 + 7 =$

132 units^2

5. $(X + 8)(2X + 2)$

6. $((5) + 8)(2(5) + 2) = (13)(12) =$

156 ft^2

$156 - 132 = 24 \text{ ft}^2$

7. $156 \text{ ft}^2 + 50 \text{ ft}^2 = 206 \text{ ft}^2$

$206 \div 9 = 22.89 \text{ yd}^2 (\text{rounded})$

23 yd^2 will need to

be purchased.

8. $23 \times 15 = \$345$

9. $23 \times 10 = \$230$

$230 + 150 = \$380$

No, the cost of installation will more than offset the per-yard cost savings.

10. $L = 2(2X^2 + 2X - 7) + 2(X^2 + 3X - 2) =$

$4X^2 + 4X - 14 + 2X^2 + 6X - 4 =$

$6X^2 + 10X - 18$

$6(2)^2 + 10(2) - 18 =$

$24 + 20 - 18 = 26 \text{ in}$

11. $P = 6(2X^2 - 4X + 1) = 12X^2 - 24X + 6$

$2(5)^2 - 4(5) + 1 = 50 - 20 + 1 = 31 \text{ units}$

Application and Enrichment Lesson 26

1. $7 - 5 = 2$

$\dfrac{2}{5} = 2 \div 5 = 0.4 = 40\% \text{ growth}$

2. $5'4" = 64" \qquad 6'1" = 73"$

$73 - 64 = 9$

$\dfrac{9}{64} = 9 \div 64 = 0.1406 =$

$14\% \text{ growth } (\text{rounded})$

3. $6,500 - 5,000 = 1,500$

$\dfrac{1,500}{5,000} = 1,500 \div 5,000 =$

$0.3 = 30\% \text{ growth}$

4. $16 - 7 = 9$

$\dfrac{9}{7} = 9 \div 7 = 1.2857 =$

$129\% \text{ growth } (\text{rounded})$

5. $5,000 - 4,000 = 1,000$

$\dfrac{1,000}{5,000} = 1,000 \div 5,000 =$

$0.2 = 20\% \text{ decrease}$

6. $6,500 - 4,000 = 2,500$

$\dfrac{2,500}{6,500} = 2,500 \div 6,500 =$

$0.3846 = 38\% \text{ decrease}$

Application and Enrichment Lesson 27

1. Prairie Dogs:

$\dfrac{65 + 71 + 35 + 10}{4} = 45.25$

Raccoons :

$\dfrac{30 + 30 + 50 + 30}{4} = 35$

Hound Dogs:

$\dfrac{22 + 71 + 89 + 80}{4} = 65.5$

The Hound Dogs had the best record.

2. median

3. median

4. game 1: 30

game 2: 71

game 3: 50

game 4: 30

$\dfrac{30 + 71 + 50 + 30}{4} = 45.25$

5. 30

6. $80 - 10 = 70$

7. game 1: $65 - 22 = 43$

game 2: $71 - 30 = 41$

game 3: $89 - 35 = 54$

$\dfrac{43 + 41 + 54 + 70}{4} = 52$

Application and Enrichment Lesson 28

1. $\dfrac{1.024 + 1.021 + 1.023 + 1.019}{4} =$

   1.022 (rounded)

2. $1.024 - 1.022 = 0.002$

   $0.002 \div 1.022 = 0.0019$ or 0.19%

3. $1.022 - 1.019 = 0.003$

   $0.003 \div 1.022 = 0.0029$ or 0.29%

4. $\dfrac{2.056 + 2.123 + 2.007}{3} = 2.062$

5. $2.123 - 2.062 = 0.061$

   $0.061 \div 2.062 = 0.0296 = 2.96\%$

6. $2.062 - 2.007 = 0.055$

   $0.055 \div 2.062 = 0.0267 = 2.67\%$

7. No, the gauge is not giving results within allowed margin of error.

Application and Enrichment Lesson 29

1. 1,000 g

   1 kg

2. $100 \times 100 \times 100 = 1,000,000$ cc

   $1,000,000 \div 1,000 = 1,000$ l

3. 2 ml

4. $160 \times 125 = 20,000$ m^2

   $20,000 \div 10,000 = 2$ ha

5. $7 \times 10,000 = 70,000$ m^2

6. $1,000 \times 1,000 = 1,000,000$ m^2 in a km^2

   $1,000,000 \div 10,000 = 100$ ha in km^2

Application and Enrichment Lesson 30

1. yes

2. rational

3. rational

4. yes

5. no

6. rational, real

Test Solutions

Test 1
1. have
2. owe
3. owe
4. $(+8)+(-7)=+1$
5. $(-10)+(-2)=-12$
6. $(-7)+(-15)=-22$
7. $(+9)+(-11)=-2$
8. $(+32)+(+96)=+128$
9. $(+4)+(-13)=-9$
10. $(-5)+(-18)=-23$
11. $(-436)+(-251)=-687$
12. $(-511)+(+709)=+198$
13. $10\div5=2$
    $2\times1=2$
14. $9\div3=3$
    $3\times2=6$
15. $32\div8=4$
    $4\times7=28$
16. $30\div2=15$
    $15\times1=15$
17. $(+25)+(-17)=+8$ paces
18. $(\$-5)+(\$-6)=\$-11$
19. $20\div5=4$
    $4\times4=16$ won prizes
20. $254\div2=127$ pages read

5. $(-36)-(+49)=$
   $(-36)+(-49)=-85$
6. $(+22)-(-30)=$
   $(+22)+(+30)=+52$
7. $(+30)+(-24)=+6$
8. $(-53)-(+10)=$
   $(-53)+(-10)=-63$
9. $(+13)+(-2)=+11$
10. $(-33)-(+2)=$
    $(-33)+(-2)=-35$
11. $(-7)+(+1)=-6$
12. $(-4)-(+18)=$
    $(-4)+(-18)=-22$
13. $\frac{1}{5}+\frac{3}{5}=\frac{4}{5}$
14. $\frac{2}{8}-\frac{1}{8}=\frac{1}{8}$
15. $\frac{5}{7}-\frac{2}{7}=\frac{3}{7}$
16. $\frac{1}{3}+\frac{1}{3}=\frac{2}{3}$
17. $\frac{2}{4}+\frac{1}{4}=\frac{3}{4}$ of the cake
18. $15\div3=5$
    $15-5=10$ dimes
19. $\$21-\$25=\$-4$
20. $(+46)+(-15)=+31$ coins

Test 2
1. $(+52)-(-23)=$
   $(+52)+(+23)=+75$
2. $(-35)-(-16)=$
   $(-35)+(+16)=-19$
3. $(+54)-(+15)=$
   $(+54)+(-15)=+39$
4. $(-7)-(+24)=$
   $(-7)+(-24)=-31$

Test 3
1. $(-20)\times(-4)=+80$
2. $(+19)\times(-3)=-57$
3. $(-30)\times(-17)=+510$
4. $(-27)\times(+8)=-216$
5. $(-9)\times(+2)=-18$
6. $(-7)\times(-29)=+203$
7. $(+33)-(-46)=$
   $(+33)+(+46)=+79$
8. $(-27)+(-10)=-37$

9. $(-41) - (-20) =$
$(-41) + (+20) = -21$

10. $24 \div 3 = 8$
$8 \times 1 = 8$

11. $15 \div 5 = 3$
$3 \times 2 = 6$

12. $28 \div 7 = 4$
$4 \times 3 = 12$

13. $\dfrac{5}{8} - \dfrac{3}{8} = \dfrac{2}{8}$

14. $\dfrac{7}{10} - \dfrac{1}{10} = \dfrac{6}{10}$

15. $\dfrac{1}{4} + \dfrac{1}{4} = \dfrac{2}{4}$

16. $\dfrac{1}{5} = \dfrac{2}{10} = \dfrac{3}{15} = \dfrac{4}{20}$

17. $\dfrac{2}{3} = \dfrac{4}{6} = \dfrac{6}{9} = \dfrac{8}{12}$

18. $\dfrac{1}{5} + \dfrac{3}{5} = \dfrac{4}{5}$ of the chores

19. $\$ -4 \times (+5) = \$ -20$

20. $(-2) \times (+6) = -12$ ft

Test 4

1. $\dfrac{-6}{-2} = +3$

2. $35 \div (-7) = -5$

3. $(-48) \div (-6) = +8$

4. $\dfrac{-28}{-4} = +7$

5. $26 \div (-2) = -13$

6. $(-56) \div 7 = -8$

7. $(-3) \times (+6) = -18$

8. $(+4) \times (-2) = -8$

9. $(-5) \times (-6) = +30$

10. $(-38) + (+12) = -26$

11. $(+47) + (-39) = +8$

12. $(-21) - (-45) =$
$(-21) + (+45) = +24$

13. $\dfrac{1}{8} + \dfrac{3}{8} = \dfrac{4}{8} = \dfrac{1}{2}$

14. $\dfrac{4}{10} + \dfrac{4}{10} = \dfrac{8}{10} = \dfrac{4}{5}$

15. $\dfrac{7}{8} - \dfrac{1}{8} = \dfrac{6}{8} = \dfrac{3}{4}$

16. $D = 6$

17. no, whole numbers are non-negative
yes

18. $20 \div 5 = 4$
$4 \times 4 = 16$ right

19. $\dfrac{1}{3} + \dfrac{1}{3} = \dfrac{2}{3}$ were either woodpeckers
or chickadees

20. $\$ -500 \div 10 = \$ -50$ per person

Test 5

1. $5^2 = (5)(5) = 25$

2. $1^6 = (1)(1)(1)(1)(1)(1) = 1$

3. $3^3 = (3)(3)(3) = 27$

4. $\left(\dfrac{1}{2}\right)^4 = \dfrac{1}{2} \cdot \dfrac{1}{2} \cdot \dfrac{1}{2} \cdot \dfrac{1}{2} = \dfrac{1}{16}$

5. 4

6. 3

7. 5

8. 5

9. 3

10. 2

11. $\dfrac{1}{3} = \dfrac{7}{21}$

12. $\dfrac{3}{4} = \dfrac{12}{16}$

13. $\dfrac{7}{10} = \dfrac{14}{20}$

14. $\dfrac{1}{3} = \dfrac{8}{24}$; $\dfrac{3}{8} = \dfrac{9}{24}$; $\dfrac{8}{24} < \dfrac{9}{24}$

15. $\dfrac{2}{5} = \dfrac{14}{35}$; $\dfrac{4}{7} = \dfrac{20}{35}$; $\dfrac{14}{35} < \dfrac{20}{35}$

16. $\dfrac{3}{4} = \dfrac{27}{36}$; $\dfrac{5}{9} = \dfrac{20}{36}$; $\dfrac{27}{36} > \dfrac{20}{36}$

17. $3 \times 3 \times 3 \times 3$; 3^4; 81

18. $12 \div 4 = 3$
$3 \times 3 = 9$ pencils given away
$12 - 9 = 3$ pencils left

19. $\$ -63 \div 7 = \$ -9$

20. $\$ -12 + \$ -7 = \$ -19$

Test 6

1. $2 \times 10^0 + 2 \times \dfrac{1}{10^1} + 4 \times \dfrac{1}{10^2} + 7 \times \dfrac{1}{10^3}$

2. $1 \times 10^2 + 5 \times 10^1 + 6 \times 10^0 + 3 \times \dfrac{1}{10^1}$

3. $3 \times 10^3 + 9 \times 10^2 + 1 \times 10^1$

4. $7 \times 10^1 + 8 \times 10^0 + 9 \times \dfrac{1}{10^2}$

5. $5,417.412$

6. 400.13

7. dollars; dimes; pennies
   $\$4.00 + \$0.00 + \$0.02 = \$4.02$

8. 8; 1; 5; $\$8.00 + \$0.10 + \$0.05 = \$8.15$

9. 3

10. 2

11. $(-33) + (-22) = -55$

12. $(-26)(10) = -260$

13. $(91) \div (-13) = -7$

14. $(19) - (-18) =$
    $(19) + (+18) = +37$

15. $\dfrac{2}{3} + \dfrac{1}{6} = \dfrac{12}{18} + \dfrac{3}{18} = \dfrac{15}{18} = \dfrac{5}{6}$

16. $\dfrac{5}{8} - \dfrac{1}{4} = \dfrac{20}{32} - \dfrac{8}{32} = \dfrac{12}{32} = \dfrac{3}{8}$

17. $\dfrac{1}{7} + \dfrac{5}{6} = \dfrac{6}{42} + \dfrac{35}{42} = \dfrac{41}{42}$

18. $48 \div 3 = 16$
    $16 \times 2 = 32$ people in blue cars
    $48 \div 8 = 6$
    $6 \times 1 = 6$ people in red cars
    $32 + 6 = 38$ people in red or blue cars

19. $\dfrac{3}{4} - \dfrac{3}{8} = \dfrac{24}{32} - \dfrac{12}{32} = \dfrac{12}{32} = \dfrac{3}{8}$ lb eaten

20. $\$0.60 + \$0.05 + \$8.00 = \$8.65$

Test 7

1. $-(5)^2 = -(5)(5) = -25$

2. $-7^2 = -(7)(7) = -49$

3. $(-10)^3 = (-10)(-10)(-10) = -1,000$

4. $\left(-\dfrac{4}{9}\right)^2 = \left(-\dfrac{4}{9}\right)\left(-\dfrac{4}{9}\right) = \dfrac{16}{81}$

5. $-2^3 = -(2)(2)(2) = -8$

6. $(-3)^2 = (-3)(-3) = 9$

7. $-(6)^2 = -(6)(6) = -36$

8. $-\left(\dfrac{1}{8}\right)^2 = -\left(\dfrac{1}{8}\right)\left(\dfrac{1}{8}\right) = -\dfrac{1}{64}$

9. $6 \times 10^0 + 1 \times \dfrac{1}{10^1} + 9 \times \dfrac{1}{10^2}$

10. 490.823

11. $(-23) + (36) = 13$

12. $(-8)(-28) = 224$

13. $(-22) \div (-2) = 11$

14. $\dfrac{1}{3} + \dfrac{2}{5} = \dfrac{5}{15} + \dfrac{6}{15} = \dfrac{11}{15}$

15. $\dfrac{5}{9} + \dfrac{1}{4} = \dfrac{20}{36} + \dfrac{9}{36} = \dfrac{29}{36}$

16. $\dfrac{3}{4} - \dfrac{2}{3} = \dfrac{9}{12} - \dfrac{8}{12} = \dfrac{1}{12}$

17. $\dfrac{3}{8} \times \dfrac{1}{\cancel{6}_2} = \dfrac{1}{16}$

18. $\dfrac{1}{2} \times \dfrac{5}{8} = \dfrac{5}{16}$

19. $\dfrac{11}{\cancel{12}_4} \times \dfrac{\cancel{3}}{7} = \dfrac{11}{28}$

20. $\dfrac{\cancel{5}}{\cancel{9}_3} \times \dfrac{\cancel{3}}{\cancel{5}} = \dfrac{1}{3}$ of the job

Unit Test I

1. $(-8) + (-17) = -25$

2. $(-9) \times (35) = -315$

3. $(13) - (-21) =$
   $(13) + (+21) = 34$

4. $(-36) \div (-6) = 6$

5. $(-10) \times (-10) = 100$

6. $(5) + (-7) = -2$

7. $(-72) \div (9) = -8$

8. $(-2) - (14) =$
   $(-2) + (-14) = -16$

9. $(50) \div (-5) = -10$

10. $(-38) - (-12) =$
    $(-38) + (+12) = -26$

11. $(-46) + (61) = 15$

12. $(7) \times (-91) = -637$

13. $-2^3 = -(2)(2)(2) = -8$

14. $-(4)^2 = -(4)(4) = -16$

15. $(-10)^2 = (-10)(-10) = 100$

16. $\left(-\dfrac{5}{8}\right)^2 = \left(-\dfrac{5}{8}\right)\left(-\dfrac{5}{8}\right) = \dfrac{25}{64}$

17. $1 \times 10^2 + 6 \times 10^1 + 5 \times 10^0 + 9 \times \dfrac{1}{10^1}$

18. $4 \times 10^0 + 3 \times \dfrac{1}{10^2} + 8 \times \dfrac{1}{10^3}$

19. $7,300.914$

20. $\dfrac{1}{7} + \dfrac{2}{5} = \dfrac{5}{35} + \dfrac{14}{35} = \dfrac{19}{35}$

21. $\dfrac{3}{10} + \dfrac{1}{5} = \dfrac{15}{50} + \dfrac{10}{50} = \dfrac{25}{50} = \dfrac{1}{2}$

22. $\dfrac{7}{8} - \dfrac{3}{4} = \dfrac{28}{32} - \dfrac{24}{32} = \dfrac{4}{32} = \dfrac{1}{8}$

23. $\dfrac{7}{8} \times \dfrac{1}{4} = \dfrac{7}{32}$

24. $\dfrac{1}{2} \times \dfrac{5}{9} = \dfrac{5}{18}$

25. $\dfrac{\cancel{2}}{3} \times \dfrac{11}{\cancel{12}_6} = \dfrac{11}{18}$

26. $36 \div 3 = 12$
$12 \times 1 = 12$

27. $40 \div 5 = 8$
$8 \times 2 = 16$

28. $21 \div 7 = 3$
$3 \times 3 = 9$

Test 8

1. $\sqrt{16} = 4$

2. $\sqrt{9} = 3$

3. $\sqrt{7^2} = 7$

4. $\sqrt{A^2} = A$

5. $\sqrt{100} = 10$

6. $\sqrt{81} = 9$

7. $\sqrt{11^2} = 11$

8. $\sqrt{X^2} = X$

9. $-3^3 = -(3)(3)(3) = -27$

10. $(-2)^2 = (-2)(-2) = 4$

11. $\left(\dfrac{1}{100}\right)^2 = \left(\dfrac{1}{100}\right)\left(\dfrac{1}{100}\right) = \dfrac{1}{10,000}$

12. $\left(\dfrac{4}{5}\right)^2 = -\left(\dfrac{4}{5}\right)\left(\dfrac{4}{5}\right) = -\dfrac{16}{25}$

13. $(+5) + (-12) = -7$

14. $(-21) \times (-15) = 315$

15. $(+36) \div (-3) = -12$

16. $\dfrac{1}{2} \div \dfrac{1}{4} = \dfrac{4}{8} \div \dfrac{2}{8} = \dfrac{4 \div 2}{8 \div 8} = \dfrac{2}{1} = 2$

17. $\dfrac{7}{8} \div \dfrac{4}{5} = \dfrac{35}{40} \div \dfrac{32}{40} = \dfrac{35 \div 32}{40 \div 40} =$
$\dfrac{35 \div 32}{1} = \dfrac{35}{32}$

18. $\dfrac{5}{8} \div \dfrac{1}{10} = \dfrac{50}{80} \div \dfrac{8}{80} = \dfrac{50 \div 8}{80 \div 80} =$
$\dfrac{50 \div 8}{1} = \dfrac{50}{8}$

19. $\dfrac{6}{8} \div \dfrac{1}{4} = \dfrac{24}{32} \div \dfrac{8}{32} = \dfrac{24 \div 8}{32 \div 32} =$
$\dfrac{3}{1} = 3$ people

20. $\sqrt{49} = 7$ ft

Test 9

1. $X - 16 = 42$
$X = 42 + 16$
$X = 58$

2. $X - 16 = 42 \Rightarrow (58) - 16 = 42$
$42 = 42$

3. $-X + 4 = -2X - 6$
$-X = -2X - 6 - 4$
$-X = -2X - 10$
$-X + 2X = -10$
$X = -10$

4. $-X + 4 = -2X - 6 \Rightarrow -(-10) + 4 = -2(-10) - 6$
$+10 + 4 = 20 - 6$
$14 = 14$

5. $4R - 4 = 3R + 10$
$4R = 3R + 10 + 4$
$4R = 3R + 14$
$4R - 3R = 14$
$R = 14$

6. $4R - 4 = 3R + 10 \Rightarrow 4(14) - 4 = 3(14) + 10$
$$56 - 4 = 42 + 10$$
$$52 = 52$$

7. $2Y - 3 = Y - 4$
$$2Y = Y - 4 + 3$$
$$2Y = Y - 1$$
$$2Y - Y = -1$$
$$Y = -1$$

8. $2Y - 3 = Y - 4 \Rightarrow 2(-1) - 3 = (-1) - 4$
$$-2 - 3 = -1 - 4$$
$$-5 = -5$$

9. $\sqrt{25^2} = 25$

10. $\sqrt{64} = 8$

11. $-(3)^2 = -(3)(3) = -9$

12. $\dfrac{1}{2}^3 = \dfrac{1}{2} \cdot \dfrac{1}{2} \cdot \dfrac{1}{2} = \dfrac{1}{8}$

13. $\dfrac{1}{4} \div \dfrac{1}{8} = \dfrac{8}{32} \div \dfrac{4}{32} = \dfrac{8 \div 4}{32 \div 32} = \dfrac{2}{1} = 2$

14. $\dfrac{2}{3} \div \dfrac{1}{2} = \dfrac{4}{6} \div \dfrac{3}{6} = \dfrac{4 \div 3}{6 \div 6} = \dfrac{4 \div 3}{1} = \dfrac{4}{3}$

15. $\dfrac{5}{12} \div \dfrac{3}{4} = \dfrac{20}{48} \div \dfrac{36}{48} = \dfrac{20 \div 36}{48 \div 48} =$
$$\dfrac{20 \div 36}{1} = \dfrac{20}{36} = \dfrac{5}{9}$$

16. $\dfrac{6}{5}; \dfrac{5}{6} \times \dfrac{6}{5} = \dfrac{30}{30} = 1$

17. $\dfrac{1}{32}; \dfrac{32}{1} \times \dfrac{1}{32} = \dfrac{32}{32} = 1$

18. $9; \dfrac{9}{1} \times \dfrac{1}{9} = \dfrac{9}{9} = 1$

19. $2X + 15 = 37 - 2$
$$2X + 15 = 35$$
$$2X = 35 - 15$$
$$2X = 20$$
$$X = 10 \text{ years old}$$

20. $10 \div 5 = 2$
$$2 \times 2 = 4 \text{ years old}$$

Test 10

1. $5^2 + 12^2 = H^2$
$$25 + 144 = H^2$$
$$169 = H^2$$
$$H = 13 \text{ in}$$

2. $12^2 + 16^2 = 20^2$
$$144 + 256 = 400$$
$$400 = 400; \text{ yes}$$

3. $L^2 + 9^2 = 15^2$
$$L^2 + 81 = 225$$
$$L^2 = 225 - 81$$
$$L^2 = 144$$
$$L = 12 \text{ ft}$$

4. $5^2 + 10^2 = 15^2$
$$25 + 100 = 225$$
$$125 \neq 225; \text{ no}$$

5. $A - 7 = -13$
$$A = -13 + 7$$
$$A = -6$$

6. $A - 7 = -13 \Rightarrow (-6) - 7 = -13$
$$-13 = -13$$

7. $10X - 8 = 9X + 8$
$$10X = 9X + 8 + 8$$
$$10X = 9X + 16$$
$$10X - 9X = 16$$
$$X = 16$$

8. $10X - 8 = 9X + 8 \Rightarrow 10(16) - 8 = 9(16) + 8$
$$160 - 8 = 144 + 8$$
$$152 = 152$$

9. $\sqrt{X^2} = X$

10. $(-8)^2 = (-8)(-8) = 64$

11. $\left(\dfrac{1}{2}\right)^3 = \dfrac{1}{2} \cdot \dfrac{1}{2} \cdot \dfrac{1}{2} = \dfrac{1}{8}$

12. $\sqrt{\dfrac{9}{16}} = \dfrac{3}{4}$

13. $\dfrac{3}{8} \div \dfrac{1}{6} = \dfrac{3}{\underset{4}{8}} \times \dfrac{\overset{3}{6}}{1} = \dfrac{9}{4}$

14. $\dfrac{5}{9} \div \dfrac{1}{4} = \dfrac{5}{9} \times \dfrac{4}{1} = \dfrac{20}{9}$

15. $\dfrac{4}{5} \div \dfrac{3}{5} = \dfrac{4}{\underset{}{5}} \times \dfrac{\overset{}{5}}{3} = \dfrac{4}{3}$

16. $(4)-(-10) =$
$(4)+(+10) = 14$

17. 358.01

18. $3^2 + 4^2 = H^2$
$9 + 16 = H^2$
$25 = H^2$
$H = 5$ miles

19. $(-2) \times (6) = -12$ ft

20. $2X + 5 = X - 5$
$2X = X - 5 - 5$
$2X = X - 10$
$2X - X = -10$
$X = -10$

Test 11

1. addition; multiplication
2. subtraction; division
3. addition; multiplication
4. subtraction; division
5. $-5X + 8X - 4 = 3X - 4$
6. $-5Y + 3 - 6Y + 2Y + 4 =$
$(-5Y) + 3 + (-6Y) + 2Y + 4 =$
$(-5Y) + (-6Y) + 2Y + 3 + 4 = -9Y + 7$
7. $6 + X - 5 + 2X + 8 =$
$6 + X + (-5) + 2X + 8 =$
$X + 2X + 6 + (-5) + 8 = 3X + 9$
8. $3B - B + 7 + 4B =$
$3B + (-B) + 7 + 4B =$
$3B + (-B) + 4B + 7 = 6B + 7$
9. $4A - 1 = 3A + 8$
$4A = 3A + 8 + 1$
$4A = 3A + 9$
$4A - 3A = 9$
$A = 9$
10. $4A - 1 = 3A + 8 \Rightarrow 4(9) - 1 = 3(9) + 8$
$36 - 1 = 27 + 8$
$35 = 35$

11. $1 + F + 3 + F = F + 5$
$2F + 4 = F + 5$
$2F = F + 5 - 4$
$2F = F + 1$
$2F - F = 1$
$F = 1$

12. $1 + F + 3 + F = F + 5 \Rightarrow 1 + (1) + 3 + (1) = (1) + 5$
$6 = 6$

13. $\frac{1}{4} \times \frac{4}{5} = \frac{1}{5}$

14. $\frac{7}{10} \div \frac{1}{2} = \frac{7}{10} \times \frac{2}{1} = \frac{7}{5} = 1\frac{2}{5}$

15. $\frac{3}{4} \div \frac{9}{10} = \frac{3}{4} \times \frac{10}{9} = \frac{5}{6}$

16. $L^2 + 9^2 = 15^2$
$L^2 + 81 = 225$
$L^2 = 225 - 81$
$L^2 = 144$
$L = 12$ ft

17. $\sqrt{81} = 9$

18. $B + 5 = 11$
$B = 11 - 5$
$B = 6$ bushels

19. $\frac{1}{2} \times \frac{2}{3} = \frac{1}{3}$ of the bushes
$12 \div 3 = 4$
$4 \times 1 = 4$ bushes had pink blooms.

20. $\frac{3}{4} \div \frac{1}{8} = \frac{3}{4} \times \frac{8}{1} = \frac{6}{1} = 6$ days

Test 12

1. $5(3X + 5Y) =$
$5(3X) + 5(5Y) = 15X + 25Y$
2. $B(7 + 8) = 7B + 8B$ or $15B$
3. $X(4C + 3D) = 4CX + 3DX$
4. $9(Q + W) = 9Q + 9W$
5. $XY(2 + 2) = 2XY + 2XY$ or $4XY$
6. $2Y(5A + 6B) =$
$2Y(5A) + 2Y(6B) = 10AY + 12BY$
7. $10X + 10Y = 10(X + Y)$
8. $8A + 11A = A(8 + 11)$ or $19A$

9. $4XY + 8XZ = 4X(Y) + 4X(2Z) = 4X(Y + 2Z)$

10. $4Q + 4R = 4(Q + R)$

11. $5A + 25B = 5(A) + 5(5B) = 5(A + 5B)$

12. $AD + CD = D(A + C)$

13. $-A - 22 = -2A - 30$
$$-A = -2A - 30 + 22$$
$$-A = -2A - 8$$
$$-A + 2A = -8$$
$$A = -8$$

14. $-A - 22 = -2A - 30 \Rightarrow$
$$-(-8) - 22 = -2(-8) - 30$$
$$8 - 22 = 16 - 30$$
$$-14 = -14$$

15. $Y - 12 = 2Y - 4$
$$Y = 2Y - 4 + 12$$
$$Y = 2Y + 8$$
$$Y - 2Y = 8$$
$$-Y = 8$$
$$0 = 8 + Y$$
$$-8 = Y$$
$$Y = -8$$

16. $Y - 12 = 2Y - 4 \Rightarrow (-8) - 12 = 2(-8) - 4$
$$-8 - 12 = -16 - 4$$
$$-20 = -20$$

17. $\frac{4}{16} = \frac{1}{4}$ inch

18. $\frac{10}{16} = \frac{5}{8}$ inch

19. $\frac{14}{16} = \frac{7}{8}$ inch

20. $7X + 5X = 11X - 15$
$$12X = 11X - 15$$
$$12X - 11X = -15$$
$$X = -15$$

Test 13

1. $4X = -124$
$$X = -31$$

2. $4X = -124 \Rightarrow 4(-31) = -124$
$$-124 = -124$$

3. $-7Q + 6 + 5Q = 15 - 7$
$$-2Q + 6 = 8$$
$$-2Q = 8 - 6$$
$$-2Q = 2$$
$$Q = -1$$

4. $-7Q + 6 + 5Q = 15 - 7 \Rightarrow$
$$-7(-1) + 6 + 5(-1) = 15 - 7$$
$$7 + 6 - 5 = 15 - 7$$
$$8 = 8$$

5. $3(P + 5) + P = 3(2 + P)$
$$3P + 15 + P = 6 + 3P$$
$$4P + 15 = 6 + 3P$$
$$4P = 6 - 15 + 3P$$
$$4P = -9 + 3P$$
$$4P - 3P = -9$$
$$P = -9$$

6. $3(P + 5) + P = 3(2 + P) \Rightarrow$
$$3((-9) + 5) + (-9) = 3(2 + (-9))$$
$$3(-4) + (-9) = 3(-7)$$
$$-12 - 9 = -21$$
$$-21 = -21$$

7. $2(A + 4) + 6A = 2(2 + 3A)$
$$2A + 8 + 6A = 4 + 6A$$
$$8A + 8 = 4 + 6A$$
$$8A = 4 - 8 + 6A$$
$$8A = -4 + 6A$$
$$8A - 6A = -4$$
$$2A = -4$$
$$A = -2$$

8. $2(A + 4) + 6A = 2(2 + 3A) \Rightarrow$
$$2((-2) + 4) + 6(-2) = 2(2 + 3(-2))$$
$$2(2) + (-12) = 2(2 + (-6))$$
$$2(2) - 12 = 2(-4)$$
$$4 - 12 = -8$$
$$-8 = -8$$

9. $\frac{7}{16}$ inch

10. $\frac{12}{16} = \frac{3}{4}$ inch

11. $\frac{8}{16} = \frac{1}{2}$ inch

12. $\frac{\cancel{7}}{\cancel{8}_4} \times \frac{\cancel{2}}{\cancel{7}} = \frac{1}{4}$

13. $\dfrac{4}{5} \div \dfrac{1}{9} = \dfrac{4}{5} \times \dfrac{9}{1} = \dfrac{36}{5} = 7\dfrac{1}{5}$

14. $\dfrac{5}{6} \div \dfrac{4}{12} = \dfrac{5}{6} \times \dfrac{\cancel{12}^{2}}{\cancel{4}} = \dfrac{5}{2} = 2\dfrac{1}{2}$

15. $1\dfrac{3}{5} + 2\dfrac{3}{4} = 1\dfrac{12}{20} + 2\dfrac{15}{20} = 3\dfrac{27}{20} =$

$3 + \dfrac{20}{20} + \dfrac{7}{20} = 3 + 1 + \dfrac{7}{20} = 4\dfrac{7}{20}$

16. $3\dfrac{1}{6} + 5\dfrac{2}{7} = 3\dfrac{7}{42} + 5\dfrac{12}{42} = 8\dfrac{19}{42}$

17. $9\dfrac{1}{2} + 4\dfrac{1}{2} = 13\dfrac{2}{2} = 13 + 1 = 14$

18. $2X + 8 = 16$

$2X = 16 - 8$

$2X = 8$

$X = 4$

19. $4^2 + L^2 = 5^2$

$16 + L^2 = 25$

$L^2 = 25 - 16$

$L^2 = 9$

$L = 3 \text{ mi}$

20. 1

Test 14

1. $7^2 + 2^2 - 5 - 4 + 3X =$

$49 + 4 - 5 - 4 + 3X = 44 + 3X$

2. $X + 32 \div 4 - 2^2 =$

$X + 32 \div 4 - 4 =$

$X + 8 - 4 = X + 4$

3. $-Y - 5 + Y + 2(2Y - Y) - 3 =$

$-Y - 5 + Y + 2(Y) - 3 = 2Y - 8$

4. $5X - 3 - X - 3(X + 1^2) =$

$5X - 3 - X - 3(X + 1) =$

$5X - 3 - X - 3X - 3 = X - 6$

5. $5(B + 3) = 4(B - 7) + 2B$

$5B + 15 = 4B - 28 + 2B$

$5B + 15 = 6B - 28$

$5B = 6B - 28 - 15$

$5B = 6B - 43$

$5B - 6B = -43$

$-B = -43$

$B = 43$

6. $5(B + 3) = 4(B - 7) + 2B \Rightarrow$

$5((43) + 3) = 4((43) - 7) + 2(43)$

$5(46) = 4(36) + 2(43)$

$230 = 144 + 86$

$230 = 230$

7. $5^3 - 10^2 = X(8 - 2) + 2X - 3X$

$5^3 - 10^2 = X(6) + 2X - 3X$

$125 - 100 = 6X + 2X - 3X$

$25 = 5X$

$X = 5$

8. $5^3 - 10^2 = X(8 - 2) + 2X - 3X \Rightarrow$

$5^3 - 10^2 = (5)(8 - 2) + 2(5) - 3(5)$

$5^3 - 10^2 = 5(6) + 2(5) - 3(5)$

$125 - 100 = 5(6) + 2(5) - 3(5)$

$125 - 100 = 30 + 10 - 15$

$25 = 25$

9. $(-3)^2 + (8 + 3^2) = 2A$

$(-3)^2 + (8 + 9) = 2A$

$(-3)^2 + (17) = 2A$

$9 + 17 = 2A$

$26 = 2A$

$A = 13$

10. $(-3)^2 + (8 + 3^2) = 2A \Rightarrow (-3)^2 + (8 + 3^2) = 2(13)$

$(-3)^2 + (8 + 9) = 2(13)$

$(-3)^2 + (17) = 2(13)$

$9 + 17 = 26$

$26 = 26$

11. $-(8)^2 = -(8)(8) = -64$

12. $2^3 = (2)(2)(2) = 8$

13. $-4^2 = -(4)(4) = -16$

14. $1^5 = (1)(1)(1)(1)(1) = 1$

15. $7\dfrac{1}{4} - 1\dfrac{3}{8} = 7\dfrac{8}{32} - 1\dfrac{12}{32} =$

$6\dfrac{40}{32} - 1\dfrac{12}{32} = 5\dfrac{28}{32} = 5\dfrac{7}{8}$

16. $9 - 2\dfrac{2}{3} = 8\dfrac{3}{3} - 2\dfrac{2}{3} = 6\dfrac{1}{3}$

17. $10\dfrac{1}{3} - 6\dfrac{5}{9} = 10\dfrac{9}{27} - 6\dfrac{15}{27} =$

$9\dfrac{36}{27} - 6\dfrac{15}{27} = 3\dfrac{21}{27} = 3\dfrac{7}{9}$

18. $5\frac{3}{4} - 3\frac{1}{2} = 5\frac{6}{8} - 3\frac{4}{8} = 2\frac{2}{8} = 2\frac{1}{4}$ in

19. $3X - 9 = 6(6) \div 4$

$3X - 9 = 36 \div 4$

$3X - 9 = 9$

$3X = 9 + 9$

$3X = 18$

$X = 6$

20. $72 \div 4 = 18$

$18 \times 3 = 54$ bought chocolate.

$72 - 54 = 18$ didn't buy chocolate.

Unit Test II

1. $\sqrt{36} = 6$

2. $\sqrt{R^2} = R$

3. $\sqrt{64} = 8$

4. $\dfrac{3}{\sqrt{25}} = \dfrac{3}{5}$

5. $4^2 - 3(5 - 2) - 25 + 6 =$

$4^2 - 3(3) - 25 + 6 =$

$16 - 9 - 25 + 6 = -12$

6. $13 + 49 \div 7 - 2^2 =$

$13 + 49 \div 7 - 4 =$

$13 + 7 - 4 = 16$

7. $\left(3 \times 6^2 - 1\right) + 11 =$

$(3 \times 36 - 1) + 11 =$

$(108 - 1) + 11 = 118$

8. $3\left(20 - 4^2\right) + 2 \times 3 =$

$3(20 - 16) + 2 \times 3 =$

$3(4) + 2 \times 3 =$

$12 + 6 = 18$

9. $3 + 10 - R + 6R = -3 + 9R + 5 - 5$

$13 + 5R = -3 + 9R$

$5R = -3 - 13 + 9R$

$5R = -16 + 9R$

$5R - 9R = -16$

$-4R = -16$

$R = 4$

10. $3 + 10 - R + 6R = -3 + 9R + 5 - 5 \Rightarrow$

$3 + 10 - (4) + 6(4) = -3 + 9(4) + 5 - 5$

$3 + 10 - 4 + 24 = -3 + 36 + 5 - 5$

$33 = 33$

11. $(-3)^2 + \left(F + 3^2\right) = 2 \cdot 4 + 6$

$(-3)^2 + (F + 9) = 2 \cdot 4 + 6$

$9 + F + 9 = 2 \cdot 4 + 6$

$9 + F + 9 = 8 + 6$

$F + 18 = 14$

$F = 14 - 18$

$F = -4$

12. $(-3)^2 + \left(F + 3^2\right) = 2 \cdot 4 + 6 \Rightarrow$

$(-3)^2 + \left((-4) + 3^2\right) = 2 \cdot 4 + 6$

$(-3)^2 + (-4 + 9) = 2 \cdot 4 + 6$

$(-3)^2 + (5) = 2 \cdot 4 + 6$

$9 + 5 = 8 + 6$

$14 = 14$

13. $-3X + 4X = 2 \times 4 - X$

$X = 8 - X$

$X + X = 8$

$2X = 8$

$X = 4$

14. $-3X + 4X = 2 \cdot 4 - X \Rightarrow$

$-3(4) + 4(4) = 2 \cdot 4 - (4)$

$-12 + 16 = 8 - 4$

$4 = 4$

15. $\dfrac{8}{16} = \dfrac{1}{2}$ inch

16. $\dfrac{3}{16}$ inch

17. $\dfrac{14}{16} = \dfrac{7}{8}$ inch

18. $12^2 + 16^2 = H^2$

$144 + 256 = H^2$

$400 = H^2$

$H = 20$ yd

19. $L^2 + 9^2 = 15^2$
$L^2 + 81 = 225$
$L^2 = 225 - 81$
$L^2 = 144$
$L = 12$ ft

20. $\dfrac{1}{2} \div \dfrac{1}{6} = \dfrac{1}{\cancel{2}} \times \dfrac{\cancel{6}^{3}}{1} = \dfrac{3}{1} = 3$

21. $\dfrac{7}{8} \div \dfrac{3}{4} = \dfrac{7}{\cancel{8}_{2}} \times \dfrac{\cancel{4}}{3} = \dfrac{7}{6} = 1\dfrac{1}{6}$

22. $\dfrac{5}{7} \div \dfrac{5}{9} = \dfrac{\cancel{5}}{7} \times \dfrac{9}{\cancel{5}} = \dfrac{9}{7} = 1\dfrac{2}{7}$

23. $3\dfrac{5}{8} + 2\dfrac{5}{6} = 3\dfrac{30}{48} + 2\dfrac{40}{48} = 5\dfrac{70}{48} =$
$5 + \dfrac{48}{48} + \dfrac{22}{48} = 5 + 1 + \dfrac{22}{48} =$
$6\dfrac{22}{48} = 6\dfrac{11}{24}$

24. $8 - 1\dfrac{1}{3} = 7\dfrac{3}{3} - 1\dfrac{1}{3} = 6\dfrac{2}{3}$

25. $9\dfrac{1}{8} - 5\dfrac{4}{5} = 9\dfrac{5}{40} - 5\dfrac{32}{40} =$
$8\dfrac{45}{40} - 5\dfrac{32}{40} = 3\dfrac{13}{40}$

26. $5^2 + 6^2 = 8^2$
$25 + 36 = 64$
$61 \neq 64$; no

27. $8N - 5 = 7N + 5$
$8N = 7N + 5 + 5$
$8N = 7N + 10$
$8N - 7N = 10$
$N = 10$

28. no

Test 15

1. $2(6 \times 6) + 2(6 \times 6) + 2(6 \times 6) =$
$2(36) + 2(36) + 2(36) =$
$72 + 72 + 72 = 216$ in^2

2. $4\left(\dfrac{1}{2} \times 3 \times 8\right) + (3 \times 3) =$
$4(12) + 9 =$
$48 + 9 = 57$ in^2

3. $2(15 \times 25) + 2(25 \times 8) + 2(8 \times 15) =$
$2(375) + 2(200) + 2(120) =$
$750 + 400 + 240 = 1{,}390$ ft^2

4. $4\left(\dfrac{1}{2} \times 10 \times 10\right) + (10 \times 10) =$
$4(50) + (100) =$
$200 + 100 = 300$ in^2

5. $5B + 10 = -10$
$5B = -10 - 10$
$5B = -20$
$B = -4$

6. $5B + 10 = -10 \Rightarrow 5(-4) + 10 = -10$
$-20 + 10 = -10$
$-10 = -10$

7. $A - 9 = -14$
$A = -14 + 9$
$A = -5$

8. $A - 9 = -14 \Rightarrow (-5) - 9 = -14$
$-14 = -14$

9. $4(J + 4) + 3J = 3(28 + J)$
$4J + 16 + 3J = 84 + 3J$
$7J + 16 = 84 + 3J$
$7J = 84 - 16 + 3J$
$7J = 68 + 3J$
$7J - 3J = 68$
$4J = 68$
$J = 17$

10. $4(J + 4) + 3J = 3(28 + J) \Rightarrow$
$4((17) + 4) + 3(17) = 3(28 + (17))$
$4(21) + 3(17) = 3(45)$
$84 + 51 = 135$
$135 = 135$

11.
$$(-5)^2 + (9 + 4^2) = 5B$$
$$(-5)^2 + (9 + 16) = 5B$$
$$(-5)^2 + (25) = 5B$$
$$25 + 25 = 5B$$
$$50 = 5B$$
$$B = 10$$

12.
$$(-5)^2 + (9 + 4^2) = 5B \Rightarrow (-5)^2 + (9 + 4^2) = 5(10)$$
$$(-5)^2 + (9 + 16) = 5(10)$$
$$(-5)^2 + (25) = 5(10)$$
$$25 + 25 = 5(10)$$
$$25 + 25 = 50$$
$$50 = 50$$

13.
$$2\frac{7}{8} \times 2\frac{6}{9} \times 1\frac{1}{2} = \frac{23}{8} \times \frac{\overset{3}{\cancel{24}}}{9} \times \frac{3}{2} =$$
$$\frac{23}{1} \times \frac{\cancel{3}}{\underset{3}{\cancel{9}}} \times \frac{3}{2} = \frac{23}{2} = 11\frac{1}{2}$$

14.
$$3\frac{1}{4} \times 6\frac{2}{3} \times 5\frac{1}{10} = \frac{13}{4} \times \frac{20}{\cancel{3}} \times \frac{\overset{17}{\cancel{51}}}{10} =$$
$$\frac{13}{\cancel{4}} \times \frac{\overset{5}{\cancel{20}}}{1} \times \frac{17}{10} = \frac{13}{1} \times \frac{\cancel{5}}{1} \times \frac{17}{\underset{2}{\cancel{10}}} =$$
$$\frac{221}{2} = 110\frac{1}{2}$$

15.
$$\frac{1}{7} \times 1\frac{2}{5} \times 4\frac{1}{5} = \frac{1}{\cancel{7}} \times \frac{\cancel{7}}{5} \times \frac{21}{5} = \frac{21}{25}$$

16.
$$2(25 \times 15) =$$
$$2(375) = 750 \text{ ft}^2$$

17. $750 \div 100 = 7.5$
7.5 rounded up = 8 squares

18.
$$2(18 \times 10) + 2(25 \times 10) + 2\left(\frac{1}{2} \times 12 \times 18\right) =$$
$$2(180) + 2(250) + 2(108) =$$
$$360 + 500 + 216 = 1,076 \text{ ft}^2$$

19. $1,076 \div 100 = 10.76$
10.76 rounded up = 11 squares

20. $24 \div 3 = 8$
$8 \div 2 = 4$ pennies

Test 16

1.
$$\overset{12}{\cancel{6}}\cancel{0} \times \frac{9}{\cancel{5}} + 32 = F$$
$$108 + 32 = F$$
$$F = 140°$$

2.
$$\overset{22}{\cancel{1}}\cancel{1}\cancel{0} \times \frac{9}{\cancel{5}} + 32 = F$$
$$198 + 32 = F$$
$$F = 230°$$

3.
$$13 \times \frac{9}{5} + 32 = F$$
$$\frac{117}{5} + 32 = F$$
$$23.4 + 32 = F$$
$$F = 55.4°$$

4.
$$42 \times \frac{9}{5} + 32 = F$$
$$\frac{378}{5} + 32 = F$$
$$75.6 + 32 = F$$
$$F = 107.6°$$

5.
$$Q + 4 = 3Q - 6$$
$$Q = 3Q - 6 - 4$$
$$Q = 3Q - 10$$
$$Q - 3Q = -10$$
$$-2Q = -10$$
$$Q = 5$$

6.
$$Q + 4 = 3Q - 6 \Rightarrow (5) + 4 = 3(5) - 6$$
$$5 + 4 = 15 - 6$$
$$9 = 9$$

7.
$$7^2 - X - 1 = 5X$$
$$49 - X - 1 = 5X$$
$$48 - X = 5X$$
$$-X = 5X - 48$$
$$-X - 5X = -48$$
$$-6X = -48$$
$$X = 8$$

8.
$$7^2 - X - 1 = 5X \Rightarrow 7^2 - (8) - 1 = 5(8)$$
$$49 - 8 - 1 = 40$$
$$40 = 40$$

9. $7.00 + 0.45 = 7.45$

10. $11.2 - 3 = 8.2$

11. $6.15 + 2.2 = 8.35$

7. $7^2 - X - 1 = 5X$
$49 - X - 1 = 5X$
$48 - X = 5X$
$-X = 5X - 48$
$-X - 5X = -48$
$-6X = -48$
$X = 8$

8. $7^2 - X - 1 = 5X \Rightarrow 7^2 - (8) - 1 = 5(8)$
$49 - 8 - 1 = 40$
$40 = 40$

9. $7.00 + 0.45 = 7.45$
10. $11.2 - 3 = 8.2$
11. $6.15 + 2.2 = 8.35$
12. $58.9 - 7.1 = 51.8$
13. $6 \times 10^1 + 1 \times 10^0 + 3 \times \dfrac{1}{10^1} + 2 \times \dfrac{1}{10^2}$
14. 0°C; 32°F
15. 100°C; 212°F
16. 37°C; 98.6°F
17. $2(12 \times 16) + 2(16 \times 2) + 2(2 \times 12) =$
$2(192) + 2(32) + 2(24) =$
$384 + 64 + 48 = 496 \text{ ft}^2$
18. $4\left(\dfrac{1}{2} \times 1 \times 2\right) + (1 \times 1) =$
$4(1) + 1 =$
$4 + 1 = 5 \text{ yd}^2$
19. $2.32 + 1.5 = 3.82 \text{ mi}$
20. $1\dfrac{3}{4} + 3\dfrac{2}{3} = 1\dfrac{9}{12} + 3\dfrac{8}{12} = 4\dfrac{17}{12} =$
$4 + \dfrac{12}{12} + \dfrac{5}{12} = 4 + 1 + \dfrac{5}{12} = 5\dfrac{5}{12} \text{ mi}$

Test 17

1. $(41 - 32)\dfrac{5}{9} = C$ or: $(41 - 32)(0.56) = C$
$(9)\dfrac{5}{9} = C$ $(9)(0.56) = C$
$C = 5°$ $C \approx 5.0°$

2. $(140 - 32)\dfrac{5}{9} = C$ or: $(140 - 32)(0.56) = C$
$(108)\dfrac{5}{9} = C$ $(108)(0.56) = C$
$C = 60°$ $C \approx 60.5°$

3. $(91 - 32)\dfrac{5}{9} = C$ or: $(91 - 32)(0.56) = C$
$(59)\dfrac{5}{9} = C$ $(59)(0.56) = C$
$\dfrac{295}{9} = C$ $C \approx 33.0°$
$C \approx 32.8°$

4. $200 \times \dfrac{9}{5} + 32 = F$
$360 + 32 = F$
$F = 392°$

5. $35 \times \dfrac{9}{5} + 32 = F$
$63 + 32 = F$
$F = 95°$

6. $56 \times \dfrac{9}{5} + 32 = F$
$\dfrac{504}{5} + 32 = F$
$100.8 + 32 = F$
$F = 132.8°$

7. $3(5) + 1 = -5X + X$
$15 + 1 = -4X$
$16 = -4X$
$X = -4$

8. $3(5) + 1 = -5X + X \Rightarrow 3(5) + 1 = -5(-4) + (-4)$
$15 + 1 = 20 - 4$
$16 = 16$

9. $D(2 - 5) - 8 = -3D - 2D + 6$
$2D - 5D - 8 = -5D + 6$
$-3D - 8 = -5D + 6$
$-3D = -5D + 6 + 8$
$-3D = -5D + 14$
$-3D + 5D = 14$
$2D = 14$
$D = 7$

10. $D(2 - 5) - 8 = -3D - 2D + 6 \Rightarrow$
$(7)(2 - 5) - 8 = -3(7) - 2(7) + 6$
$14 - 35 - 8 = -21 - 14 + 6$
$-29 = -29$

11. $2.8 \times 0.31 = 0.868$

12. $0.456 \times 0.2 = 0.0912$

13. $0.78 \times 0.59 = 0.4602$

14. $1\frac{1}{6} \times 5\frac{3}{8} = \frac{7}{6} \times \frac{43}{8} = \frac{301}{48} = 6\frac{13}{48}$

15. $2\frac{4}{5} \times 2\frac{1}{2} = \frac{14}{5} \times \frac{5}{2} = \frac{14}{1} \times \frac{1}{2} = \frac{7}{1} = 7$

16. $3\frac{1}{10} \times 3\frac{1}{3} = \frac{31}{10} \times \frac{10}{3} = \frac{31}{3} = 10\frac{1}{3}$

17. $\$10.00 - \$4.37 = \$5.63$

18. $\$8.98 \times 3 = \$26.94$

19. above; water freezes at 0°C.

20. $2(12 \times 8) + 2(11 \times 8) =$
$2(96) + 2(88) =$
$192 + 176 = 368 \text{ ft}^2$

Test 18

1. $(8-4)^2 \times |4-8| =$
$(4)^2 \times |-4| =$
$16 \times 4 = 64$

2. $11 + |3 + 2^2| =$
$11 + |3 + 4| =$
$11 + |7| =$
$11 + 7 = 18$

3. $|1 - 3^2| + |-5| =$
$|1 - 9| + |-5| =$
$|-8| + |-5| =$
$8 + 5 = 13$

4. $2Y - 2 = 3Y - |6 - 12|$
$2Y - 2 = 3Y - |-6|$
$2Y - 2 = 3Y - 6$
$2Y = 3Y - 6 + 2$
$2Y = 3Y - 4$
$2Y - 3Y = -4$
$-Y = -4$
$Y = 4$

5. $2Y - 2 = 3Y - |6 - 12| \Rightarrow$
$2(4) - 2 = 3(4) - |6 - 12|$
$8 - 2 = 12 - |-6|$
$8 - 2 = 12 - 6$
$6 = 6$

6. $-2X + |6 + 1| + 3X - 4 = 10 - 1$
$-2X + |7| + 3X - 4 = 10 - 1$
$X + 3 = 9$
$X = 9 - 3$
$X = 6$

7. $-2X + |6 + 1| + 3X - 4 = 10 - 1 \Rightarrow$
$-2(6) + |6 + 1| + 3(6) - 4 = 10 - 1$
$-12 + |7| + 18 - 4 = 10 - 1$
$9 = 9$

8. $4A + 5 - 3 = 2A + (2)^2(2)$
$4A + 5 - 3 = 2A + 4(2)$
$4A + 5 - 3 = 2A + 8$
$4A + 2 = 2A + 8$
$4A = 2A + 8 - 2$
$4A = 2A + 6$
$4A - 2A = 6$
$2A = 6$
$A = 3$

9. $4A + 5 - 3 = 2A + (2)^2(2) \Rightarrow$
$4(3) + 5 - 3 = 2(3) + (2)^2(2)$
$4(3) + 5 - 3 = 2(3) + 4(2)$
$12 + 5 - 3 = 6 + 8$
$14 = 14$

10. $-2B + 3 + 5B + 1^2 = 2(3 + 2) + 3^2$
$-2B + 3 + 5B + 1^2 = 2(5) + 3^2$
$-2B + 3 + 5B + 1 = 2(5) + 9$
$-2B + 3 + 5B + 1 = 10 + 9$
$3B + 4 = 19$
$3B = 19 - 4$
$3B = 15$
$B = 5$

11. $-2B+3+5B+I^2 = 2(3+2)+3^2 \Rightarrow$

$\qquad -2(5)+3+5(5)+I^2 = 2(3+2)+3^2$

$\qquad -2(5)+3+5(5)+I^2 = 2(5)+3^2$

$\qquad -2(5)+3+5(5)+1 = 2(5)+9$

$\qquad -10+3+25+1 = 10+9$

$\qquad\qquad 19 = 19$

12. $\overset{7}{\cancel{3}}\cancel{5} \times \dfrac{9}{\cancel{5}} + 32 = F$

$\qquad 63+32 = F$

$\qquad\qquad F = 95°$

13. $(72-32)\dfrac{5}{9} = C$ or: $(72-32)(0.56) = C$

$\qquad (40)\dfrac{5}{9} = C \qquad\qquad (40)(0.56) = C$

$\qquad\qquad\qquad\qquad\qquad\qquad C \approx 22.4°$

$\qquad \dfrac{200}{9} = C$

$\qquad\qquad C \approx 22.2°$

14.
```
  0.87
5)4.35
  40
  35
  35
   0
```

15.
```
   2.88
8)23.04
  16
   70
   64
   64
   64
    0
```

16.
```
  4.5
4)18.0
  16
   20
   20
    0
```

17.
```
  0.72
3)2.16
  21
   06
    6
    0
```

18. $100°C$

19. $2(5.2 \times 5.2) + 2(5.2 \times 5.2) + 2(5.2 \times 5.2) =$

$\qquad 2(27.04) + 2(27.04) + 2(27.04) =$

$\qquad\qquad 54.08 + 54.08 + 54.08 = 162.24 \text{ in}^2$

or

$\qquad 6(5.2 \times 5.2) =$

$\qquad 6(27.04) = 162.24 \text{ in}^2$

20. $3.6 \text{ lb} \div 9 \text{ people} = 0.4 \text{ lb}$
of meat per person

Test 19

1. $\dfrac{7}{12} = \dfrac{A}{60}$

$\qquad 12A = 7(60)$

$\qquad 12A = 420$

$\qquad A = 35$

2. $\dfrac{9}{10} = \dfrac{81}{T}$

$\qquad 9T = 10(81)$

$\qquad 9T = 810$

$\qquad T = 90$

3. $\dfrac{6}{G} = \dfrac{8}{16}$

$\qquad 8G = 6(16)$

$\qquad 8G = 96$

$\qquad G = 12$

4. $\dfrac{3}{5} = \dfrac{Q}{25}$

$\qquad 5Q = 3(25)$

$\qquad 5Q = 75$

$\qquad Q = 15$

5. $\dfrac{6}{11} = \dfrac{36}{X}$

$\qquad 6X = 11(36)$

$\qquad 6X = 396$

$\qquad X = 66$

6. $\dfrac{2}{D} = \dfrac{3}{9}$

$\qquad 3D = 2(9)$

$\qquad 3D = 18$

$\qquad D = 6$

7. $2(2X+1) = 2(X+4)$

$\qquad 4X+2 = 2X+8$

$\qquad 4X = 2X+8-2$

$\qquad 4X = 2X+6$

$\qquad 4X-2X = 6$

$\qquad 2X = 6$

$\qquad X = 3$

8. $2(2X+1) = 2(X+4) \Rightarrow 2(2(3)+1) = 2((3)+4)$

$\qquad\qquad\qquad\qquad\qquad\qquad 2(6+1) = 2((3)+4)$

$\qquad\qquad\qquad\qquad\qquad\qquad 2(7) = 2(7)$

$\qquad\qquad\qquad\qquad\qquad\qquad 14 = 14$

9. $11 + Q = \left| (-3^2) \right| + 4^2 - 5$

    $11 + Q = \left| (-(3)(3)) \right| + 4^2 - 5$

    $11 + Q = |-9| + 4^2 - 5$

    $11 + Q = 9 + 16 - 5$

    $11 + Q = 20$

    $\quad\quad Q = 20 - 11$

    $\quad\quad Q = 9$

10. $11 + Q = \left| (-3^2) \right| + 4^2 - 5 \Rightarrow$

    $11 + (9) = \left| (-3^2) \right| + 4^2 - 5$

    $11 + 9 = |-9| + 4^2 - 5$

    $11 + 9 = 9 + 16 - 5$

    $\quad 20 = 20$

11. $1.8 \times 0.23 = 0.414$

12. $0.169 + 1.5 = 1.669$

13. $17.2 - 3.4 = 13.8$

14.
$$\begin{array}{r} 3 \\ 5\overline{)15} \\ \underline{15} \\ 0 \end{array}$$

15.
$$\begin{array}{r} 1,200 \\ 8\overline{)9,600} \\ \underline{8} \\ 16 \\ \underline{16} \\ 0 \end{array}$$

16.
$$\begin{array}{r} 2 \\ 6\overline{)12} \\ \underline{12} \\ 0 \end{array}$$

17. sunny to total $= \dfrac{5}{30} = \dfrac{1}{6}$

    sunny to cloudy $= \dfrac{5}{25} = \dfrac{1}{5}$

18. $\dfrac{3}{5}$ of 365:

    $365 \div 5 = 73$

    $73 \times 3 = 219$ sunny days

19. $75 \div 1.5 = 50$ pieces

20. $32 \div 2 = 16$ stayed

    $\dfrac{3}{4}$ of 16:

    $16 \div 4 = 4$

    $4 \times 3 = 12$ bored

    16 stayed $- 12$ bored $= 4$ not bored

Test 20

1. $\dfrac{X}{12} = \dfrac{5}{15}$

    $15X = 12(5)$

    $15X = 60$

    $\quad X = 4$ units

2. $\dfrac{R}{13} = \dfrac{28}{13}$

    $13R = 13(28)$

    $\quad R = 28$ units

3. $\dfrac{V}{36} = \dfrac{10}{18}$

    $18V = 36(10)$

    $18V = 360$

    $\quad V = 20$ units

4. $\dfrac{X}{2} = \dfrac{144}{6}$

    $6X = 2(144)$

    $6X = 288$

    $\quad X = 48$ units

5. $\dfrac{5}{6} = \dfrac{45}{Y}$

    $5Y = 6(45)$

    $5Y = 270$

    $\quad Y = 54$

6. $\dfrac{9}{G} = \dfrac{63}{70}$

    $63G = 9(70)$

    $63G = 630$

    $\quad G = 10$

7. $\dfrac{3}{8} = 3 \div 8 = 0.375$

8. $\dfrac{10}{100} = 10 \div 100 = 0.1$

9. $\dfrac{5}{10} = 5 \div 10 = 0.5$

10. $\dfrac{1}{4} = 1 \div 4 = 0.25$

11. $0.45 = \dfrac{45}{100} = \dfrac{9}{20}$

12. $0.8 = \dfrac{8}{10} = \dfrac{4}{5}$

13. $0.66 = \dfrac{66}{100} = \dfrac{33}{50}$

14. $0.4 = \dfrac{4}{10} = \dfrac{2}{5}$

15. $3X - 9 + 7X - 10 = 9X - 5X + 5$
$10X - 19 = 4X + 5$
$10X = 4X + 5 + 19$
$10X = 4X + 24$
$10X - 4X = 24$
$6X = 24$
$X = 4$

16. $3X - 9 + 7X - 10 = 9X - 5X + 5 \Rightarrow$
$3(4) - 9 + 7(4) - 10 = 9(4) - 5(4) + 5$
$12 - 9 + 28 - 10 = 36 - 20 + 5$
$21 = 21$

17. $5^2 \div 5 + 3(X + 7) = 2X + 27$
$5^2 \div 5 + 3X + 21 = 2X + 27$
$25 \div 5 + 3X + 21 = 2X + 27$
$5 + 3X + 21 = 2X + 27$
$3X + 26 = 2X + 27$
$3X = 2X + 27 - 26$
$3X = 2X + 1$
$3X - 2X = 1$
$X = 1$

18. $5^2 \div 5 + 3(X + 7) = 2X + 27 \Rightarrow$
$5^2 \div 5 + 3((1) + 7) = 2(1) + 27$
$5^2 \div 5 + 3(8) = 2 + 27$
$25 \div 5 + 3(8) = 2 + 27$
$5 + 24 = 2 + 27$
$29 = 29$

19. 1 pie $- 0.75$ pie $= 0.25$ of a pie left
$0.25 = \dfrac{25}{100} = \dfrac{1}{4}$ of a pie left

20. $40.5 - 36.2 = 4.3$ ft

Test 21

1. $18 = 2 \times 3 \times 3$

2. $25 = 5 \times 5$

3. $7 = 1 \times 7$; 1 is not a prime number, so the most accurate answer is 7.

4. $32 = 2 \times 2 \times 2 \times 2 \times 2$
$64 = 2 \times 2 \times 2 \times 2 \times 2 \times 2$
$LCM = 2 \times 2 \times 2 \times 2 \times 2 \times 2 = 64$

5. $5 = 1 \times 5$; $7 = 1 \times 7$
$LCM = 5 \times 7 = 35$

6. $25 = 5 \times 5$; $10 = 2 \times 5$
$LCM = 2 \times 5 \times 5 = 50$

7. $11 = 1 \times 11$; $12 = 2 \times 2 \times 3$
$LCM = 2 \times 2 \times 3 \times 11 = 132$

8. $8 = 2 \times 2 \times 2$; $12 = 2 \times 2 \times 3$
$LCM = 2 \times 2 \times 2 \times 3 = 24$

9. $4 = 2 \times 2$; $6 = 2 \times 3$
$LCM = 2 \times 2 \times 3 = 12$

10. $\dfrac{R}{3.1} = \dfrac{8.4}{2.8}$
$2.8R = 3.1(8.4)$
$2.8R = 26.04$
$R = 9.3$ units

11. $\dfrac{E}{0.7} = \dfrac{8}{0.5}$
$0.5E = 0.7(8)$
$0.5E = 5.6$
$E = 11.2$ units

12. $\dfrac{1}{4} = 1 \div 4 = 0.25 = 25\%$

13. $\dfrac{2}{3} = 2 \div 3 = 0.66\dfrac{2}{3} = 66\dfrac{2}{3}\%$

14. $\dfrac{3}{100} = 3 \div 100 = 0.03 = 3\%$

15. $\dfrac{7}{8} = 7 \div 8 = 0.87\dfrac{1}{2} = 87\dfrac{1}{2}\%$

16. $\dfrac{1}{2} = 1 \div 2 = 0.5 = 50\%$

17. $\dfrac{40}{100} = 40 \div 100 = 0.40 = 40\%$

18. $3^2 + 5^2 = 7^2$
$9 + 25 = 49$
$34 \neq 49$; no

19. $5\dfrac{1}{2} - 1\dfrac{3}{8} = 5\dfrac{4}{8} - 1\dfrac{3}{8} = 4\dfrac{1}{8}$ bushels

20. $S + 2S = 9$
$3S = 9$
$S = 3$ songs

Test 22

1. $24 = \underline{2} \times 2 \times 2 \times \underline{3}$; $42 = \underline{2} \times \underline{3} \times 7$
   $GCF = 2 \times 3 = 6$

2. $21 = 3 \times \underline{7}$; $14 = 2 \times \underline{7}$
   $GCF = 7$

3. $44 = \underline{2} \times \underline{2} \times 11$; $36 = \underline{2} \times \underline{2} \times 3 \times 3$
   $GCF = 2 \times 2 = 4$

4. $12 = \underline{2} \times \underline{2} \times \underline{3}$; $48 = \underline{2} \times \underline{2} \times 2 \times 2 \times \underline{3}$
   $GCF = 2 \times 2 \times 3 = 12$

5. $3 = 1 \times \underline{3}$; $9 = \underline{3} \times 3$
   $GCF = 3$

6. $10 = \underline{2} \times \underline{5}$; $60 = \underline{2} \times 2 \times 3 \times \underline{5}$
   $GCF = 2 \times 5 = 10$

7. $3 = 1 \times 3$; $6 = 2 \times 3$
   $LCM = 2 \times 3 = 6$

8. $8 = 2 \times 2 \times 2$; $12 = 2 \times 2 \times 3$
   $LCM = 2 \times 2 \times 2 \times 3 = 24$

9. $5 = 1 \times 5$; $15 = 3 \times 5$
   $LCM = 3 \times 5 = 15$

10. $(-13) + (-45) = -58$

11. $(-32) \times 19 = -608$

12. $-1 - 2 = -3$

13. $0.35 \times 100 = 35$

14. $0.02 \times 4.6 = 0.092$

15. $0.10 \times 0.95 = 0.095$

16. $25 \div 5 + 3(X + 7) = 2X + 3^3$
   $25 \div 5 + 3X + 21 = 2X + 3^3$
   $25 \div 5 + 3X + 21 = 2X + 27$
   $5 + 3X + 21 = 2X + 27$
   $3X + 26 = 2X + 27$
   $3X = 2X + 27 - 26$
   $3X = 2X + 1$
   $3X - 2X = 1$
   $X = 1$

17. $4X + 3^2 - 9 + 17 = 27 - X$
   $4X + 9 - 9 + 17 = 27 - X$
   $4X + 17 = 27 - X$
   $4X = 27 - 17 - X$
   $4X = 10 - X$
   $4X + X = 10$
   $5X = 10$
   $X = 2$

18. $\$12.45 + \$35.95 = \$48.40$ spent
   $\$50.00 - \$48.40 = \$1.60$ left

19. $0.20 \times 50 = 10$ found Monday
   $0.10 \times 50 = 5$ found Tuesday
   $5 + 10 = 15$ found altogether
   $50 - 15 = 35$ left to find

20. $\dfrac{1}{4} = 1 \div 4 = 0.25 = 25\%$ of the pie

Unit Test III

1. $\dfrac{1}{3} = \dfrac{25}{Y}$
   $1Y = 3(25)$
   $Y = 75$

2. $\dfrac{14}{32} = \dfrac{A}{16}$
   $32A = 14(16)$
   $32A = 224$
   $A = 7$

3. $\dfrac{7}{G} = \dfrac{7}{8}$
   $7G = 7(8)$
   $G = 8$

4. $\dfrac{R}{4} = \dfrac{9}{3}$
   $3R = 4(9)$
   $3R = 36$
   $R = 12$ units

5. $\dfrac{E}{0.6} = \dfrac{10}{0.2}$
   $0.2E = 0.6(10)$
   $0.2E = 6$
   $E = 30$ units

6. $4 = 2 \times 2$; $5 = 1 \times 5$
   $LCM = 2 \times 2 \times 5 = 20$

7. $15 = 3 \times 5$; $20 = 2 \times 2 \times 5$
   $LCM = 2 \times 2 \times 3 \times 5 = 60$

8. $12 = 2 \times 2 \times 3$; $32 = 2 \times 2 \times 2 \times 2 \times 2$
   $LCM = 2 \times 2 \times 2 \times 2 \times 2 \times 3 = 96$

9. $11 = 1 \times \underline{11}$; $22 = 2 \times \underline{11}$
   $GCF = 11$

10. $16 = \underline{2} \times 2 \times 2 \times 2$; $18 = \underline{2} \times 3 \times 3$
    GCF $= 2$

11. $10 = 2 \times \underline{5}$; $25 = \underline{5} \times 5$
    GCF $= 5$

12. $2(15 \times 13) + 2(10 \times 13) + 2(15 \times 10) =$
    $2(195) + 2(130) + 2(150) =$
    $390 + 260 + 300 = 950 \, \text{ft}^2$

13. $4\left(\dfrac{1}{2} \times 10 \times 5\right) + (10 \times 10) =$
    $4(25) + (100) =$
    $100 + 100 = 200 \, \text{ft}^2$

14. $^{6}\cancel{3}0 \times \dfrac{9}{\cancel{5}} + 32 = F$
    $54 + 32 = F$
    $F = 86°$; summer

15. $(41 - 32)\dfrac{5}{9} = C$ or: $(41 - 32)(0.56) = C$
    $(\cancel{9})\dfrac{5}{9} = C$ $(9)(0.56) = C$
    $C = 5°$ $C \approx 5.0°$

16. $|2 - 4^2| + |5| =$
    $|2 - 16| + |5| =$
    $|-14| + |5| =$
    $14 + 5 = 19$

17. $8.5 + 2.0 = 10.5$

18. $5.28 - 3.04 = 2.24$

19. $4.00 + 2.69 = 6.69$

20. $6.0 - 1.7 = 4.3$

21. $5.9 \times 0.4 = 2.36$

22. $0.006 \times 0.36 = 0.00216$

23. $7.8 \times 3.1 = 24.18$

24. $2.8 \div 7 = 0.4$

25. $75 \div 0.05 = 1,500$

26. $0.06 \div 0.03 = 2$

27. $\dfrac{4}{5} = 4 \div 5 = 0.80 = 80\%$

28. $\dfrac{1}{3} = 1 \div 3 = 0.33\dfrac{1}{3} = 33\dfrac{1}{3}\%$

29. $\dfrac{16}{100} = 16 \div 100 = 0.16 = 16\%$

30. $\$80.00 \times 0.05 = \$4.00$ should be saved

Test 23

1. $7X^2 - 4X - 1$
   $\underline{+ \;\; -5X^2 + 2X + 4}$
        $2X^2 - 2X + 3$

2. $3X^2 + 3X + 6$
   $\underline{+ \;\; 1X^2 + 6X - 1}$
        $4X^2 + 9X + 5$

3. $-X^2 - 2X - 2$
   $\underline{+ \;\; -4X^2 + \;X + 9}$
        $-5X^2 - \;X + 7$

4. $9X^2 - 6X - 3$
   $\underline{+ \;\; -6X^2 + 3X + 8}$
        $3X^2 - 3X + 5$

5. $5X^2 + \;4X + 7$
   $\underline{+ \;\; 3X^2 + \;8X - 2}$
        $8X^2 + 12X + 5$

6. $-7X^2 - \;X - 5$
   $\underline{+ \;\; -X^2 + 5X + 1}$
        $-8X^2 + 4X - 4$

7. $\dfrac{2}{5} = \dfrac{R}{100}$
   $5R = 2(100)$
   $5R = 200$
   $R = 40$

8. $\dfrac{46}{X} = \dfrac{2}{7}$
   $2X = 46(7)$
   $2X = 322$
   $X = 161$

9. $\dfrac{11}{50} = \dfrac{22}{Q}$
   $11Q = 50(22)$
   $11Q = 1,100$
   $Q = 100$

10. $5.00 \times 3.4 = 17$

11. $2.50 \times 78.2 = 195.5$

12. $1.20 \times 40 = 48$

13. $6 = 2 \times 3$; $8 = 2 \times 2 \times 2$
    LCM $= 2 \times 2 \times 2 \times 3 = 24$

14. $6 = 2 \times \underline{3}$; $21 = \underline{3} \times 7$
    GCF = 3

15. $2(7 \times 10.1) + 2(10.1 \times 8) + 2(8 \times 7) =$
    $2(70.7) + 2(80.8) + 2(56) =$
    $141.4 + 161.6 + 112 = 415$ in^2

16. $4\left(\dfrac{1}{2} \times 0.5 \times 0.12\right) + (0.5 \times 0.5) =$
    $4(0.03) + (0.25) =$
    $0.12 + 0.25 = 0.37$ ft^2

17. 212°F

18. $\dfrac{3}{4} + 1\dfrac{1}{2} + 2\dfrac{1}{4} = \dfrac{3}{4} + 1\dfrac{2}{4} + 2\dfrac{1}{4} = 3\dfrac{6}{4} =$
    $3 + \dfrac{4}{4} + \dfrac{2}{4} = 3 + 1 + \dfrac{2}{4} = 4\dfrac{2}{4} = 4\dfrac{1}{2}$ lb

19. $1.70 \times \$1,500 = \$2,550$ was raised

20. $3X + 9 = -2X + 44$
    $3X = -2X + 44 - 9$
    $3X = -2X + 35$
    $3X + 2X = 35$
    $5X = 35$
    $X = 7$

Test 24

1. $V \approx 3.14(4)^2(10) \approx 502.4$ in^3

2. $V \approx 3.14(3)^2(13) \approx 367.4$ yd^3

3. $V \approx 3.14(3)^2(5) \approx 141.3$ ft^3

4. $V \approx 3.14(5)^2(6) \approx 471$ ft^3

5. $\quad 5X^2 + 4X + 9$
   $+ \ 3X^2 + 8X + 1$
   $\overline{\quad 8X^2 + 12X + 10}$

6. $\quad 8X^2 + 9X + 6$
   $+ \ {-5X^2} - 3X - 2$
   $\overline{\quad 3X^2 + 6X + 4}$

7. $\quad 3X^2 + 7X + 8$
   $+ \ {-2X^2} + 3X - 9$
   $\overline{\quad X^2 + 10X - 1}$

8. $10 = 2 \times 5$; $15 = 3 \times 5$
   LCM $= 2 \times 3 \times 5 = 30$

9. $12 = \underline{2} \times \underline{2} \times \underline{3}$; $48 = \underline{2} \times \underline{2} \times 2 \times 2 \times \underline{3}$
   GCF $= 2 \times 2 \times 3 = 12$

10. $800\% = \dfrac{800}{100} = 8$

11. $375\% = \dfrac{375}{100} = 3\dfrac{75}{100} = 3\dfrac{3}{4}$

12. $240\% = \dfrac{240}{100} = 2\dfrac{40}{100} = 2\dfrac{2}{5}$

13. line segment

14. plane

15. point

16. line

17. 37°C

18. $\dfrac{1}{3} = 33\dfrac{1}{3}\%$
    $33\dfrac{1}{3}\% > 25\%$

19. $3\dfrac{1}{2} \times 50 = \dfrac{7}{\cancel{2}} \times \dfrac{\cancel{50}^{\,25}}{1} = \dfrac{7}{1} \times \dfrac{25}{1} = \dfrac{175}{1} = 175$ mi

20. $\dfrac{18}{20} = 18 \div 20 = 0.90 = 90\%$

Test 25

1. $(X + 3)(X + 1) = X^2 + 4X + 3$

2. $(X + 7)(X + 2) = X^2 + 9X + 14$

3. $(2X + 1)(X + 3) = 2X^2 + 7X + 3$

4. $(2X + 3)(X + 4) = 2X^2 + 11X + 12$

5. $(X + 5)(X + 7) = X^2 + 12X + 35$

6. $(2X + 2)(X + 5) = 2X^2 + 12X + 10$

7. $\quad 5X^2 + 4X + 1$
   $+ \ 8X^2 + 4X - 2$
   $\overline{\quad 13X^2 + 8X - 1}$

8. $\quad 3X^2 - \ X - 8$
   $+ \ 7X^2 - 9X - 1$
   $\overline{\quad 10X^2 - 10X - 9}$

9. $\quad {-4X^2} + 7X - 6$
   $+ \ 9X^2 - 3X + 5$
   $\overline{\quad 5X^2 + 4X - 1}$

10. $8(4+45) =$
$8(4)+8(45) =$
$32+360 = 392$

11. $6(Q+3) =$
$6Q+6(3) = 6Q+18$

12. $X(X+11) = X^2+11X$

13. $6^2+3\cdot5-4+|-1| =$
$6^2+3\cdot5-4+1 =$
$36+3\cdot5-4+1 =$
$36+15-4+1 = 48$

14. $|20\div5+(5)(6)-3| =$
$|4+30-3| =$
$|31| = 31$

15. $(4^2-8)+14-7 =$
$(16-8)+14-7 =$
$8+14-7 = 15$

16. $\dfrac{W}{27} = \dfrac{12}{36}$
$36W = 27(12)$
$36W = 324$
$W = 9$ units

17. $\dfrac{X}{2} = \dfrac{5.1}{3.4}$
$3.4X = 2(5.1)$
$3.4X = 10.2$
$X = 3$ units

18. $\$80.00 \times 0.40 = \$32.00$ off
$\$80.00 - \$32.00 = \$48.00$ is new price
$\$48.00 \times 1.06 = \$50.88$ is price with tax
$\$50.00 < \$50.88$; no

19. angle

20. line segment

Test 26

1.
$$
\begin{array}{r}
4\overset{2}{\cancel{3}}\overset{1}{2} \\
-\ 1:1\ 7 \\
\hline
3:1\ 5
\end{array}
$$

2. $5:48 + :10 = 5:58$
$-\ 2:50 + :10 = 3:00$
$\overline{\hphantom{-\ 2:50 + :10 = }2:58}$

3.
$$
\begin{array}{r}
{}^14:16 \\
+\ 2:50 \\
\hline
7:06
\end{array}
$$

4.
$$
\begin{array}{r}
{}^13:29 \\
+\ 2:48 \\
\hline
6:17
\end{array}
$$

5.
$$
\begin{array}{r}
{}^16:37 \\
+\ 5:29 \\
\hline
12:06
\end{array}
$$

6. $10:10 + :24 = 10:34$
$-\ 2:36 + :24 = \ 3:00$
$\overline{\hphantom{-\ 2:36 + :24 = }7:34}$

7. $(X+7)(X+8) = X^2+15X+56$

8. $(X+1)(X+3) = X^2+4X+3$

9. $(2X+2)(X+6) = 2X^2+14X+12$

10. $0.8 = \dfrac{8}{10} = \dfrac{4}{5}$

11. $0.45 = \dfrac{45}{100} = \dfrac{9}{20}$

12. $0.70 = \dfrac{70}{100} = \dfrac{7}{10}$

13. $\dfrac{9}{10} = 9\div10 = 0.9$

14. $\dfrac{1}{6} = 1\div6 \approx 0.17$

15. $\dfrac{5}{7} = 5\div7 \approx 0.71$

16. obtuse

17. acute

18. $90°$

19. $180°$

20. $6:15 + 5:40 = 11:55$

Test 27

1. $V \approx \dfrac{1}{3}(8 \times 8 \times 10)$

   $V \approx \dfrac{1}{3}(640) \approx 213.33 \text{ ft}^3$

2. $V \approx \dfrac{1}{3}(3.14)(4)^2(7)$

   $V \approx \dfrac{1}{3}(3.14)(16)(7)$

   $V \approx \dfrac{1}{3}(351.68) \approx 117.23 \text{ in}^3$

3. $\begin{array}{r} {}^{1}4:\!28 \\ +\ \ 6:33 \\ \hline 11:01 \end{array}$

4. $\begin{array}{r} 6:43+0:09 = 6:52 \\ -\ 2:51+0:09 = 3:00 \\ \hline 3:52 \end{array}$

5. $\begin{array}{r} {}^{1}7:35 \\ +\ \ 2:30 \\ \hline 10:05 \end{array}$

6. $(X+6)(X+6) = X^2 + 12X + 36$

7. $(2X+4)(X+1) = 2X^2 + 6X + 4$

8. $(A)(A+8) = A^2 + 8A$

9. $\dfrac{7}{8} - \dfrac{1}{4} = \dfrac{7}{8} - \dfrac{2}{8} = \dfrac{5}{8}$

10. $\dfrac{\cancel{3}}{4} \times \dfrac{5}{\underset{6}{\cancel{18}}} = \dfrac{5}{24}$

11. $\dfrac{4}{5} \div \dfrac{1}{2} = \dfrac{4}{5} \times \dfrac{2}{1} = \dfrac{8}{5} = \dfrac{5}{5} + \dfrac{3}{5} = 1\dfrac{3}{5}$

12. mode

13. median

14. mean

15. acute

16. length or width

17. $\sqrt{36} = 6$

18. $\text{mean} = \dfrac{21+36+42}{3} = \dfrac{99}{3} = 33$

19. $180°$

20. $(X+5)(X+5) = X^2 + 10X + 25 \text{ units}^2$

Test 28

1. $8:11+12:00 = 2011$

2. 0200

3. 0000

4. 0530

5. $1536 - 12:00 = 3:36 \text{ p.m.}$

6. $2045 - 12:00 = 8:45 \text{ p.m.}$

7. $3:05 \text{ AM}$

8. $12:00 \text{ p.m. (noon)}$

9. $\begin{array}{r} {}^{1}0550 \\ +\ \ 1730 \\ \hline 2320 \end{array}$

10. $\begin{array}{r} 2{}^{1}030 \\ +\ \ 0145 \\ \hline 2215 \end{array}$

11. $\begin{array}{r} 1619+32 = 1651 \\ -\ 1128+32 = 1200 \\ \hline 0451 \end{array}$

12. $\begin{array}{r} 0643+10 = 0653 \\ -\ 0350+10 = 0400 \\ \hline 0253 \end{array}$

13. $V \approx \dfrac{1}{3}(2.3 \times 2.3 \times 4)$

   $V \approx \dfrac{1}{3}(21.16) \approx 7.05 \text{ in}^3$

14. $V \approx \dfrac{1}{3}(3.14)(0.5)^2(2)$

   $V \approx \dfrac{1}{3}(3.14)(0.25)(2) \approx 0.52 \text{ yd}^3$

15. $V \approx (3.14)(3)^2(9.4)$

   $V \approx (3.14)(9)(9.4) \approx 265.64 \text{ ft}^3$

16. $\dfrac{5}{5,000} = \dfrac{1}{1,000}$

17. $21 = 3 \times \underline{7} \ ; \ 49 = \underline{7} \times 7$

    $\text{GCF} = 7$

18. 4

19. $1230 + 0245 = 1515$

20. $\dfrac{24}{32} = \dfrac{3}{4} = 3 \div 4 = 0.75 = 75\%$

Test 29

1.
$$\begin{array}{r} 29'\ 6" \\ +\quad 5'\ 5" \\ \hline 34'11" \end{array}$$

2.
$$\begin{array}{r} 4\ \text{yd}\ 2\ \text{ft} \\ +\quad 8\ \text{yd}\ 2\ \text{ft} \\ \hline 12\ \text{yd}\ 4\ \text{ft} \end{array} = 13\ \text{yd}\ 1\ \text{ft}$$

3.
$$\begin{array}{r} 9\ \text{lb}\ 10\ \text{oz} \\ +\quad 6\ \text{lb}\ 11\ \text{oz} \\ \hline 15\ \text{lb}\ 21\ \text{oz} \end{array} = 16\ \text{lb}\ 5\ \text{oz}$$

4.
$$\begin{array}{r} 46'15" \\ -\quad 18'\ 4" \\ \hline 28'11" \end{array}$$

5.
$$\begin{array}{r} 4\ \text{yd}\ 3\ \text{ft} \\ -\quad 2\ \text{yd}\ 2\ \text{ft} \\ \hline 2\ \text{yd}\ 1\ \text{ft} \end{array}$$

6.
$$\begin{array}{r} 6\ \text{lb}\ 25\ \text{oz} \\ -\quad 1\ \text{lb}\ 13\ \text{oz} \\ \hline 5\ \text{lb}\ 12\ \text{oz} \end{array}$$

7. 0113

8. $2:45+12:00=1445$

9. 10:30 a.m.

10. $2107-12:00=9:07$ p.m.

11. $3\frac{5}{8} \div 1\frac{1}{4} = \frac{29}{8} \div \frac{5}{4} = \frac{29}{\cancel{8}_2} \times \frac{\cancel{4}}{5} = \frac{29}{10} = 2\frac{9}{10}$

12. $5\frac{7}{9} \times 4\frac{2}{3} = \frac{52}{9} \times \frac{14}{3} = \frac{728}{27} = 26\frac{26}{27}$

13. $6\frac{1}{7} + 3\frac{3}{14} = 6\frac{2}{14} + 3\frac{3}{14} = 9\frac{5}{14}$

14. kilogram

15. centimeter

16. $\frac{1}{1,000}$

17. $\frac{4}{55}$

18. mean $= \frac{80+85+90+95+100}{5} = \frac{450}{5} = 90$

19. 90

20. $V \approx \frac{1}{3}(3.14)(10)^2(10)$

$V \approx \frac{1}{3}(3.14)(100)(10) \approx 1,046.67\ \text{ft}^3$

Test 30

1. rational
2. rational
3. irrational
4. irrational
5. rational
6. rational

7.
$$\begin{array}{r} 7'15" \\ -\quad 5'\ 5" \\ \hline 2'10" \end{array}$$

8.
$$\begin{array}{r} 6\ \text{yd}\ 1\ \text{ft} \\ +\quad 4\ \text{yd}\ 2\ \text{ft} \\ \hline 10\ \text{yd}\ 3\ \text{ft} \end{array} = 11\ \text{yd}$$

9.
$$\begin{array}{r} 19\ \text{lb}\quad 2\ \text{oz}+5\ \text{oz} = 19\ \text{lb}\ 7\ \text{oz} \\ -\quad 7\ \text{lb}\ 11\ \text{oz}+5\ \text{oz} =\ 8\ \text{lb}\ 0\ \text{oz} \\ \hline 11\ \text{lb}\ 7\ \text{oz} \end{array}$$

10. $10\times10=100$ millimeters

11. $23\div1,000=0.023$ kilometers

12. $5\times100=500$ centiliters

13. $2,000\div1,000=2$ grams

14. yes

15. $\left|-\left(4^2\right)\right| =$
$\left|-(16)\right| = 16$

16. obtuse

17. $V \approx (3.14)(7)^2(5)$
$V \approx (3.14)(49)(5) \approx 769.3\ \text{in}^3$

18. $250+50=300$ total buttons
$\frac{50}{300} = \frac{1}{6}$

19. $0.692 = \frac{692}{1,000} = \frac{173}{250}$

20. 33

Unit Test IV

1.
$$\begin{array}{r} 4X^2 + 3X + 1 \\ +\quad 7X^2 + 2X - 3 \\ \hline 11X^2 + 5X - 2 \end{array}$$

2.
$$\begin{array}{r} 2X^2 - X - 6 \\ + \quad 8X^2 - X - 2 \\ \hline 10X^2 - 2X - 8 \end{array}$$

3.
$$\begin{array}{r} -5X^2 + 8X - 7 \\ + \quad 6X^2 - 4X + 6 \\ \hline X^2 + 4X - 1 \end{array}$$

4. $(3X + 4)(X + 5) = 3X^2 + 19X + 20$

5. $(X + 6)(X + 8) = X^2 + 14X + 48$

6. $(2X + 3)(X + 6) = 2X^2 + 15X + 18$

7. $(X + 2)(X + 1) = X^2 + 3X + 2$

8. $(X + 6)(X + 4) = X^2 + 10X + 24$

9. $(2X + 2)(X + 2) = 2X^2 + 6X + 4$

10. $V = \dfrac{1}{3}(7 \times 7 \times 5)$

$V = \dfrac{1}{3}(245) \approx 81.67 \text{ in}^3$

11. $V = \dfrac{1}{3}(3.14)(0.3)^2(1)$

$V = \dfrac{1}{3}(3.14)(0.09)(1)$

$V = \dfrac{1}{3}(.2826) \approx 0.09 \text{ yd}^3$

12. $V = (3.14)(6)^2(11.2)$

$V = (3.14)(36)(11.2) \approx 1,266.05 \text{ ft}^3$

13.
$$\begin{array}{r} 3:21 \\ - \quad 1:06 \\ \hline 2:15 \end{array}$$

14.
$$\begin{array}{r} 4:37 + :11 = 4:48 \\ - \quad 1:49 + :11 = 2:00 \\ \hline 2:48 \end{array}$$

15.
$$\begin{array}{r} 5:28 \\ + \quad 7:52 \\ \hline 12:80 \quad = 13:20 \end{array}$$

16. $10:19 + 12:00 = 2219$

17. 0400

18. $5:30 + 12:00 = 1730$

19. $12:17$ a.m.

20. $3:30$ a.m.

21. $1945 - 12:00 = 7:45$ p.m.

22.
$$\begin{array}{r} 0540 \\ + \quad 1720 \\ \hline 2260 \quad = 2300 \end{array}$$

23.
$$\begin{array}{r} 2010 \\ + \quad 0235 \\ \hline 2245 \end{array}$$

24.
$$\begin{array}{r} 1012 + 18 = 1030 \\ - \quad 0642 + 18 = 0700 \\ \hline 0330 \end{array}$$

25.
$$\begin{array}{r} 9'2" + 6" = 9'8" \\ - \quad 4'6" + 6" = 5'0" \\ \hline 4'8" \end{array}$$

26.
$$\begin{array}{r} 5 \text{ yd } 2 \text{ ft} \\ + \quad 8 \text{ yd } 2 \text{ ft} \\ \hline 13 \text{ yd } 4 \text{ ft} \quad = 14 \text{ yd } 1 \text{ ft} \end{array}$$

27.
$$\begin{array}{r} 12 \text{ lb } 4 \text{ oz} + 9 \text{ oz} = 12 \text{ lb } 13 \text{ oz} \\ - \quad 5 \text{ lb } 7 \text{ oz} + 9 \text{ oz} = 6 \text{ lb } 0 \text{ oz} \\ \hline 6 \text{ lb } 13 \text{ oz} \end{array}$$

28. irrational

29. rational

30. irrational

31. $250\% = 2.50$

$2.50 \times 34 = 85$

32. $90°$

33. mean $= \dfrac{4 + 6 + 6 + 7 + 12}{5} = \dfrac{35}{5} = 7$

34. $\dfrac{6}{432} = \dfrac{1}{72}$

35. kilometer

Final Test

1. $(-8) + (-25) = -33$

2. $(-7) \times (-15) = 105$

3. $(11) - (-6) =$
$(11) + (+6) = 17$

4. $(-45) \div (9) = -5$

5. $-1^3 = -(1)(1)(1) = -1$

6. $-(5)^2 = -(5)(5) = -25$

7. $(-8)^2 = (-8)(-8) = 64$

8. $\left(-\dfrac{2}{3}\right)^2 = \left(-\dfrac{2}{3}\right)\left(-\dfrac{2}{3}\right) = \dfrac{4}{9}$

9. $9 \times 10^1 + 5 \times 10^0 + 2 \times \dfrac{1}{10^1} + 1 \times \dfrac{1}{10^2} + 4 \times \dfrac{1}{10^3}$

10. $1,825.6$

11. $\sqrt{100} = 10$

12. $\sqrt{Y^2} = Y$

13. $8 \cdot 2 + 5^2 - Y = 2(Y+1) + 6$
$8 \cdot 2 + 25 - Y = 2Y + 2 + 6$
$16 + 25 - Y = 2Y + 2 + 6$
$41 - Y = 2Y + 8$
$-Y = 2Y + 8 - 41$
$-Y = 2Y - 33$
$-Y - 2Y = -33$
$-3Y = -33$
$Y = 11$

14. $8 \cdot 2 + 5^2 - Y = 2(Y+1) + 6 \Rightarrow$
$8 \cdot 2 + 5^2 - (11) = 2((11)+1) + 6$
$8 \cdot 2 + 5^2 - 11 = 2(12) + 6$
$8 \cdot 2 + 25 - 11 = 2(12) + 6$
$16 + 25 - 11 = 24 + 6$
$30 = 30$

15. $8M - 4M - 6 - 3 + 5M = 8^2 - 1$
$8M - 4M - 6 - 3 + 5M = 64 - 1$
$9M - 9 = 63$
$9M = 63 + 9$
$9M = 72$
$M = 8$

16. $8M - 4M - 6 - 3 + 5M = 8^2 - 1 \Rightarrow$
$8(8) - 4(8) - 6 - 3 + 5(8) = 8^2 - 1$
$8(8) - 4(8) - 6 - 3 + 5(8) = 64 - 1$
$64 - 32 - 6 - 3 + 40 = 64 - 1$
$63 = 63$

17. $(-3)^2 \div 9 + 6 = D$
$9 \div 9 + 6 = D$
$1 + 6 = D$
$D = 7$

18. $(-3)^2 \div 9 + 6 = D \Rightarrow$ $(-3)^2 \div 9 + 6 = (7)$
$9 \div 9 + 6 = 7$
$1 + 6 = 7$
$7 = 7$

19. $\dfrac{1}{8} = \dfrac{7}{Y}$
$1Y = 8(7)$
$Y = 56$

20. $\dfrac{11}{12} = \dfrac{A}{48}$
$12A = 11(48)$
$12A = 528$
$A = 44$

21. $\dfrac{R}{8} = \dfrac{9}{6}$
$6R = 8(9)$
$6R = 72$
$R = 12 \text{ units}$

22. $3 = 1 \times 3;\ 4 = 2 \times 2$
$LCM = 2 \times 2 \times 3 = 12$

23. $6 = 2 \times 3;\ 9 = 3 \times 3$
$LCM = 2 \times 3 \times 3 = 18$

24. $24 = 2 \times 2 \times 2 \times 3;\ 40 = 2 \times 2 \times 2 \times 5$
$GCF = 2 \times 2 \times 2 = 8$

25. $15 = 3 \times 5;\ 35 = 5 \times 7$
$GCF = 5$

26. $\begin{aligned} &5X^2 + 4X + 2 \\ &\underline{+\ \ 8X^2 + 3X - 4} \\ &13X^2 + 7X - 2 \end{aligned}$

27. $\begin{aligned} &7X^2 - X - 3 \\ &\underline{+\ \ 6X^2 - 2X - 5} \\ &13X^2 - 3X - 8 \end{aligned}$

28. $\begin{aligned} &{-4X^2} + 9X - 8 \\ &\underline{+\ \ 2X^2 - 6X + 1} \\ &{-2X^2} + 3X - 7 \end{aligned}$

29. $2(7 \times 5) + 2(5 \times 6) + 2(6 \times 7) =$
$2(35) + 2(30) + 2(42) =$
$70 + 60 + 84 = 214 \text{ in}^2$

30. $4\left(\dfrac{1}{2} \times 11 \times 9\right) + (9 \times 9) =$
$4(49.5) + (81) =$
$198 + 81 = 279 \text{ ft}^2$

31. $9^2 + 12^2 = H^2$

    $81 + 144 = H^2$

    $225 = H^2$

    $H = 15$ ft

32. $(2X + 1)(X + 6) = 2X^2 + 13X + 6$

33. $(X + 7)(X + 9) = X^2 + 16X + 63$

34. $(2X + 4)(X + 5) = 2X^2 + 14X + 20$

35. $V \approx \dfrac{1}{3}(3 \times 3)(4)$

    $V \approx \dfrac{1}{3}(9)(4) \approx 12$ in^3

36. $V \approx \dfrac{1}{3}(3.14)(5)^2(7)$

    $V \approx \dfrac{1}{3}(3.14)(25)(7) \approx 183.17$ yd^3

37. $V \approx (3.14)(6)^2(10)$

    $V \approx (3.14)(36)(10) \approx 1{,}130.4$ ft^3

38.   7 : 18
    $-$ 3 : 05
    ⎯⎯⎯⎯⎯
         4 : 13

39.   2 : 44
    $+$ 1 : 59
    ⎯⎯⎯⎯⎯
         3 : 103 $= 4 : 43$

40.   0136
    $+$ 0438
    ⎯⎯⎯⎯⎯
         0574 $= 0614$

41. $2120 - 12:00 = 9:20$ p.m.

42. $1611 - 12:00 = 4:11$ p.m.

43. $3:45$ a.m.

44. $8'3'' + 7'' = 8'10''$
    $-$ $5'5'' + 7'' = 6'\ 0''$
    ⎯⎯⎯⎯⎯⎯⎯⎯⎯⎯
                     $2'10''$

45.   6 yd 1 ft
    $+$ 9 yd 1 ft
    ⎯⎯⎯⎯⎯⎯⎯
         15 yd 2 ft

46.   25 lb 8 oz + 6 oz = 25 lb 14 oz
    $-$ 15 lb 10 oz + 6 oz = 16 lb 0 oz
    ⎯⎯⎯⎯⎯⎯⎯⎯⎯⎯⎯⎯⎯⎯⎯
                             9 lb 14 oz

47. irrational

48. rational

Symbols and Tables

PERIMETER
Add the lengths of all the sides.

AREA
rectangle, square = bh (base times height)
triangle = 1/2 bh
circle = πr^2

SURFACE AREA
Add the area of each face.

VOLUME
rectangular solid = area of base (B) x h
cylinder = area of base (B) x h
pyramid and cone = 1/3 area of base (B) x h

TEMPERATURE
freezing point of water = 0°C or 32°F
normal body temperature = 37°C
    or 98.6°F
boiling point of water = 100°C
    or 212°F

Celsius to Fahrenheit
    (C x 9/5) + 32° = F
Fahrenheit to Celsius
    (F – 32°) x 5/9 = C

PYTHAGOREAN THEOREM
    $leg^2 + leg^2 = hypotenuse^2$

U.S. CUSTOMARY MEASURE
3 teaspoons (tsp) = 1 tablespoon (Tbsp)
2 pints (pt) = 1 quart (qt)
8 pints = 1 gallon (gal)
4 quarts = 1 gallon
12 inches (in) = 1 foot (ft) or 12" = 1'
3 feet or 36 in= 1 yard (yd)
5,280 feet = 1 mile (mi)
16 ounces (oz) = 1 pound (lb)
2,000 pounds = 1 ton

METRIC TO U.S. CUSTOMARY
1 meter (39.37") ≈ 1.1 yards
1 centimeter ≈ 0.4 inches
1 liter ≈ 1.06 quarts
1 kilogram ≈ 2.2 pounds
1 kilometer ≈ 0.6 miles
1 liter ≈ 1.06 quarts

METRIC MEASURE
1,000 millimeters (mm) = 1 meter (m)
100 centimeters (cm) = 1 meter
10 decimeters (dm) = 1 meter
10 meters = 1 dekameter (dam)
100 meters = 1 hectometer (hm)
1,000 meters = 1 kilometer (km)

Replace meter with liter or gram
for volume and weight equivalents.

U.S. CUSTOMARY TO METRIC
1 inch = 2.54 centimeters
1 ounce ≈ 28 grams

SYMBOLS

=	equals
≠	is not equal to
~	similar
≈	approximately equals
<	less than
>	greater than
±	plus or minus
| |	absolute value
%	percent
π	pi (commonly rounded to 22/7 or 3.14)
r^2	r squared or r · r
$\sqrt{}$	square root
∞	infinity
↔	line
.	point
→	ray
∠	angle
▱	plane
⌐	right angle

A - C

absolute value- the value of a number without its sign, or the difference between a number and zero expressed as a positive number

acute angle- an angle with a measure greater than 0° and less than 90°

additive inverse- the number that, when added to another, results in a sum of 0

algebra- a branch of mathematics that deals with numbers, which may be represented by letters or symbols

altitude- the perpendicular distance from a vertex to the opposite side of a polygon or from the base to the apex of a cone or pyramid

angle- a geometric figure formed by two rays joined at their origins

apex- the point farthest from the base in a geometric figure

area- the measure of the space covered by a plane shape, expressed in square units

Associative Property- a property that states that the way terms are grouped in an addition or multiplication expression does not affect the result

average- a measure of center in a set of numbers; could be measured using a mean, median, or mode, but usually refers to the mean

base- 1. a particular side or face of a geometric figure used to calculate area or volume; 2. a number that is raised to a power; 3. the number that is the foundation in a given number system; for example, the decimal system describes numbers in relation to powers of 10, such as 100, 101, 102, and so on

base unit- a metric unit of measurement that can be modified by adding a prefix to represent fractions or multiples

binomial- an algebraic expression with two terms

Celsius- a scale for measuring temperature where the freezing point of water is 0° and the boiling point of water is 100°

centi- in the metric system, the Latin prefix representing one hundredth of the base unit

coefficient- a quantity placed before and multiplying the variable in an algebraic expression

common factor- a number or algebraic expression that divides evenly into each of a group of numbers or expressions

common multiple- a number or algebraic expression that is a multiple of each of a group of numbers or expressions

Commutative Property- a property that states that the order in which numbers are added or multiplied does not affect the result

cone- a solid with a circular base and a curved surface that rises to a point

congruent- having exactly the same size and shape

counting numbers- whole numbers from 1 to infinity; also called *natural numbers*

cube- 1. a solid with six congruent square faces that meet at right angles; 2. a number multiplied by itself three times

cylinder- a solid with one curved surface and two congruent circular bases

D - E

deci- in the metric system, the Latin prefix representing one tenth of the base unit

decimal (fraction)- a fraction written using a decimal point and place value

decimal system- a number system based on ten, also called *base ten*

decompose- to separate a number into parts

degree- a measure of temperature

deka- also *deca-*; in the metric system, the Greek prefix representing ten of the base unit

denominator- the bottom number in a fraction, which shows the number of parts in the whole

diameter- a straight line passing through the center of a circle and touching both sides

dimension- a measurement in a particular direction (length, width, height, depth)

Distributive Property of Multiplication over Addition- a property for multiplying a sum by a given factor

dividend- the number being divided

divisor- a number that is being divided into another

equation- a mathematical statement that uses an equal sign to show that two expressions have the same value

equivalent- having the same value

expanded notation- a way of writing numbers by showing each digit multiplied by its place value

exponent- a raised number that indicates the number of times a factor is multiplied by itself; also called *power*

exponential notation- a form of expanded notation where each place value is indicated by 10 with an exponent

F - H

face- one of the flat surfaces of a solid

factor- (n.) a whole number that multiplies with another to form a product; (v.) to find the factors of a given product

factor tree- a diagram used to find the prime factors of a composite number

Fahrenheit- a scale for measuring temperature where the freezing point of water is 32° and the boiling point of water is 212°

fraction- a number indicating part of a whole

geometry- a branch of mathematics that deals with figures in space

gram- the basic unit of mass in the metric system

greatest common factor (GCF)- the greatest number that will divide evenly into two or more numbers

hectare- a metric unit of area, defined as 10,000 square meters

hecto- in the metric system, the Greek prefix representing 100 of the base unit

height- the perpendicular distance from the base to the top of a figure

hypotenuse- the side opposite the right angle in a right triangle

I - L

improper fraction- a fraction with a numerator greater than its denominator

infinite- without end; unable to be counted or measured

integer- a non-fractional number that can be positive, negative, or zero

inverse- opposite or reverse

irrational numbers- numbers that cannot be written as fractions and form non-repeating, non-terminating decimals

kilo- In the metric system, the Greek prefix representing 1,000 of the base unit

least common multiple (LCM)- the least number that is a multiple of two or more other numbers

leg- in a right triangle, one of the two sides that make up the right angle

line- in geometry, a set of connected points that extends infinitely in two directions

line segment- a section of a line bounded by two endpoints

liter- the basic unit of liquid volume in the metric system

M - O

mean - also called *average*; a measure of center found by dividing the sum of a set of values by the number of values

median- the middle value in a list of numbers when they are arranged in order from least to greatest

meter- the basic unit of linear measure in the metric system

metric system- a system of measurement based on ten

military time- a system of measuring time based on the full 24 hours in a day, rather than using a.m. and p.m.

milli- in the metric system, the Latin prefix representing one thousandth of the base unit

mixed number- a number written as a whole number and a fraction

mode- in a data set, the item that appears the most often

monomial- a mathematical expression with only one term

multiple- the product of a given number and another whole number

multiplicative inverse- the number that, when multiplied by a given number, has a product of 1; also called *reciprocal*

natural numbers- whole numbers from 1 to infinity; also called *counting numbers*

negative number- a number less than zero

number line- a line on which every point corresponds to a real number

numerator- the top number in a fraction, which shows the number of parts being considered

obtuse angle- an angle with a measure greater than 90° and less than 180°

P

pi- the Greek letter π, which represents an irrational number with an approximate value of 22/7 or 3.14

place value- the position of a digit which indicates its assigned value

place value notation- a way of writing numbers that shows the place value of each digit

plane- a flat, two-dimensional surface that extends infinitely in all directions

point- a defined position in space that has no dimensions; represented with a dot

polygon- a closed plane shape having three or more straight sides that do not cross

polynomial- an algebraic expression with more than one term

positive numbers- numbers greater than zero

power- another name for an *exponent*; indicates the number of times a factor is multiplied by itself

prime factorization- renaming a number as a product of two or more prime numbers

prime factors- all the factors of a number that are prime numbers

prime number- a number that has only two factors: one and itself

probability- the likelihood of getting a desired outcome, given all possible outcomes

proper fraction- a fraction with a numerator less than its denominator

proportion- two ratios that are equal to each other

pyramid- a solid with a polygonal base and triangular faces that rise to a point

Pythagorean theorem- states that the square of the length of the hypotenuse of a right triangle is equal to the sum of the squares of the lengths of the other sides

R-S

radical- an expression containing a root

radius- the distance from the center of a circle to its edge; in a regular polygon, the distance from the center to any vertex; in a sphere, the distance from the center to any point on the surface; plural is *radii*

ratio- the relationship between two values; can be written in fractional form

rational numbers- numbers that can be written as ratios or fractions, including decimals

ray- a geometric figure that starts at a definite point (called the origin) and extends infinitely in one direction

real numbers- numbers that can be written as decimals, including rational and irrational numbers

reciprocal- the number that, when multiplied by a given number, has a product of 1; also called *multiplicative inverse*

rectangle- a quadrilateral with two pairs of opposite parallel sides and four right angles

rectangular solid- a solid with six rectangular faces that meet at right angles

repeated division- a method for finding the prime factors of a number

right angle- an angle measuring 90°

right triangle- a triangle with one right angle

root- a number that can be multiplied by itself a given number of times to form a specified product

"Rule of Four"- a Math-U-See method for finding the common denominator of two fractions

"same difference theorem"- a Math-U-See method for subtraction that adds the same value to minuend and subtrahend to avoid regrouping

similar- having the same shape but different sizes

slant height- the height of a triangle forming a face of a pyramid

square- 1. a quadrilateral in which the four sides are perpendicular and congruent; 2. a number multiplied by itself

square root- a number that can be multiplied by itself to form a specified product

standard form- the usual way of writing a number, with each digit representing a different place; also called *standard notation*

straight angle- an angle with a measure of 180°

surface area- the sum of the areas of all the faces of a solid

T - Z

term- a part of an algebraic expression which may be a number, a variable, or a product

trapezoid- a four-sided polygon with a set of parallel sides

trinomial- an algebraic expression with three terms

unit- 1. the place in a place-value system representing numbers less than the base; 2. a quantity used as a standard of measure

unknown- a specific quantity that has not yet been determined

variable- a value that is not fixed or determined, often representing a range of possible values

vertex- the endpoint shared by two rays, line segments, or edges; plural is *vertices*

volume- the number of cubic units that can be contained in a solid

whole numbers- counting numbers from zero to infinity, excluding fractions

Secondary Levels Master Index

This index lists the levels at which main topics are presented in the instruction manuals for *Pre-Algebra* through *PreCalculus*. For more detail, see the description of each level at <u>MathUSee.com</u>.

Absolute value .... Pre-Algebra, Algebra 1
Additive inverse .... Pre-Algebra
Age problems .... Algebra 2
Angles .... Geometry, PreCalculus
Angles of elevation and depression............
    PreCalculus
Arc functions .... PreCalculus
Area.. Geometry
Associative Property................ Pre-Algebra,
    Algebra 1
Axioms .... Geometry
Bases other than 10 .... Algebra 1
Binomial theorem...................... Algebra 2
Boat in current problems............. Algebra 2
Chemical mixtures .... Algebra 2
Circumference .... Geometry
Circumscribed figures .... Geometry
Cofunctions .... PreCalculus
Coin problems .... Algebra 1 and 2
Commutative Property .... Pre-Algebra,
    Algebra 1
Completing the square................ Algebra 2
Complex numbers .... Algebra 2
Congruency .... Pre-Algebra, Geometry
Conic sections .... Algebra 1 and 2
Conjugate numbers .... Algebra 2
Consecutive integers.......... Algebra 1 and 2
Determinants............................. Algebra 2
Difference of two squares .. Algebra 1 and 2
Discriminants............................. Algebra 2
Distance formula .... Algebra 2
Distance problems .... Algebra 2
Distributive Property .... Pre-Algebra,
    Algebra 1
Expanded notation.................... Pre-Algebra
Exponents
    fractional.................... Algebra 1 and 2
    multiply and divide ...... Algebra 1 and 2
    negative .... Algebra 1 and 2
    notation .... Pre-Algebra
    raised to a power........ Algebra 1 and 2
Factoring polynomials....... Algebra 1 and 2
Functions.............................. PreCalculus
Graphing
    Cartesian coordinates .... Algebra 1

circle .... Algebra 1 and 2
ellipse .... Algebra 1 and 2
hyperbola .... Algebra 1 and 2
inequality .... Algebra 1 and 2
line.............................. Algebra 1 and 2
parabola..................... Algebra 1 and 2
polar .... PreCalculus
trig functions..................... PreCalculus
Greatest common factor........... Pre-Algebra
Identities .... PreCalculus
Imaginary numbers .... Algebra 2
Inequalities .....Algebra 1 and 2, PreCalculus
Inscribed figures .... Geometry
Integers .... Pre-Algebra
Interpolation .... PreCalculus
Irrational numbers .... Pre-Algebra
Latitude and longitude .... Geometry
Least common multiple............. Pre-Algebra
Limits .... PreCalculus
Line
    graphing..................... Algebra 1 and 2
    properties of .... Geometry
Logarithms .... PreCalculus
Matrices..................................... Algebra 2
Maxima and minima.................... Algebra 2
Measurement, add and subtract.................
    Pre-Algebra
Metric-customary conversions ..... Algebra 1,
    Algebra 2
Midpoint formula .... Algebra 2
Motion problems......................... Algebra 2
Multiplicative inverse .... Pre-Algebra
Natural logarithms .... PreCalculus
Navigation .... PreCalculus
Negative numbers .... Pre-Algebra
Number line............. Pre-Algebra, Algebra 1
Order of operations .... Pre-Algebra,
    Algebra 1
Parallel and perpendicular lines
    graphing..................... Algebra 1 and 2
    properties of .... Geometry
Pascal's triangle .... Algebra 2
Percent problems .... Algebra 2
Perimeter.................................... Geometry
Pi .... Pre-Algebra, Geometry

Place value .... Pre-Algebra
Plane .... Geometry
Points, lines, rays .... Geometry
Polar coordinates and graphs .... PreCalculus
Polygons
      area .... Geometry
      similar .... Pre-Algebra, Geometry
Polynomials
      add .... Pre-Algebra, Algebra 1
      divide .... Algebra 1
      factor .... Algebra 1 and 2
      multiply .... Pre-Algebra, Algebra 1 and 2
Postulates .... Geometry
Prime factorization .... Pre-Algebra
Proofs .... Geometry
Pythagorean theorem .... Pre-Algebra,
    Geometry, PreCalculus
Quadratic formula .... Algebra 2
Radians .... PreCalculus
Radicals .... Pre-Algebra,
    Algebra 1 and 2, Geometry
Ratio and proportion .... Pre-Algebra,
    Geometry, Algebra 2
Rational expressions .... Algebra 2
Real numbers .... Pre-Algebra
Reference angles .... PreCalculus
Same difference theorem .... Pre-Algebra
Scientific notation .... Algebra 1 and 2
Sequences and series .... PreCalculus
Sets .... Geometry
Significant digits .... Algebra 1
Similar polygons ...... Pre-Algebra, Geometry
Simultaneous equations ..... Algebra 1 and 2
Sine, cosine, and tangent, laws of ....
PreCalculus
Slope-intercept formula ...... Algebra 1 and 2
Solve for unknown .... Pre-Algebra,
    Algebra 1 and 2
Special triangles .... Geometry, PreCalculus
Square roots ... Pre-Algebra, Algebra 1 and 2
Surface area .... Pre-Algebra, Geometry
Temperature conversions .... Pre-Algebra
Time
      add and subtract .... Pre-Algebra
      military .... Pre-Algebra
Time and distance problems .... Algebra 2
Transformations .... Geometry
Transversals .... Geometry
Triangles .... Geometry, PreCalculus
Trig ratios .... Geometry, PreCalculus
Unit multipliers .... Algebra 1 and 2

Vectors .... Algebra 2, PreCalculus
Volume .... Pre-Algebra, Geometry
Whole numbers .... Pre-Algebra

Pre-Algebra Index

Topic	Lesson
Absolute value	18
Additive inverse	9, 13
Algebra	23
Algebra-decimal inserts	9, 23, 25
Altitude	27
Angles	Student 25D, 26D
Area	3H, 4H, 5H, 12H, 23H, 25H
circle	24
rectangle	Student 3A note; 15
square	Student 3E #20
triangle	15
Associative Property	11
Average	Student 27D
Base	24, 27
Base ten	6, 23
Base X	23
Binomial	23, 25
Categories of numbers	30H
Celsius	16, 17
Centigrade	16
Circle	24, 3H, 4H, 5H, 6H
Coefficient	11, 13
Common factor	12
Commutative Property	11
Counting numbers	30H
Cone	27
Congruent	20
Cube	
exponent	5
surface area	15
volume	24
Cylinder	24
Decimal system	6
Decimals	
add	Student 16D
change to fraction	Student 20D
change to percent	Student 21D
divide	Student 18D, 19D
multiply	Student 17D
subtract	Student 16D
Diameter	24
Distributive Property	12
Dividend	Student 19D
Divisibility rules	22H
Divisor	Student 19D

Topic	Lesson
Dodecahedron	16H
Equations, solving	9, 13, 14, 16, 17, 19
Expanded notation	6
Exponential growth	5H
Exponential notation	6
Exponents	5, 7
Face	15
Factor	5, 8, 12, 21, 22
Factoring	22H
Factor tree	21
Fahrenheit	16, 17
Fibonacci numbers	7H
Fractions	
add	Student 2D, 6D, 13D
common denominator	Student 5D
compare	Student 5D
divide	Student 8D, 10D
equivalent	Student 3D; 19
improper	Student 11D, 13D
measurement	Student 12D
mixed numbers	Student 11D, 13-15D
multiply	Student 7D, 15D
of a number	Student 1D
reciprocal	Student 9D, 10D
rule of four	Student 5D, 6D, 8D
simplify	Student 4D
subtract	Student 2D, 6D, 14D
to decimals	Student 20D
to percents	Student 21D
Graphs	2H
Geometry	Student 24D, 25D, 26D
Greatest common factor (GCF)	22
Hectare	29H
Height	24, 27
Hexahedron	16H
Homeowner problems	12H, 13H, 15H, 23H, 24H, 25H
Hypotenuse	10, 20
Icosahedron	16H
Integers	4, 30H
Inverse	4, 8, 9, 13
Inverse ratio	19H
Irrational numbers	30 , 30H

Topic	Lesson
Key words	8H
Least common multiple (LCM)	21
Line	Student 24D
Line segment	Student 24D
Maps	20H
Mean	Student 27D
Measurement conversion problems	24H, 25H
Measurement	
add	29
metric	Student 29D, 30D
rulers	Student 12D
subtract	29
Median	Student 27D, 27H
Metric measurement	Student 29D, 30D
Metric system	29H
Mixed numbers	Student 11D, 13D, 14D, 15D
Mode	Student 27D, 27H
Money	6
Monomial	23
Multiplicative inverse	13
Natural numbers	30H
Negative numbers	2H
add	1
divide	4
exponents	7
multiply	3
subtract	2
Number line	4, 30, 11H
Octahedron	16H
Order of operations	14
Parallelogram	4H
Parentheses	1, 7, 14
Patterns	
numerical	7H, 11H
Pascal's triangle	17H, 21H
three-dimensional	7H
visual	6H, 11H
word problems	8H
Pascal's triangle	17H, 21H
Percent	
decimal or fraction to percent	Student 21D
greater than 100	Student 23D
of a number	Student 22D
to a decimal	Student 22D
Percent of change	26H
Percent of error	28H
Perimeter	3H, 12H, 23H
Pi	24, 30
Place value	6
Plane	Student 24D
Point	Student 24D
Polygon	20, 16H
Polyhedron	16H
Polynomials	23H, 25H
add	23
multiply	25
Power of a number	5, 6
Prime factorization	21, 22
Prime numbers	21, 22
Probability	Student 28D
Product	3, 8
Proportion	19, 20, 19H, 20H
Pyramid	15, 27
Pythagorean theorem	10, 10H, 13H
Radius	24
Ratio	19, 20, 19H, 20H
Rational numbers	19; Stud 20D; 30, 30H
Ray	Student 25D
Real numbers	30, 30H
Reciprocal	Student 9D; 13
Rectangle	Student 3A; 15
Rectangular solid	15
Repeated division	21
Right angle	10; Student 26D
Roots	8
Rule of four	Student 5D, 6D, 8D
Rulers	Student 12D
Same difference theorem	26, 29
Similar	20
Similarity	20H
Slant height	27
Solving for the unknown	
in word problems	14H, 18H, 19H, 20H, 21H
Square of a number	5
Square roots	8, 30, 13H
Surface area	15, 15H, 16H
Tetrahedron	16H
Temperature conversions	16, 17
Time	
add	26, 28
military	28
subtract	26, 28
Trapezoid	5H

Topic ....**Lesson**

Triangle 10, 15, 20
Trinomial 23
Vertex...27, 16H
Variable 9, 13
Volume 24H
      cone 27
      cube 24
      cylinder 24
      pyramid.. 27
Whole numbers.. 4
Word problems—help in solving............. 8H